A DYNAMIC STRATIGRAPHY
OF THE BRITISH ISLES

TITLES OF RELATED INTEREST

Cathodoluminescence of geological materials
D. J. Marshall

Chemical fundamentals of geology
R. Gill

The dark side of the Earth
R. Muir Wood

Deep marine environments
K. Pickering *et al.*

Experiments in physical sedimentology
J. R. L. Allen

Heavy minerals in colour
M. Mange & H. Maurer

Image interpretation in geology
S. Drury

Karst geomorphology and hydrology
D. C. Ford & P. W. Williams

Marine geochemistry
R. Chester

Mathematics in geology
J. Ferguson

Perspectives on a dynamic Earth
T. R. Paton

Petroleum geology
F. K. North

Petrology of the sedimentary rocks
J. T. Greensmith

A practical approach to sedimentology
R. C. Lindholm

Principles of physical sedimentology
J. R. L. Allen

Rocks and landforms
J. Gerrard

Rutley's elements of mineralogy
C. D. Gribble

Sedimentary structures
J. Collinson & D. Thompson

Sedimentology: process and product
M. R. Leeder

Soils of the past
G. Retallack

Statistical methods in geology
R. F. Cheeney

Trace fossils
R. Bromley

Volcanic successions
R. A. F. Cas & J. V. Wright

The young Earth
E. G. Nisbet

A Dynamic Stratigraphy of the British Isles

A study in crustal evolution

R. ANDERTON
Department of Applied Geology, University of Strathclyde

P. H. BRIDGES
Division of Geology, Derby Lonsdale College of Higher Education

M. R. LEEDER
Department of Earth Sciences, University of Leeds

B. W. SELLWOOD
Department of Geology, University of Reading

CHAPMAN & HALL
London · New York · Tokyo · Melbourne · Madras

**Published by Chapman & Hall, 2-6 Boundary Row,
London SE1 8HN, UK**

Chapman & Hall, 2-6 Boundary Row, London SE1 8HN, UK

Blackie Academic & Professional, Wester Cleddens Road,
Bishopbriggs, Glasgow G64 2NZ, UK

Chapman & Hall Inc., One Penn Plaza, 41st Floor,New York
NY 10119, USA

Chapman & Hall Japan, Thomson Publishing Japan, Hirakawacho
Nemoto Building, 6F, 1-7-11 Hirakawa-cho, Chiyoda-ku, Tokyo 102,
Japan

Chapman & Hall Australia, Thomas Nelson Australia, 102 Dodds
Street, South Melbourne, Victoria 3205, Australia

Chapman & Hall India, R. Seshadri, 32 Second Main Road, CIT East,
Madras 600 035, India

First edition 1979
Reprinted 1990, 1992, 1993

© 1979 R.Anderton, P.Bridges, M.Leeder and B.W.Sellwood

Typeset in 10 on 12 point Baskerville by George Over Limited, Rugby
and London
Printed in Great Britain at The Alden Press, Oxford.

ISBN 0 412 445107 (PB)

A catalogue record for this book is available from the British Library
Library of Congress Cataloging-in-Publication Data available

Preface

This book is in essence a story; an illustrated description of the evolution of the British Isles. Although this in itself is not new the authors believe that there is a need for a change in the approach to the subject. The correlation of rocks is the most fundamental aspect of stratigraphy, but it is necessary to utilise other geological disciplines in order to discover the origin of rocks and unravel the evolution of a terrain. Geochemistry, geophysics, igneous and metamorphic petrology, palaeobiology, sedimentology and structural geology must all play their part together.

This text focuses on the geological events that led to the evolution of the small but complex part of the Earth's crust now called the British Isles. The level of treatment assumes a minimum background knowledge equivalent to a first-year course in geology at university. The evidence is drawn not only from the rocks of the British Isles but also from the adjacent terrains of the present and of the past. The book is divided into four parts. Although they are in chronological order the emphasis is on major events. Thus first, in Part 1, we consider the evolution of the early crust. This is followed in Part 2 by a study of the building of the Caledonides. In Part 3 we trace the development of the Hercynides, and in Part 4 we consider the post-Palaeozoic evolution of the British Isles on the passive eastern margin of the N. Atlantic Ocean.

Since the development of the theory of plate tectonics there have been numerous hypotheses and speculations concerning the evolution of the British Isles. Although there continues to be a substantial outflow of new data it is generally difficult to confirm or totally to refute these hypotheses. This means that our story, especially in the Palaeozoic and Precambrian, must be followed with caution. For this reason we have provided essential data in the form of diagrams and tables. It is hoped that readers will compare the interpretations and views expressed in the text with data located in the figures. Selected references are cited in the text, and key references, those considered most useful, are indicated by an asterisk.

Acknowledgements

Our book owes its origin to the inspiration of Professor P. Allen FRS. His enthusiasm and energy, expressed in his writings on the lower Cretaceous of the Weald, have greatly influenced the authors. We are very grateful to Professor E. K. Walton, Dr G. E. Bowes, Professor E. H. Francis, Dr J. M. Hancock, Dr J. D. Hudson, Dr W. E. A. Phillips, Dr H. G. Reading, Mr J. Rose, Dr A. D. Stewart and Professor J. V. Watson for their encouragement and their numerous constructive criticisms of the draft manuscript. Dr J. Collinson, Dr M. E. Tucker and Mr I. Tunbridge have generously provided unpublished results of their recent researches. Thanks are due to Mr A. Cross for drafting many of the diagrams in Part 4, and to our colleagues at Derby, Leeds, Reading and Strathclyde for their interest and helpful advice. Our wives Annette, Madeleine, Sue and Jan deserve a special thank-you for their patience and encouragement. Finally we are greatly indebted to Mr Roger Jones of Allen and Unwin who has carefully guided this work to completion.

Contents

Part One
THE CRUSTAL FOUNDATIONS

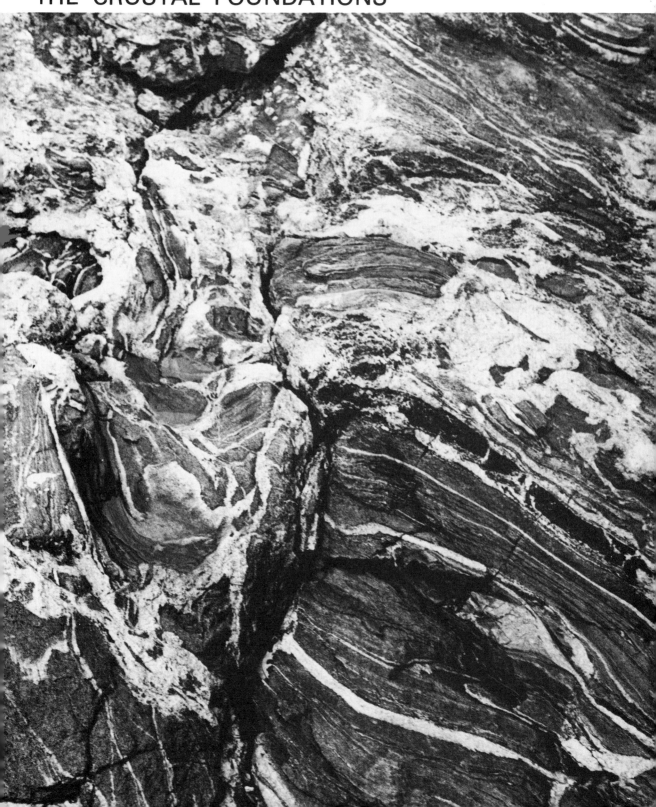

Theme

The most complete record of the early evolution of the Earth is preserved in the Precambrian shields that form the nuclei of the present continents. In the British Isles, which lie between the shield areas of N. America and the Baltic, early Precambrian or Archaean rocks only crop out over a small area, although a basement composed of Archaean gneisses probably forms the crustal foundation underlying most of the British Isles. The recognition and interpretation of Archaean events rely heavily on our understanding of structural geology, geochemistry and geochronology in addition to such disciplines as palaeobiology, sedimentology and stratigraphy, which play a more significant role in our comprehension of the later parts of geological history. Also, since the Earth underwent some of its most profound evolutionary changes during the early Precambrian, it is probable that some of the major geological processes acting then were very different from those of today.

Chapter 1 traces the early evolution of the Earth as recorded in the Archaean crust of the shield areas and then applies this information to the limited Archaean rock record in the British Isles. The further development of this early crust during the later Precambrian or Proterozoic is treated in Chapter 2. Here again, information from the shield areas is useful in establishing the significance of some of the features seen in the British Isles, because major time breaks within the rather fragmentary Proterozoic rock sequence here prevent us from obtaining a clear picture of Proterozoic evolution from British rocks alone. It is only from late Precambrian times onwards, i.e. after 800–700 Ma ago, that the stratigraphic record in the British Isles becomes reasonably continuous. That part of the geological record will be considered later in Parts 2, 3 and 4.

Plate 1 Lewisian gneiss on the northern shore of Loch Torridon, NW Highlands of Scotland. The photograph shows dark, basic gneiss, probably of igneous origin and formed during the Scourian, folded and migmatised, i.e. intruded by a network of quartz and feldspar veinlets.

1 The Archaean crust and early Precambrian evolution

1a Introduction

About 4600 Ma (million years) ago the Earth condensed from a cloud of dust and gas. Ever since it has been changing and evolving. It is the purpose of this book to chart, in some detail, the evolution of that small part of the Earth's crust now occupied by the British Isles.

Only those parts of Earth history that have left their record in the rocks can be deduced. As with human history, the record generally becomes more fragmentary and difficult to decipher as we go back in time. Later geological processes tend to obscure or destroy the evidence of earlier events. Fortunately the location of the areas of intense geological activity, the orogenic belts, has changed through time. As a result the geological map of the Earth's continental crust shows not a haphazard pattern, but a series of tectonic provinces (Fig. 1.1), each of which developed its major geological characteristics during a different part of geological history.

The areas of ancient Precambrian rocks called shields (large crustal blocks that have not undergone tectonic disturbance since the Precambrian) or cratons (a more general term which also includes shields with an undisturbed cover of later sediments) record early Precambrian history in detail and have been largely unaffected by later events. These are surrounded by younger belts which record more recent geological episodes although they may incorporate an ancient basement (Hurley & Rand 1969). The British Isles lie astride belts of

Figure 1.1 Geological map of the world showing major tectonic provinces.

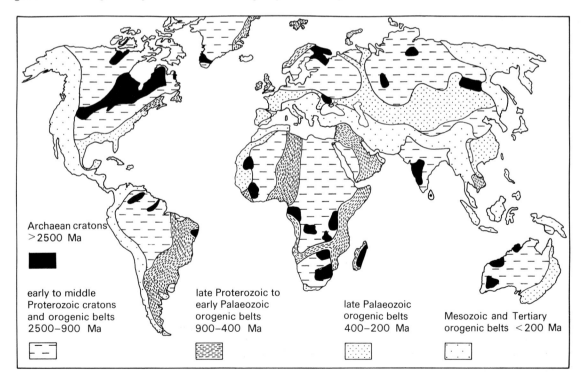

Archaean cratons
>2500 Ma

early to middle
Proterozoic cratons
and orogenic belts
2500–900 Ma

late Proterozoic to
early Palaeozoic
orogenic belts
900–400 Ma

late Palaeozoic
orogenic belts
400–200 Ma

Mesozoic and Tertiary
orogenic belts <200 Ma

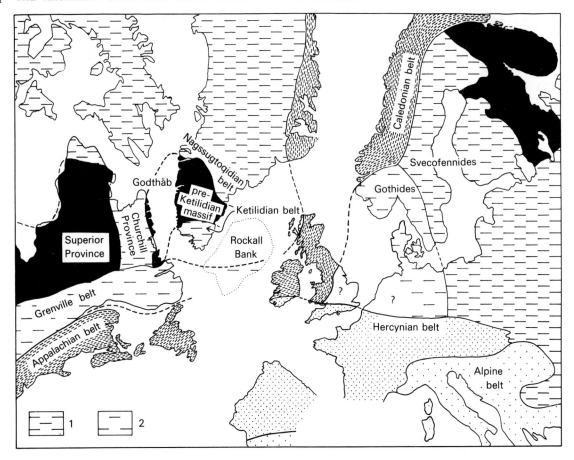

Figure 1.2 Geological map of the N. Atlantic region on a Permian predrift reconstruction. Ornament is as for Fig 1.1 except that 1 = 2500–1700 Ma and 2 = 1300–900 Ma orogenic belts.

many ages (Fig. 1.2), and their rocks therefore record a large span of geological history. They have remained in an area of change throughout a large portion of geological time. The British Isles lie at a geological crossroads and, in comparison with other areas of the world of a similar small size, they record a large and varied part of the Earth's history.

Figure 1.3 shows how well each part of the Earth's history is represented in the British Isles in comparison with other areas of the world. Clearly, although the Phanerozoic record is fairly detailed, Precambrian, and especially early Precambrian, rocks are rather sparse. So although the Phanerozoic history of the British Isles is reasonably complete, it will be necessary to refer to evidence of early Precambrian events from other parts of the world to get even an outline of their early history.

1b The first 800 million years

There is no direct evidence about the Earth's early evolution. Our ideas about it come from the construction of theoretical models, extrapolation back from more recent periods, and evidence from outside the Earth. Briefly, it is imagined that, as the Earth condensed, it was heated by the release of radiogenic and gravitational energy. Convection currents were initiated and the Earth started to differentiate, the heavy components sinking towards the centre to form the core and the lighter materials rising towards the surface to form the crust. Initially the Earth's surface would have been very hot, possibly partly molten, with intense volcanic activity and the continuous fracturing and foundering of crustal material. Although the initial,

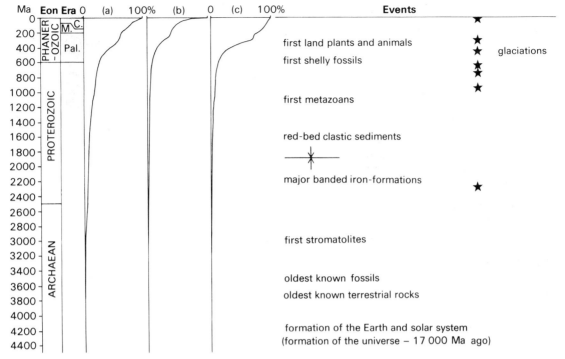

Figure 1.3 The dates of important events in the earth's history and estimates of the age of the rocks at the earth's surface. Cumulative curves show the percentages of the total land area of (a) the whole world, (b) the USA and (c) the British Isles that are older than the corresponding dates on the vertical time scale. Pal. = Palaeozoic, M. = Mesozoic, C. = Cainozoic.

probably basic, crust may have been repeatedly renewed by the eruption of basic volcanics followed by their sinking and resorption back into the mantle, any light acid rocks probably tended to accumulate at the surface, being too buoyant to sink. The acid material thus progressively increased in volume to form the earliest continental crust. This crust may initially have formed a uniform layer over the Earth, its thickness being limited by the shallow depth at which melting took place; but as the Earth cooled and the crust was allowed to thicken, it may have been dragged towards the sites of descending mantle convection currents to form one or several large masses of granitic crust, the first continents.

As well as being very hot the Earth's surface at this time would have been dry and without life. Not until the Earth's surface had cooled below 100°C would the oceans start to condense, and without water there would not have been even the possibility of life. Some impression of the nature of the early Earth can be gained by looking at other planetary bodies such as the moon which, because of their small size, cooled rapidly after their formation and fossilised those stages of planetary evolution that have been obliterated on the Earth.

1c The early continental crust

The record age for the oldest crustal rocks so far dated is c. 3750 Ma, held by the quartzo-feldspathic Amitsôq Gneiss of the Godthåb area of W. Greenland (although this, like any other record, is likely to be broken!). Therefore, although it is only possible to speculate about the earliest phase of Earth history that predates the oldest-known rocks, we know that after 800 Ma the 'boiling jam-pot' Earth had cooled sufficiently for at least parts of its surface to have become a permanent crust floating on denser mantle. Several questions can be raised about this very early or primordial continental crust. Did it have a similar composition and thickness to that of today? Did it cover as large an area as the present continents or has continental crust been added throughout geological time? Was there one large, or many small, early continents?

The earliest Precambrian rocks, such as the Amitsôq Gneiss and the associated Isua supra-crustals (metamorphosed sediments and volcanics), include a variety of lithologies such as granitic to dioritic gneisses, basic and ultrabasic volcanics, quartzites and banded iron-formations (Moorbath 1975*). Although some of the lithologies are not common in the Phanerozoic (e.g. the banded iron-formations) and some show chemical differences from today's equivalents, they are broadly similar to those of the present continental crust. Although rocks over 3500 Ma old are rare, rocks in the age range 3500−2500 Ma are found on all the continents (Fig. 1.1). Condie (1973) has shown that the thickness of the early crust can be estimated from several indicators, including the presence of mineral assemblages in granulite facies rocks, which can form only under pressures equivalent to crustal depths of 20−30 km. Extensive areas of granulites are found only in rocks younger than about 2900 Ma, but not in older rocks which are characterised by the high temperature but lower pressure rocks of the amphibolite facies. This suggests that the primordial crust may have been too thin for granulites to form, but had thickened to near present crustal thicknesses by c. 2500 Ma ago, the end of the period of Earth history known as the Archaean (from the Greek for 'ancient'). The initial $^{87}Sr/^{86}Sr$ ratios of many, although not all, Archaean gneisses are low, at about 0.701 (Moorbath 1975*). This indicates that they were formed from mantle-derived material (either directly or via subducted oceanic lithosphere) intruded not more than a few hundred million years before their date of metamorphism. Such gneisses were formed by the addition, rapidly followed by the metamorphism, of new crustal material; they were not formed by the metamorphism of much older crust. We may surmise that new continental crust was continuously being added during the Archaean, and it is thought that at least half of the present continental crust had formed by the end of the Archaean. It has probably increased in volume subsequently, but at a slower rate.

To summarise: continental crust had certainly started to form before 3800 Ma ago by the differentiation of the mantle. After this the volume and thickness of the continental crust increased rapidly and by 2500 Ma ago this crust occupied at least half the area of, and had a thickness comparable with, the present continents. To consider the geometry and motions of the Archaean continent(s), it is now necessary to look in more detail at Archaean tectonics.

1d   Archaean tectonics

There is no global stratigraphic classification scheme for the Precambrian. Difficulties of inter-regional correlation within the Precambrian have allowed local terminologies to evolve in isolation from each other. As a result there is a plethora of different names for time and rock units. In the absence of general agreement about the subdivision of the Precambrian, the vaguely defined terms 'Archaean' and 'Proterozoic' are in most common use for the older and younger parts of the Precambrian respectively (Fig. 1.3). Their meaning has changed several times since their introduction about a century ago and, although there is a wide difference of opinion as to the age of their common boundary, a date of c. 2500 Ma ago is now the most widely accepted.

There are two types of Archaean terrain. The first comprises narrow synclinal belts of sediments, basic and ultrabasic volcanics metamorphosed to green-schist facies and called greenstone belts, which are surrounded by granite batholiths. Clearly the greenstone belts were formed at the Earth's surface and have never been deeply buried. The contacts between the granites and the greenstone belts are often tectonically complex, and there is therefore some debate about their mutual relationships. One school of thought (e.g. Windley 1973*) considers that the greenstone belts were deposited on granitic crust which was later remobilised to form granite diapirs which rose up through the overlying sediments and volcanics, pinching them into tight synclines (Fig. 1.4). These belts may have been initiated as deep troughs or basins in incipient rifts over rising mantle convection-currents. An alternative hypothesis is that the greenstone belts are fragments of primitive island arc to ocean floor environments trapped between colliding microcontinents (e.g. Anhaeusser 1973). Although field relations and

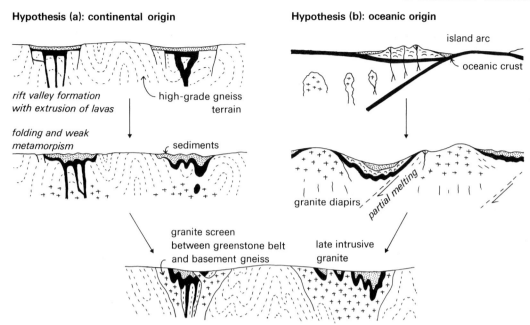

Hypothesis (a): continental origin

rift valley formation
with extrusion of lavas

high-grade gneiss
terrain

folding and weak
metamorpism

sediments

granite screen
between greenstone belt
and basement gneiss

Hypothesis (b): oceanic origin

island arc

oceanic crust

granite diapirs

partial melting

late intrusive
granite

Figure 1.4 Hypotheses for the origin of greenstone belts (continental origin after Windley 1973*, oceanic origin after Anhaeusser 1973).

radiometric dating have confirmed the former theory in certain areas (Hawkesworth *et al.* 1975), the calc-alkaline nature of the volcanics is consistent with the latter.

The second type of area consists of high-grade amphibolite- and granulite-facies granite gneisses (using 'granite' in the widest sense), sometimes finely interbanded with similarly metamorphosed supracrustals. The granulite facies rocks must have formed at considerable pressures in the lower crust. Fyfe (1973) has proposed that intrusions of granite into the lowest crust could have suffered partial remelting, the resulting melt removing the more volatile components and rising diapirically to form high level plutons, leaving behind a dry residuum exhibiting granulite facies metamorphism. Windley (1973*) has pointed out that dense basic volcanics are particularly well represented in the infolded supracrustal assemblages within the gneisses. The sinking of these supracrustals to the depths of 20 km or more necessary for granulite facies metamorphism may have been concomitant with the rise of granite diapirs.

Archaean gneiss terrains are characterised by complex structural patterns, including interference folds, domes and basins, without dominant linear trends. Such patterns are produced by vertical movements rather than horizontal compression. These terrains may have been characterised by much convective vertical mixing within a relatively plastic crust. The two components of the Archaean cratons, the granite–greenstone belts and the high-grade gneiss belts, seem to have been formed in different but contemporaneous environments, the upper crust and lower crust respectively, and brought together by subsequent tectonic movements.

If intense small-scale convection characterised the early mantle, the primordial granitic crust would have accumulated over downgoing convection currents at many points on the Earth, and an ocean-floor-spreading mechanism could have formed the greenstone belts. However, large scale convection would have produced several large continents, or a single supercontinent; and the Archaean deformation and igneous activity would then have been produced within, rather than at the margins of, the continents by mechanisms unlike those operating in post-Archaean times. Palaeomagnetic studies have as yet been unable to demonstrate any large relative horizontal movements within the Archaean continents. Although there is no consensus of opinion regarding Archaean tectonics, the balance of the evidence at the moment seems to favour a driving force different from the present-day ocean floor spreading mechanism. This would be likely if the Archaean Earth had higher thermal gradients and a thinner crust than the present. Perhaps the volume of the Archaean continents increased more by vertical accretion (i.e. direct addition of granitic rock to the crust from the mantle) than by lateral accretion (i.e. addition via subducted oceanic lithosphere). Orogenic belts may have formed over rising mantle convection-currents that rifted the continents without being strong enough to separate them. Horizontal movements within the continents were probably limited, and deformation was probably dominated by convection within a still-plastic crust.

1e   The atmosphere, hydrosphere and sedimentation

The Earth's atmosphere contains very small proportions of the noble gases, such as xenon and neon, in comparison with their cosmic abundances. It is therefore thought that the primordial atmosphere was stripped off, very early in the Earth's history, and that the present atmosphere and hydrosphere have formed largely by the addition of fluids from volcanic activity. As the juvenile gases erupted from modern volcanoes consist largely of water vapour and carbon dioxide, the Archaean atmosphere may have consisted mainly of carbon dioxide, with volcanic water vapour condensing to form the oceans when the Earth's surface had cooled sufficiently.

As the oldest known sediments, the Isua supracrustals, contain banded iron-formations (see below), a lithology thought to be precipitated in an aqueous environment, the oceans must have started to form before 3750 Ma ago. Most of the minor volcanic gases would not have accumulated in the atmosphere because they are either too soluble and chemically reactive (e.g. hydrogen sulphide) or too light to be retained by the Earth's gravity field (e.g. hydrogen). However, since it does not form any geologically stable sediments, the small proportion of nitrogen exhaled by volcanoes would have accumulated to form the present nitrogen-dominated atmosphere. The volume of the atmosphere and hydrosphere must have increased through time, and it is most likely that the hydrosphere grew rapidly during the Archaean and at a decelerating rate thereafter, thereby keeping the depth to which the growing continents were flooded fairly constant throughout geological history.

By the end of the early Archaean, then, there were rivers and oceans and presumably winds and storms. The physical processes that control sedimentation (from mechanical weathering to transport and deposition) would have been much the same as they are today. However, as the composition of the atmosphere and oceans would have been different, so too would have been the chemical controls.

The best example of this difference in chemistry is the presence in Archaean greenstone belts and early Proterozoic shelf sequences of extensive banded iron-formations (Fig. 1.3). These consist of alternating centimetre-thick layers of chert and iron minerals (siderite, magnetite, haematite, iron silicates and pyrite) which were chemically or biochemically precipitated. These deposits form a major part of the world's economic iron reserves and are exploited on a very large scale in Canada and Australia. The major banded iron-formations only occur before 1900 Ma ago, a date that roughly corresponds with the first appearance of terrestrial red-bed (i.e. haematite-bearing) facies. Another difference is the presence in early Proterozoic fluvial sediments of detrital pyrite and uraninite, minerals that rapidly oxidise under present-day surface conditions. All these facts are consistent with a gradual change from a reducing to an increasingly oxidising atmosphere and hydrosphere during the Precambrian (but see Dimroth & Kimberley 1976 for a contrary view). Free oxygen would have formed in the early atmosphere only by the photodissociation of water vapour. Later, with the appearance of photosynthesising organisms, increasing amounts of oxygen would be released. Under the early reducing conditions iron derived from rock weathering and volcanic exhalations would be easily transported by rivers as a ferrous solution but could be precipitated by biochemical oxidation on reaching the sea. A change to general oxidising conditions would drastically cut down the mobility of iron and bring to an end the deposition of extensive banded iron-formations (Garrels et al. 1973*).

1f Life

The oldest-known 'fossils' are microscopic spheres and filaments preserved in cherts dated at 3555 Ma. These are found in the Onverwacht Group of the Swaziland sequence, part of a South African greenstone belt. Simple micro-organisms may well have existed for a long time before this, as the essentials for life probably existed very early in the Earth's evolution, soon after the formation of the oceans.

Chemical reactions produced by geologically likely energy sources, such as ultraviolet radiation or electrical discharges, can produce complex organic compounds (e.g. hydrocarbons and amino acids) from mixtures of simple gases such as hydrogen, carbon dioxide, water vapour and ammonia. Under Archaean reducing conditions such compounds would not be oxidised away, as they would today, but would accumulate, react with each other and become more complex. When these compounds collected together to form 'cells' and became organised by the presence of nucleic acids so that they could reproduce, life began (Sylvester-Bradley 1972*). The first organisms would have obtained energy by the assimilation of 'food' composed of organic compounds. However, organisms capable of obtaining energy directly from the sun by photosynthesis must have evolved quickly since stromatolites, presumably formed by blue-green algae (the most primitive plants), are found in the 3100 Ma old Bulawayan System in Rhodesia. This shows not only that life had begun by the Archaean but also that it was already having a direct effect on the sedimentary environment. However, Archaean life must have been restricted to very simple and tolerant organisms such as bacteria and blue-green algae.

The effect of organisms on the evolution of the atmosphere and hydrosphere was mentioned above (Ch. 1e). If the hypothesis of a reducing early atmosphere is accepted, then early life must have been anaerobic. The oxygen produced both by early photosynthetic organisms and by photodissociation would be taken up by the reservoir of reduced elements in the oceans, notably iron, resulting in its local precipitation as banded iron-formations. Not until this reservoir was exhausted would free oxygen accumulate in significant amounts in the atmosphere, and this, Cloud (1972) has suggested, began to happen c. 1900 Ma ago.

1g The Archaean in Britain

Britain straddles the late Precambrian to Lower Palaeozoic Caledonian belt. Northwest and southeast of this belt lie Precambrian cratons which have cores of Archaean rocks (Fig. 1.2). Although these cores do not extend as far as Britain, small areas of Archaean rocks are preserved in Britain within post-Archaean orogenic belts. A small area of Archaean rocks is found in the Channel Isles (the Icart and Perelle gneisses) and a much larger area in NW Scotland (the Scourian gneisses, part of the Lewisian Complex; Fig. 1.5). Preliminary isotopic data suggest that the Rosslare Complex of SE Ireland may also have a history stretching back into the Archaean.

The Icart and Perelle gneisses of the Channel Isles, dated at 2620 ± 50 Ma (Adams 1976), form part of what is known as the Pentevrian basement, the oldest part of a small inlier of Precambrian rocks within the Hercynian belt. It is not known whether these gneisses are continuous at depth with rocks of the Baltic Shield, or are separated from them by the later Grenville and/or Caledonian orogenic belts passing southeastwards through Europe. Either way, it is possible that SE Britain is underlain at depth by rocks that date back into the Archaean.

The situation in Scotland is far clearer. The Scourian gneisses can be thought of as lying on the edge of the N. Atlantic Craton, an area that contains the Archaean rocks of coastal Labrador and Greenland, including the very ancient Amitsôq Gneiss (Fig. 1.2). The oldest Scourian rocks so far dated are c. 2900 Ma old, and $^{87}Sr/^{86}Sr$ ratios show that they were formed from only slightly older igneous precursors, not from primordial crust (Moorbath et al. 1975). Gneisses of this age are found on many cratons, and the Scourian crust may have formed during the major worldwide phase of crustal accretion that continued until the end of the Archaean. The Scourian gneisses have been extensively reworked by later orogenic events (Laxfordian, Grenvillian, Caledonian; Ch. 2c), but it is possible that a large part of Scotland and the northern part of Ireland are underlain at depth by a gneissic basement that has been in existence since the Archaean.

Where unaffected by later reworking the typical Scourian gneisses are coarsely banded acid to intermediate gneisses, some of which show granulite

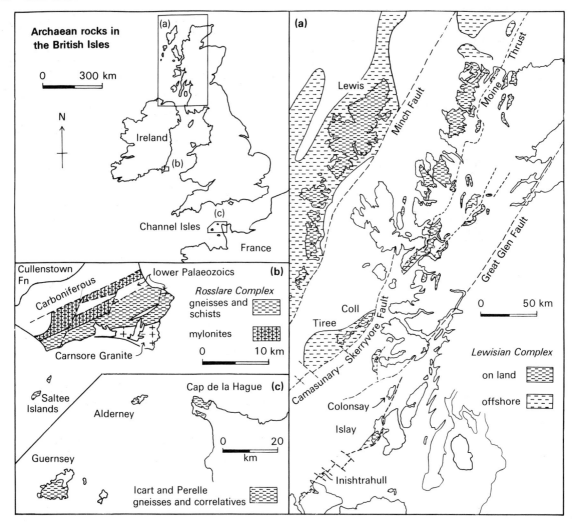

Figure 1.5   Location of Archaean rocks in the British Isles ((a) after Binns *et al.* 1974, Dobson *et al.* 1975 and Watson 1975*; (b) after Max 1975; (c) after Adams 1976).

facies metamorphism. They are typical of the Archaean high-grade gneiss terrains. However, greenstone belts are absent. In fact, the N. Atlantic Craton as a whole contains a lower proportion of supracrustal rocks than many other Archaean cratons.

Watson (1975*) has estimated that 80% of the Scourian consists of acid to intermediate gneiss. Although there is no direct evidence for their origin, these rocks are most likely to be of igneous parentage. Interbanded with these gneisses are rocks of volcanic and metasedimentary composition, and strips of ultrabasic, basic and anorthositic rocks which may show a relict mineralogical banding

characteristic of layered igneous intrusions (Davies 1974). The metasediments are well represented in the Outer Hebrides (Fig. 1.6) where they include quartzites, quartz–magnetite rocks, marbles, calc-silicate rocks and highly aluminous rocks (i.e. those rich in biotite, kyanite and garnet). No original sedimentary textures are preserved in any of the Scourian metasediments; the interpretation of their origin is based largely on their chemistry. Thus, although quartzo-feldspathic gneisses can have a variety of origins (e.g. as granites, migmatites, arkoses), such lithologies as quartzite, for example, are reasonably attributed to a sedimentary origin. Although there are no greenstone belts, the

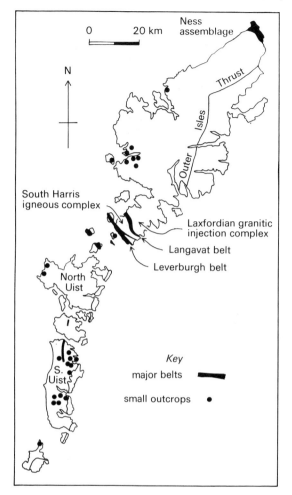

Figure 1.6 Location of Scourian metasediments in the Lewisian rocks of the Outer Hebrides (after Coward *et al.* 1969).

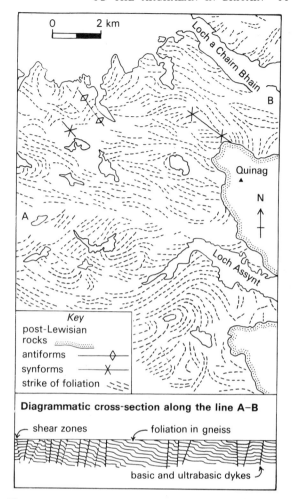

Figure 1.7 Structural pattern of the Scourian gneiss in the Loch Assynt area of Sutherland (after Sheraton *et al.* 1973).

Scourian supracrustal assemblage includes representatives of many of the greenstone belt lithologies, and it is interesting to speculate whether the quartz–magnetite rocks could be highly metamorphosed banded iron-formations.

The gneisses, metasedimentary and meta-igneous rocks all suffered high to intermediate-pressure granulite- or amphibolite-facies metamorphism 2900–2700 Ma ago, an episode known as the Badcallian (Fig. 2.2). They are characterised by low concentrations of water, uranium, thorium, rubidium and potassium, components that may have been expelled during metamorphism. Although the gneisses are composed largely of quartz and plagioclase, it is the presence of pyroxene as the dominant mafic mineral that indicates their granulite metamorphic facies. The Badcallian deformation produced a near horizontal foliation with a disordered fold pattern (Fig. 1.7). Although there is a slight NW–SE grain to the structural pattern, these rocks have a generally chaotic and 'stirred-up' appearance, on both a small and a large scale.

The small areas of Archaean rocks in Britain are similar in many ways to those of the large Archaean cratons such as the Superior Province of the Canadian Shield (Fig. 1.2). The outcrop of the Channel Isles rocks is too small to make many deductions, but the Scourian gneisses of NW Scotland are similar in their lithology, metamorphism, deformation and geochemistry to the high-grade gneisses of other Archaean cratons. The Scourian rocks are not as old as the oldest rocks in Greenland and Africa and, although no remnants of the original Scottish continental crust have been positively identified, it is thought that the Scourian continental crust had formed by approximately 2900 Ma ago. Sediments and volcanics were deposited on this crust, and magmas intruded into it forming layered igneous intrusions. Within 200 Ma of the initial formation of this crust the whole assemblage was carried down to a considerable depth within the crust and metamorphosed under hot, high pressure, but dry conditions to the granulite and amphibolite facies. The deformation and metamorphism were roughly contemporaneous with similar episodes in other Archaean cratons, such as the Kenoran Orogeny of the Superior Province.

Although the Scourian gneisses form only a small part of the Lewisian Complex, radiometric and geochemical data suggest that much of the complex is of Archaean origin and has subsequently been reworked (Ch. 2c). Geophysical work (Bamford *et al.* 1977) shows that the present crustal structure of NW Scotland is similar to that of much of the rest of Britain. Therefore it is possible that Archaean rocks, although much reworked, form the basement of much of Britain; they are the crustal foundation on which the rest of British geology rests.

1h Summary

The early history of the Earth is obscure and speculative. However, it is known that, within 800 Ma of the Earth's formation, continental crust had become a permanent feature of the Earth's surface, a wide variety of rock types were being produced and therefore many geological processes such as volcanism, metamorphism, weathering and sedimentation were already operating. During the next 1000 Ma the continents rapidly increased in volume, probably reaching their present thickness and half their present area by the end of the Archaean. These Archaean rocks form the nuclei of the present continents and possibly form the basement of NW and SE Britain. Formed at a time of great crustal mobility (thin crust, high heat flow), the Archaean rocks have evolved in a different tectonic environment from that of later rocks and show evidence of vertical mixing rather than a tectonic grain due to horizontal compression. Archaean rocks vary from granulite- and amphibolite-facies gneisses of various origins to belts of greenschist-facies metasediments and metavolcanics. In Britain the former are predominant, recording for the most part events within the crust. On other continents, especially Africa, the belts of low grade rocks (greenstone belts) record the early history of the Earth's surface, including the origin of life.

2 Proterozoic developments: Sedimentation, orogeny and magmatism

2a Introduction

Although there was no sudden change in Earth history at the Archaean/Proterozoic boundary *c.* 2500 Ma ago, no single process either starts or stops at this date, the Archaean can be thought of as a period of rapid geological evolution with rapid accretion of continental crust, high heat flow, great crustal mobility and rapid outgassing of the mantle. By the early Proterozoic the rate of change had decreased, large areas of continental crust had become stable and, although there are many features of the early Proterozoic that are not found in the Phanerozoic, it is much easier to visualise the Proterozoic in terms of modern processes. Thus while the Archaean is dominated by metamorphic rocks, with low grade sediments and volcanics being restricted to infolded tracts between areas of high grade basement, undeformed Proterozoic rocks including extensive areas of flat-lying sediments are common.

The irregular tectonic pattern of the Archaean shields is cut by a network of Proterozoic orogenic belts, each showing a roughly linear tectonic grain. In many cases these have formed by the deformation and metamorphism of an igneous and sedimentary cover together with its Archaean gneiss basement. There is no evidence for the existence of oceanic crust in most of these rocks; deformation seems often to be related to transcurrent fault movement, producing shear zones, and the metamorphism to the rise of fluids along these zones. Thus by the end of the Archaean the continental crust seems to have become sufficiently rigid to have responded to mantle convection currents by being broken into a series of fairly small continental blocks between which there were only limited horizontal displacements. Some Proterozoic sequences, however, such as those that formed in the Corona-

tion Geosyncline of NW Canada *c.* 2000 Ma ago, show a depositional and deformational history similar to that of many Phanerozoic orogenic belts, suggesting an origin related to ocean floor spreading.

The response of the Earth's crust to mantle convection has probably changed slowly through time. It is unlikely that the processes of ocean floor spreading, as we know them today, took over suddenly from more 'primitive' tectonic regimes. Perhaps there was a gradual evolution from the small-scale plastic deformation characteristic of the Archaean, to limited and largely transcurrent movement between small blocks, to the formation of small rifts and then small ocean basins, and finally to the opening and closing of major oceans between large rigid continental plates. Some or all of these processes may have operated simultaneously during the Proterozoic, although their relative importance may have changed with time. Palaeomagnetic evidence suggests that most of the continental crust was lumped into one supercontinent during the early Proterozoic and that major sea-floor spreading was initiated *c.* 1100 Ma ago (J. D. A. Piper 1974*). If this is confirmed, it would imply that prior to *c.* 1100 Ma ago most orogenic episodes were produced by deformation, metamorphism and plutonism in ensialic mobile belts within this continent, although subduction zones may have existed around its margins. Thus although some form of ocean floor spreading may have been taking place locally since the early Proterozoic or even earlier, transcurrent faulting and minor rifting dominated Proterozoic tectonics.

The widespread occurrence of stromatolites in Proterozoic sediments shows that algae remained the dominant organism, although they probably became more diverse and complex. However, more significant evolutionary changes must have been taking place, since burrows suggestive of fairly advanced metazoans (multicelled animals) are found in sediments as old as 1000 Ma in Zambia (Clemmey 1975) and a varied assemblage of soft-bodied metazoan fossils, known as the Ediacara fauna, is found in the latest Proterozoic sediments of Australia and Africa (Fig. 1.3). Two members of this assemblage, *Charnia* and *Charnodiscus*, have also been found in the Proterozoic sediments of Charnwood Forest, England (Ch. 6d; Fig. 6.1). The early Proterozoic saw the end of the major phase of deposition of banded iron-formations and the appearance of terrestrial red-beds. These changes may have been due to the accumulation of photosynthetically produced oxygen in the hydrosphere and atmosphere, as discussed in Chapter 1e. The increased pace of biological evolution was probably also related to the accumulation of oxygen, as the more advanced organisms that multiply mainly by sexual reproduction can only exist under aerobic conditions. Further build-up of atmospheric oxygen in the late Proterozoic may have aided the development of metazoans, since oxygen ionised to ozone in the upper atmosphere provides an effective screen against the ultraviolet radiation that is harmful to the higher forms of life.

The Proterozoic eon ends at the base of the Cambrian. However, as the late Proterozoic to Lower Palaeozoic history of the British Isles is related to the evolution of the Iapetus Ocean, this chapter deals only with the earlier part of the Proterozoic, prior to the probable formation of this ocean.

2b The Proterozoic: dates and divisions

The solution to the related problems of correlation and dating is central to the understanding of any segment of the Earth's history. These problems are particularly acute in the Precambrian, where (a) palaeontological zoning techniques are not generally applicable (but see below), and (b) the geological record is very patchy because of the small exposed area of rocks preserved in comparison with the vast amount of time available for their formation (Fig. 1.3).

The different subdivisions of the Proterozoic used in Canada and Russia, two of the major areas of exposed Proterozoic rocks, illustrate how different correlation techniques have evolved in different geological environments. The Russians, working on well-preserved sedimentary sequences in Siberia and the Urals, have used the presence of different forms of stromatolites to divide the later part of the Proterozoic, known as the Riphean, into four intervals (Fig. 2.1). This scheme has proved useful for correlation within Eurasia. The four intervals have been further subdivided on the basis of stromatolites for local correlation, and attempts have been made to apply the Russian scheme to the Proterozoic of other continents, e.g. Australia and N. Africa, although this work is still in its infancy.

In Canada, research has concentrated on the metamorphic and structural history of the Precambrian rocks. Gastil (1960) has noted that radiometric mineral ages from the Canadian Shield are concentrated within certain time intervals which he has interpreted as the duration of major orogenies. The termination of these orogenies — the Kenoran, Hudsonian and Grenvillian — has been used to divide the Proterozoic into three eras — the Aphebian, Helikian and Hadrynian — and to define the end of the Archaean. The dates of the boundaries of these eras have subsequently been revised (Fig. 2.1) in the light of more recent U−Pb dates and Rb−Sr whole-rock isochrons. These give better estimates of the date of the metamorphic peak of each orogeny and the dates of crystallisation of post-tectonic intrusions, than do mineral ages, which only date some point during the subsequent cooling history of the rocks. These dates can be related to the sedimentary rock sequence because the duration of orogenies is usually marked by unconformities within the local stratigraphic succession. The Canadian and Russian classifications are therefore similar in that they are based on geological events, the ages of which can be estimated by radiometric dating, although the dates themselves do not define the stratigraphic boundaries.

Britain lies between Canada and Russia, and neither scheme is directly applicable to Britain. Proterozoic stromatolites are found in Britain (Chs 4d and 6c) but not sufficiently widely for correlation with the Riphean intervals. The Proterozoic history of N. Britain has, however, some links with the Canadian Shield, and the Canadian terms could therefore be used.

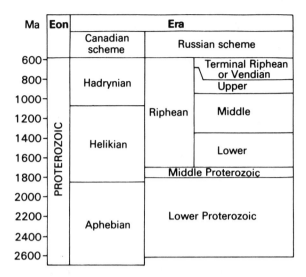

Figure 2.1 Proterozoic dates and divisions (after Stockwell 1972 and Krylov & Semikhatov 1976).

So far Proterozoic classification and correlation have been discussed on a continental scale. On a local scale, and in the absence of palaeontological data, Precambrian correlation relies heavily on radiometric dating, an expensive and lengthy procedure by comparison with the identification of zone fossils in the Phanerozoic. In most cases, Precambrian rocks can only be correlated if they can be mapped laterally into each other or if radiometric dating demonstrates their contemporaneity. Lithological similarity or similarity in stratigraphic relationships is not usually an adequate justification for correlation, although unusual and distinctive horizons, such as tillites, can be correlated over large areas. In the absence of comprehensive radiometric dating cover for the Precambrian there still remain many rock units that, although their origin can be interpreted, this interpretation can not be fitted into the regional geological history.

2c Lewisian developments

The Lewisian Complex of NW Scotland (Fig. 1.5) includes the oldest rocks in the British Isles. The Archaean origins of these rocks were discussed in Ch. 1g. However, the story does not end there because the Lewisian rocks record a whole series of events from 2900 Ma ago down to less than 1700 Ma ago.

Prior to the development of radiometric dating, an important time marker was recognised within the complex. This is a suite of NW−SE trending tholeiitic dolerite dykes, ranging in width from a few centimetres to over 100 m. In some areas these dykes cut sharply across the gneisses; in others they are deformed and metamorphosed to the amphibolite facies with the surrounding gneisses. Clearly the dykes were intruded after the last episode of deformation and metamorphism in some areas, whereas in others a still later orogenic episode affected the dykes and gneisses alike (Fig. 2.2). Sutton and Watson (1951) have christened the early and later episodes the Scourian and Laxfordian respectively, the dykes now being known as Scourie dykes. Subsequently the unmetamorphosed parts of the dyke suite have been dated at around 2200 Ma.

The main, and some of the local, Lewisian events are tabulated in Figure 2.3. It can be seen that the main episode of Scourian metamorphism, the Badcallian, was an Archaean event closely following the formation of the parent rocks. However, metamorphism, preceded by the formation of pegmatites and associated with minor folding, lingered

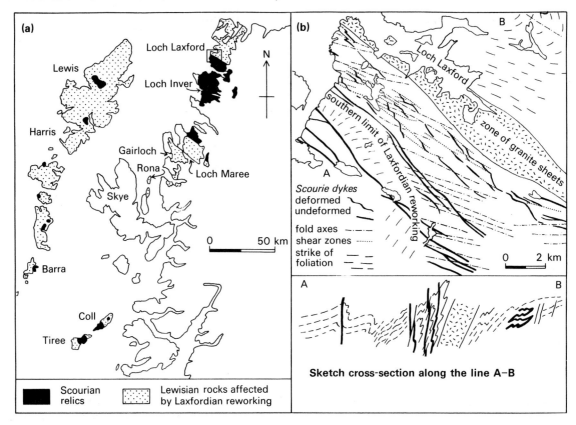

Figure 2.2 (a) Distribution of Laxfordian reworking in the Lewisian outcrops on land west of the Moine Thrust. (b) Map and section of the Scourian/Laxfordian boundary in the Loch Laxford area, the Laxford Front, showing the northeastward increase in Laxfordian deformation (after Beach *et al.* 1974).

on in a small area around Loch Inver (Fig. 2.2) until *c.* 2200 Ma ago, an episode known as the Inverian (Evans & Lambert 1974).

Laxfordian history following the intrusion of the Scourie dykes can be simplified into four phases: (1) the deposition of sediments, (2) deformation and metamorphism, (3) intrusion of igneous rocks, and (4) uplift and cooling. This is a similar sequence of events to that found in many Phanerozoic orogenic belts. However, with the exception of the small volume of sedimentary rocks involved, the rocks metamorphosed by the Laxfordian event had largely been in existence since the Badcallian gneiss-forming metamorphism. This is demonstrated by Rb−Sr whole-rock isochrons and U−Pb zircon ages on Laxfordian gneisses which give dates of *c.* 2700 Ma, although mineral ages and K−Ar dates record Laxfordian reheating *c.* 1900−1700 Ma ago. In other words the effect of the Laxfordian

metamorphism was largely to remetamorphose and deform 'old' gneisses rather than to form gneisses from 'new' igneous and sedimentary rocks. This process is known as reworking and is fairly common in Precambrian gneiss terrains.

Some Scourian (Badcallian) gneisses exhibit pyroxene granulite-facies metamorphism dominated by anhydrous quartz−feldspar−pyroxene assemblages. Laxfordian metamorphism was of the amphibolite facies, giving the hydrous mafic minerals hornblende and biotite. In other words, the reworking of the Badcallian granulitic gneisses during the Laxfordian was characterised by the retrogression and hydration of the pyroxene to hornblende and biotite. The distribution of Scourian and Laxfordian gneisses (Fig. 2.2) shows that the Laxfordian reworking was fairly pervasive. With the exception of a large central belt on the mainland between Loch Laxford and Loch Maree, only small

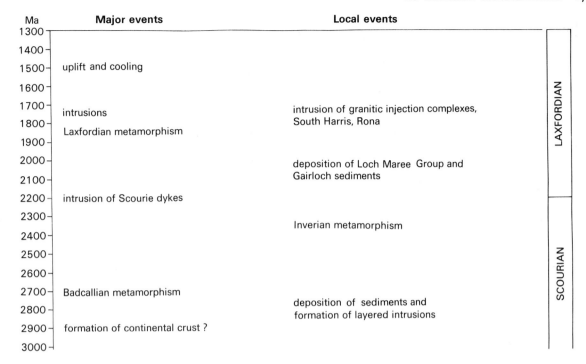

Figure 2.3 Dates of Lewisian events.

relics of unreworked Scourian occur within the Laxfordian. Geophysical work suggests that older Scourian gneiss may underlie the younger Laxfordian in the Loch Laxford area. The large Scourian relics may therefore have been roofed originally by reworked Laxfordian rocks.

The intensity of the Laxfordian deformation appears to be related to the retrogressive metamorphism. For example, parts of some Scourie dykes that have suffered metamorphism to amphibolites are intensely deformed, but adjacent parts of the same dykes are unaffected and show their original mineralogy. Watson (1973) has suggested that both aspects of reworking, retrogressive metamorphism and deformation, are related to the influx of water. Therefore the pattern of reworking reflects the distribution of ductile shear zones and fractures along which the water could gain access. The boundaries of the large Scourian relics are fairly straight (e.g. the Laxford Front; Fig. 2.2), although the smaller Scourian masses have an irregular outline. Watson believes that this may reflect reworking at different crustal levels. In the middle crust reworking may have been restricted to vertical fracture belts along which volatiles expelled from the

lower crust rose, the deformation being produced by movement between the flanking rigid blocks (Fig. 2.4). At shallow crustal levels the volatiles may have spread out laterally, producing more pervasive reworking with smaller intervening relics. Although the relics may have been heated to the same temperature as the surrounding reworked gneisses, they were unchanged because they remained impervious to water.

The Laxfordian structures are near vertical with NW–SE trends when adjacent to or between older Scourian relics. The more extensive (?high-level) Laxfordian tracts show structures with more variable orientations. The influence of mineralogy on deformation is clear even on a small scale within Laxfordian tracts, biotite- and hornblende-bearing rocks suffering more than the more competent quartzo-feldspathic gneisses.

Although most of the rocks involved in the Laxfordian metamorphism were 'old' Scourian gneisses, some post-Scourian sediments were also affected. The Loch Maree Group and Gairloch sediments lie in the southern part of the mainland Lewisian outcrop (Fig. 2.2). They include such distinctive metasediments as mica schist, graphite

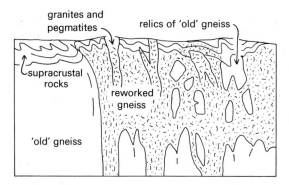

Figure 2.4 Sketch section through the middle and upper crust illustrating suggested relationships between reworked and 'old' gneisses (after Watson 1973).

schist, quartz-magnetite schist and marble. Bikerman *et al.* (1975*) have dated the metamorphism of these metasediments at 1975 ± 75 Ma, with initial $^{87}Sr/^{86}Sr$ ratios suggesting that they were deposited after 2200 Ma ago. Although their field relations with the over- and underlying quartzo-feldspathic gneisses are ambiguous, isotopic data suggest that they were deposited on a Scourian basement of gneiss with which they were subsequently interleaved after the main Laxfordian metamorphism. However, some of the adjacent gneisses may have been formed from the metasediments themselves since gradations from metasedimentary schists to feldspar-rich gneisses are found. Small strips of quartzites and calc-silicate rocks are also found north of Loch Laxford among the Laxfordian quartzo-feldspathic gneisses and amphibolites, although these also may be reworked Scourian metasediments.

The intrusion of small granite sheets and extensive pegmatite veins followed the Laxfordian reworking. The former are often found in the zones of most intense Laxfordian deformation. The largest plutonic complex, the granitic injection complex of Harris (Fig. 1.6), includes granite, migmatite and pegmatite, dated at 1750 ± 34 Ma, giving a minimum age for the local Laxfordian deformation (van Breeman *et al.* 1971). The formation of these intrusions *c.* 1700–1800 Ma ago was probably the last major episode in the Lewisian story, although some minor folding may have followed. A large number of K–Ar mineral ages in the range 1700–1400 Ma have been recorded from the Lewisian, and these probably record dates during

uplift by which time the rocks had cooled below the temperature at which radiogenic isotopes become mobile.

East of the Moine Thrust, Lewisian rocks are exposed as inliers in the Moine (Fig. 2.7). They have a long and complex history, including the Scourian and Laxfordian episodes of deformation, the intrusion of basic rocks and the deposition of sedimentary assemblages. They were also deformed together with the younger Moine metasediments during the Grenvillian Orogeny *c.* 1050 Ma ago (Ch. 2f). Further south, on Inishtrahull and Islay (Fig. 1.5), another Lewisian inlier was caught up in the Caledonian Orogeny. The distribution of these rocks suggests that much of N. Britain may be underlain by Lewisian basement, formed during the Scourian, reworked during the Laxfordian and possibly reworked again by later orogenic events. Marine sampling has shown that Lewisian rocks crop out over much of the continental shelf west and north of Scotland and on the northern part of the Rockall Bank (Fig. 1.5). Prior to the opening of the Atlantic, NW Scotland would have been continuous with SE Greenland. The Laxfordian is a continuation of the Proterozoic belts of Greenland (the Ketilidian and the Nagssugtoqidian), and the Scourian relics are equivalent to the large central Archaean core of S. Greenland known as the pre-Ketilidian massif (Fig. 1.2). The Laxfordian was roughly contemporaneous with the Hudsonian and Svecofennian Orogenies of the Canadian and Baltic Shields respectively. These episodes were all typical of Proterozoic ensialic mobile belts in which deformation, metamorphism and plutonism resulted from limited movements between crustal blocks, producing shear belts through which mantle heat and fluids were channelled.

2d Proterozoic basement in S. Britain

Compared with the extensive outcrops of Archaean and Proterozoic gneissic basement in the N. Atlantic Craton, of which NW Scotland is a part, our knowledge of the basement under S. Britain is only very sketchy. The Pentevrian gneisses of the Channel Isles, with a history stretching back into the Archaean (Ch. 1g), may have been remetamorphosed *c.* 1950 Ma ago, and similar gneisses to the south on the French mainland record thermal events *c.* 1100–900 Ma ago (Adams 1976).

The Rosslare Complex of SE Ireland (Fig. 1.5) may have an Archaean origin, although most of its development is of Proterozoic age (Max 1975). The oldest part of the complex consists of mica-rich metasedimentary and amphibolitic gneisses. These have been intruded by gabbro, granite and several generations of basic dykes and repeatedly deformed and metamorphosed to produce a gneiss complex similar in character to the Lewisian. Along the strike to the northeast, gneisses in the Mona Complex of Anglesey have been correlated with the Rosslare rocks although they may be much younger (Ch. 6c).

With the exception of the Channel Isles Pentevrian basement and the Rosslare Complex there are no exposures of basement in S. Britain that are definitely of comparable antiquity to the Lewisian of NW Scotland. Although there are several small exposures of metamorphic rocks in S. Wales and the English Midlands, they may all be younger than 1000 Ma (Chs 6d and 6e). There is a similar lack of Proterozoic basement in the other areas on the southern side of the Caledonian belt in W. Europe and the Avalon Peninsula of Newfoundland. The Baltic Shield is the nearest extensive tract of Proterozoic basement gneiss on the southern side of the Caledonian orogen.

2e Correlation and dating in Scotland and N. Ireland

The problems of Precambrian correlation, discussed in Chapter 2b, are nicely illustrated by the Proterozoic rocks of N. Britain. Two major dislocations, the Moine Thrust (a Caledonian structure that thrust the rocks on its southeastern side northwestwards during the Lower Palaeozoic) and the Great Glen Fault (a major transcurrent fault that has been active at least since the Devonian), divide the region into three areas, each with a rather different stratigraphic sequence (Fig. 2.5). A splay of the Great Glen Fault further divides the southern area, isolating yet another distinct sequence on the islands between Colonsay and Inishtrahull. As it is not possible to trace any rock units across these faults, the relationships between the rocks in differ-

ent areas have long been the subject of debate. Prior to the development of radiometric dating, correlations were proposed usually, and quite unjustifiably, on the basis of similarities in lithology and stratigraphic position only. Modern correlations rely heavily on radiometric dating, although detailed structural, geochemical and sedimentological studies can provide confirmatory evidence.

The last major event recorded in the Lewisian is the Laxfordian Orogeny which terminated *c.* 1700 Ma ago. Geochronological, geochemical and structural studies show that the Lewisian basement exposed in areas 2 and 3 of Figure 2.5 records a similar history and that the small outcrops in area 1 are probably of similar antiquity. There are no more outcrops of Lewisian to the southeast, but geophysical work suggests that a Lewisian-type basement underlies N. Britain, at least as far southeast as the Southern Uplands Fault.

The post-1700 Ma Proterozoic history of the N. Atlantic region outside the British Isles is dominated by two major orogenies: (a) one at *c.* 1100 Ma known as the Grenvillian in N. America and the Gothic in Scandinavia, and (b) the Caledonian-Appalachian Orogeny which began at the end of the Proterozoic and continued into the Palaeozoic. In Britain the Caledonian belt is well known and is the subject of Part 2 of this book. However, if the Grenville–Gothide belt is projected across Britain, it appears to lie completely within the area of the later Caledonides (Fig. 1.2). Therefore, overlying the Lewisian basement in N. Britain we may expect to find sedimentary rocks (a) deposited before *c.* 1000 Ma ago and subsequently deformed by both the Grenvillian and Caledonian Orogenies, (b) deposited after *c.* 1000 Ma ago and affected only by the Caledonian deformation, and (c) deposited during the 1700–600 Ma interval outside either orogenic belt and hence remaining undeformed. Although our understanding of the Proterozoic history of N. Britain is still based on many speculative correlations, a picture consistent with the history of other parts of the N. Atlantic region is emerging. At least some of the Moine northwest of the Great Glen Fault has suffered both Caledonian and pre-Caledonian deformation and may be the local representative of the Grenville belt. It is discussed in Chapter 2f. Rocks affected only by the Caledonian Orogeny, including the Dalradian and some of the

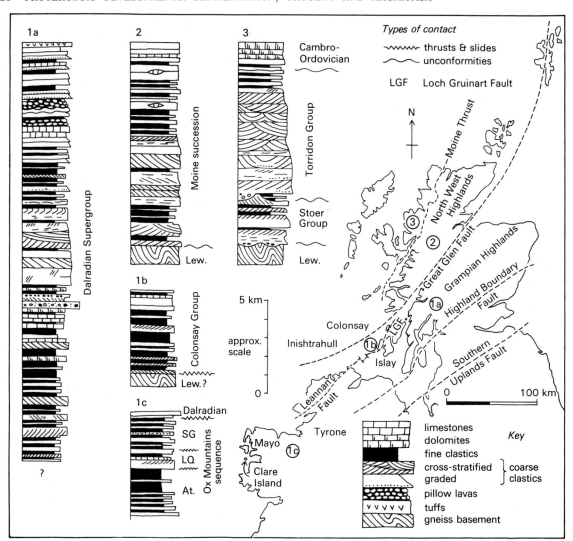

Figure 2.5 Proterozoic rocks of N. Britain. Logs show a typical example of the stratigraphic sequence to be found in each of the areas indicated on the map.
At. = Attymas Group; Lew. = Lewisian; LQ = Leckee Quartzitic Group; SG = Slieve Gamph Group.

'Moine-like' rocks of area 1 (Fig. 2.5), are included with the Caledonides (Ch. 4). The non-metamorphosed Torridonian, deposited outside either orogenic belt, is discussed here (Ch. 2g).

2f Sedimentation and orogeny: the Moine metasediments

The thick sequence of metasediments known as the Moine succession, lying between the Great Glen Fault and the Moine Thrust, was deposited uncon-

formably on Lewisian basement which is now preserved at the base of thrust slices and in fold cores. In the southwestern parts of the outcrop, around Morar where the intensity of deformation is least, the stratigraphic scheme summarised in Figure 2.6 has been worked out. The nature of the junctions between the major divisions of this scheme are unknown and may prove to be tectonic rather than sedimentary, although they have been traced almost to the northern coast of Scotland (Fig. 2.7). The metasediments were originally deposited

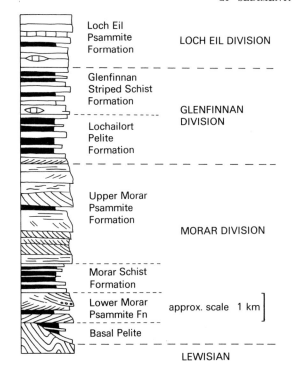

Figure 2.6 Stratigraphic log for the Moine of Morar. Note: the boundaries between the divisions may be tectonic rather than stratigraphic.

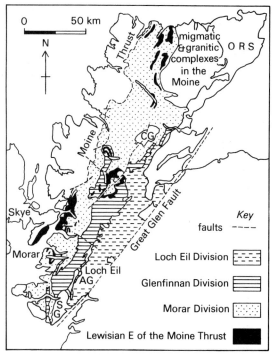

Figure 2.7 The Moine rocks of the NW Highlands of Scotland (after Johnstone 1975).
AG = Ardgour gneiss, a Grenvillian migmatite; SG and CG = Strontian Granite and Carn Chuinneag Granite, Caledonian 'Newer Granites'; ORS = Old Red Sandstone.

mainly as sands and muds. Sedimentary structures are locally preserved in the two thick psammite (i.e. metamorphosed sandstone) units of the Morar division. Palaeocurrents deduced from cross-bedding measurements show a flow towards the north and northeast at widely separated localities. Although no detailed sedimentological work has yet been published on these psammites, their great lateral persistence, the presence of broad shallow channels, wave ripples and heavy mineral bands and the absence of obvious cyclicity are consistent with a high-energy shallow marine shelf to beach environment of deposition. The pelite and striped schist formations were originally muds or thin sands interbedded with muds. They have been intensely deformed in most places, and sedimentary structures are rare. Their depositional environment can only be guessed at; a shelf origin is quite possible, but other possibilities cannot be excluded.

The deformation and metamorphism of the Moine has until recently been assumed to be of

Caledonian age, because the thrusts along its northwestern boundary cut Cambro-Ordovician rocks. However, an anomalous group of radiometric dates of $c.$ 740 Ma on pegmatites cutting the Morar Moines have for a long time suggested a much longer history for at least part of the Moine, a suggestion subsequently confirmed by Rb–Sr whole-rock isochrons on the Morar Schist Formation (Fig. 2.6) and the Ardgour Gneiss (a metasomatic derivative of the Moine metasediments) of 1024 ± 96 and 1050 ± 46 Ma respectively (Brook *et al.* 1977). These isochrons are thought to date the main (or M2) metamorphism of the area. Detailed work shows that in the southwestern part of the outcrop a low grade metamorphism (M1) and the first phase of deformation (D1), which formed major isoclinal folds with Lewisian cores (Fig. 2.8), some minor folds and schistosity (D. Powell 1974*), predate a later deformational phase (D2), which was contemporaneous with the main M2 amphibolite-facies (garnet grade) metamorphism. All these events are

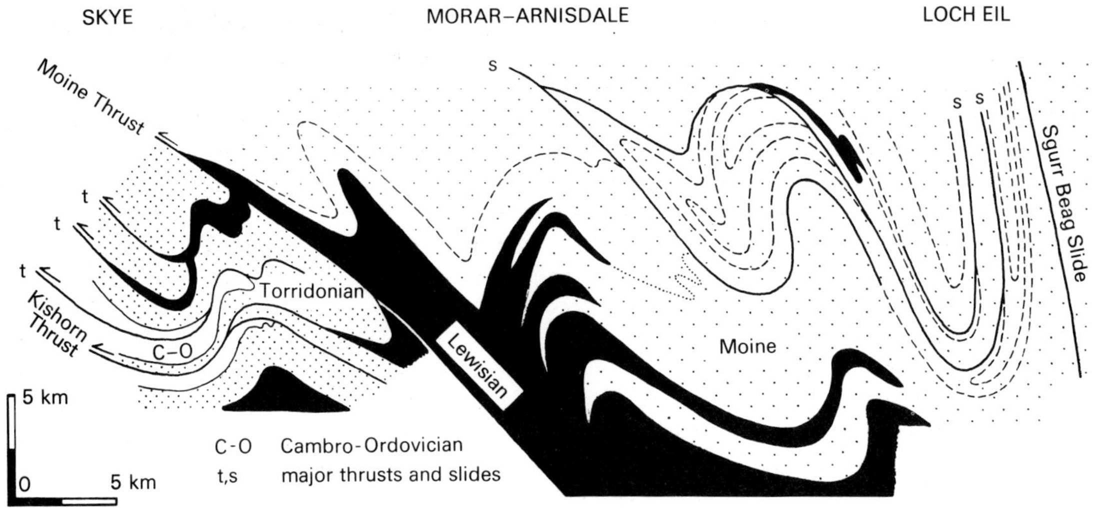

Figure 2.8 Structural section across the southwestern part of the Moine outcrop (see Fig. 2.7) (after D. Powell 1974*).

therefore pre-Caledonian, and the later D3 and D4 deformational phases and a mild retrogressive metamorphism M3 are of Caledonian age. Thus at least part of the Moine was deposited, deformed and metamorphosed before 1000 Ma ago and was subsequently caught up in the Caledonian orogen to be deformed and metamorphosed once again during the Lower Palaeozoic Caledonian Orogeny.

As the dates for the pre-Caledonian metamorphism and deformation of the Moine are very similar to those from the Grenville belt of N. America, and as evidence of Grenville rocks is expected in N. Britain, these Moinian events can be ascribed to the Grenvillian Orogeny. Although pre-Caledonian deformation has only been demonstrated in the rocks of the southwestern part of the Moine outcrop, often known as the Morar Moines, most of the Moine northwest of the Great Glen Fault may be of the same antiquity. Many dates of between 800 and 700 Ma have also been recorded from Moine pegmatites in the Morar area and attributed to a 'Morarian event'. It is uncertain at the moment whether this was an orogenic event or just a phase of pegmatite formation.

Southeast of the Great Glen Fault the Central Highland Granulites, previously included with the Moine, are a downward continuation of the late Precambrian to lower Ordovician Dalradian Supergroup and are therefore much younger than the

Moines to the northwest (Ch. 4b). The presumably unconformable junction between the Moine and Dalradian successions may lie in uninvestigated ground adjacent to the Great Glen Fault or may have been concealed by displacements along it. Indeed, as the detailed pre-Devonian history of this major transcurrent fault is not known, it cannot be assumed that the rocks now exposed on either side of it were deposited during the late Proterozoic in anything like their present relative positions (Ch. 3g).

Other pre-Dalradian rocks that show no evidence of a pre-Caledonian deformation but have in the past been called 'Moine' are found on Islay (the Bowmore Group) and in NW Mayo (the Erris Group). These are probably all much younger than the type Moine of the NW Highlands and are best considered together with the Dalradian (Ch. 4). The affinities of a strip of metamorphic rocks which crops out from Clare Island to Tyrone in Ireland and includes the Ox Mountain sequence and Deer Park Complex are also open to debate (Fig. 4.1). These include greenschist- to amphibolite-facies pelites and psammites (metamorphosed mudstones and sandstones respectively) with rare limestones and amphibolites, possibly of both volcanic and intrusive origin. Phillips et al. (1975) have suggested that some of these rocks have a pre-Caledonian history analogous to the Morar Moines, but Long and Max (1977) have correlated them with the

Dalradian and attribute all their deformation to the Caledonian.

In N. Britain, the Morar Moines northwest of the Great Glen Fault are the only rocks that have definitely suffered a post-Lewisian pre-Caledonian orogenic history. Although as yet there are few confirmatory radiometric dates, these rocks may prove to be parts of the extensive Grenville—Gothide orogenic belt caught up within the later Caledonides. What can be deduced about the tectonic environment of these rocks?

The style of deformation and grade of metamorphism produced by the Grenvillian event in Britain are similar to those seen in many younger orogenic belts but different from those of such areas as the Lewisian. Clearly the Moine rocks were deformed at a higher crustal level than the older gneiss complexes, but were they subjected to a new and different type of orogeny? If the Moine rocks are considered as constituents of the Grenville—Gothide orogen, then they form part of a linear belt. As will be seen in Part 2, late Precambrian to Lower Palaeozoic history can be explained in terms of processes related to the closure of an ocean basin, the Iapetus Ocean, which divided the British Isles (Fig. 3.5). It has been suggested that this ocean started to open much earlier in the Precambrian and that the Grenvillian Orogeny was produced by subduction along its northwestern margin (Wright 1976). Alternatively, the Grenville—Gothide belt might have been formed by continental collision resulting from the closure of an entirely earlier ocean (M. R. W. Johnson 1975). There is no direct evidence in this belt for the existence of ocean crust and subduction zones (cf. Ch. 3a), although, as palaeomagnetic work (J. D. A. Piper 1974*) suggests that the c. 1150 Ma orogenic belts, of which the Grenville—Gothide is a part, form a single broad arc across the Proterozoic supercontinent, a different tectonic regime from that responsible for the short, curved and irregularly disposed older belts is likely. The Grenvillian Orogeny therefore probably does represent a new type of tectonic regime different from that seen previously in the N. Atlantic region, although whether it resulted from the closure of a Grenville ocean, subduction along the northwestern side of the Iapetus Ocean or an incipient rifting that never developed into true sea-floor spreading, is not yet known.

2g Landscapes and sediments in the Torridonian of NW Scotland

Unconformably overlying the Lewisian basement northwest of the Moine Thrust (area 3 of Fig. 2.5) is a thick sequence of unmetamorphosed, predominantly fluvial, red-bed clastic sediments known collectively as the Torridonian, although a major unconformity within the sequence shows that these sediments form two quite distinct, if superficially similar, rock units now known as the Stoer and Torridon Groups (Stewart 1969). Rb—Sr whole-rock isochrons give their ages as 995 ± 24 and 810 ± 17 Ma respectively, so that they are significantly younger than the Morar Moines with which they were for a long time correlated.

The Stoer Group, occupying a relatively small area between Gairloch and Stoer (Fig. 2.9), com-

Figure 2.9 Distribution of the Torridonian in NW Scotland with palaeocurrents from the lowest few hundred metres of the Applecross Formation (after Stewart 1969, G. E. Williams 1969 and others). Note: the Cailleach Head Formation of the Torridon Group occurs only over a few square kilometres at Cailleach Head.

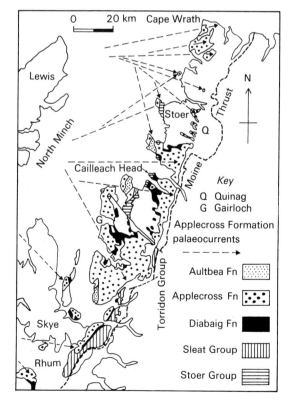

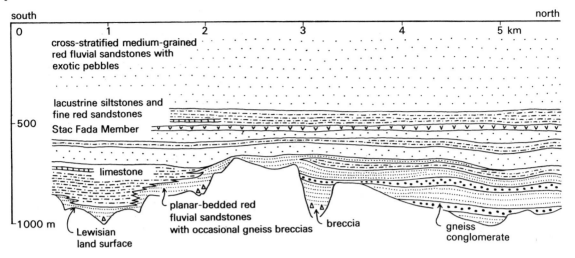

Figure 2.10 Stratigraphic profile of the Stoer Group across the peninsula at Stoer (See Fig. 2.9) (after Stewart 1969).

prises about 2 km of terrestrial clastics, which over-lie a hilly fossil landsurface, with a relief of up to 400 m, cut in the Lewisian gneiss. This surface was buried under locally eroded detritus, deposited in fluvial and lacustrine environments, which was sub-sequently blanketed by fluvial sandstones contain-ing farther-travelled sediment including quartzite pebbles unknown in the nearby basement gneisses (Fig. 2.10). Palaeocurrents in the latter sediments indicate flow from both the west and southeast (Stewart 1975*). An interesting volcanic mudflow or ashflow – the Stac Fada Member, containing pumice, shards and accretionary lapilli – provides a useful time plane which can be traced for 60 km, nearly the whole length of the Stoer Group outcrop. A hot arid climate is indicated by the general red colouration and by the alternation of fluvial with lacustrine sediments, containing desiccation cracks, ripple marks and algal limestones, which Stewart (1969) has interpreted as ephemeral playa-lake deposits. This climatic interpretation is consistent with a palaeomagnetically determined palaeolatitude for these rocks of 15°N.

The Stoer Group sediments were tilted by up to 30° to the northwest, presumably by fault move-ments, before the deposition of the overlying Torri-don Group nearly 200 Ma later. In the north the coarse fluvial clastics at the base of the Torridon Group (the Applecross Formation) directly overlie another fossil landscape. South of Quinag, however, this landscape is buried by the southward-

thickening Diabaig Formation deposited in both fluvial and marine or lacustrine environments. Further south still, the Diabaig Formation con-formably overlies the deltaic Sleat Group. If it is assumed that the lower and upper boundaries of the Diabaig Formation are time planes, it appears that there is a northward onlap reflecting more rapid subsidence in the south and a northward transgres-sion that reached as far as Quinag (Fig. 2.11). An alternative, if less likely, interpretation is that the Sleat Group, Diabaig Formation and Applecross Formation are laterally equivalent diachronous units lying above a roughly horizontal Lewisian surface.

Apart from some basal, locally derived breccia, the Sleat Group (up to 3.5 km thick on Skye) is composed mainly of cross-bedded grey sandstones and shales. Various marine to marginal-marine depositional environments have been suggested, and Stewart (1975*) favours a deltaic origin. The Diabaig Formation, which overlies the Sleat Group on Skye, directly overlies the Lewisian landscape further north. It includes locally derived breccias and conglomerates which pass outwards from the Lewisian valley sides and palaeohills into grey shales of marine or lacustrine origin. The palaeoen-vironment can be envisaged as a partially drowned hilly terrain in which screes and small alluvial fans on the hillslopes are building out into fairly shel-tered drowned valleys. Although the marginal marine or lacustrine sediments are usually fine

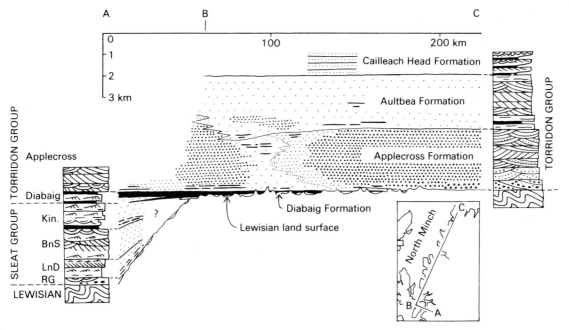

Figure 2.11 Stratigraphic profile, along the line shown on the inset map, and logs for the Sleat and Torridon Groups (after Stewart 1975* and others).
RG = Rubha Guaih Formation; LnD = Loch na Dal Formation; BnS = Beinn na Seamraig Formation; Kin. = Kinloch Formation.

grained, wave action was strong enough to cut a spectacular cliff, which is overlain by a boulder bed produced by the failure of this cliff, where the transgression intersected the (then) soft Stoer Group sandstones (Lawson 1976).

The palaeogeography of Diabaig times was abruptly terminated by the influx of vast amounts of coarse fluvial clastics of the Applecross and Aultbea Formations. The palaeocurrent pattern and sedimentary structures of the Applecross Formation (Fig. 2.9) suggest that it was deposited in braided streams on two large alluvial fans building southeastwards from a fault scarp in about the position of the present-day North Minch (G. E. Williams 1966). The varied pebble suite of the Applecross includes local acid gneiss, quartz–mica schist, pegmatite and quartzite clasts together with exotic clasts such as quartz–tourmaline rock, porphyritic volcanics and chert. Allen *et al.* (1974) have reviewed the source of the exotic clasts and suggested that it lay within the area between NW Scotland and SE Greenland. Radiometric dates of between *c.* 3000 and *c.* 800 Ma on Applecross clasts are consistent with such a source area composed of Scourian (pre-Ketilidian), Laxfordian (Ketilidian)

and possibly Grenvillian rocks. Near Cape Wrath the Lewisian directly beneath the Applecross Formation is weathered down to depths of 3–6 m and, as G. E. Williams (1969) has shown that this weathering predates the deposition of the Applecross, this is claimed to be the oldest soil profile recorded in Britain.

The Aultbea Formation, although finer grained, probably had a similar origin to the Applecross. The youngest Torridonian unit, the Cailleach Head Formation, consists of many coarsening-upwards cycles, each approximately 20 m thick. These may represent the progradation of fluvial sediments into a lake or sea.

The depositional environments of the Torridonian are better known than those of any other part of the British Precambrian. However, it is less clear how to fit these well-dated slices of sedimentological interpretation into the regional geological history of the area. As the Stoer Group, dated at *c.* 995 Ma, is only slightly younger than the *c.* 1050 Ma date for the metamorphism of the Morar Moines (Ch. 2f), it is tempting to suggest that the exotic sediments of the Stoer Group were derived from the erosion of a Grenvillian mountain chain, i.e. that they were a local Grenvillian molasse. The palaeocurrents suggest that the Stoer Group was deposited in a fairly small fault-bounded basin. If the molasse interpretation is correct, this basin was tiny, as molasse basins go, and could have been part of a more extensive deposit.

Further fault movement before deposition of the Torridon Group is suggested by the tilting of the Stoer Group sediments. The sub-Torridon Group landscape was cut before Torridon Group times and, as it declines in relief northwestwards, the eroded stump of a mountain chain might still have existed to the southeast. Subsidence then brought in a marine transgression from the south which completely drowned the landscape south of Loch Torridon but probably left the hills to the north standing as islands. The Sleat Group and possibly the Diabaig Formation prograded into this sea, depositing locally derived sediment. Major uplift of a mature quartzo-feldspathic source area along faults to the northwest then completely buried this varied palaeogeography under a vast pile of alluvial fan sediments to form the Applecross and Aultbea Formations. Clearly, this major uplift had a much more dramatic effect than the preceding subsidence. Large alluvial fans, which nearly always build out from fault scarps, are often found in molasse basins, i.e. basins filled predominantly by fluvial sediments shed from young fold mountains undergoing late orogenic uplift (cf. the Old Red Sandstone of the Midland Valley; Ch. 9). However, there is no evidence for a young fold-mountain belt northwest of Scotland at this time. Alluvial fans are also found during the early stages of continental rifting prior to, and during the formation of, new ocean crust. During the initial updoming, fluvial palaeocurrents flow away from the rift centre although later they flow towards the new continental margin. The Torridon Group could fit into a post-rifting tectonic framework, suggesting that the Iapetus Ocean was initiated before c. 810 Ma ago, but this still would not explain the major uplift to the northwest, the most important feature of Torridon Group times.

The red-bed fluvial clastics of the Stoer and Torridon Groups show many similarities to the fluvial facies of the Old Red Sandstone and Permo-Triassic in the British Isles. The sedimentary and climatic environments were broadly similar in each case, and as a result there has until recently been some doubt regarding to which of these units some of the red sandstone outliers in the Highlands should be allocated.

2h Summary

The Proterozoic was a time of less rapid geological evolution than the Archaean. It saw a change in the style of tectonism, from that of limited movement between small crustal blocks producing ensialic mobile belts in which pre-existing gneisses were reworked, to that of continental rifting, the formation of new oceans and subduction and collision tectonics producing linear orogenic belts. By the middle Proterozoic, compositional changes in the atmosphere and hydrosphere had led to the deposition of sedimentary rocks similar in most aspects to those of today and, by the upper Proterozoic, the higher forms of life were evolving. In the British Proterozoic, tectonic reworking, together with igneous and sedimentary processes, is well exhibited by the Lewisian rocks of Scotland, which are overlain by upper Proterozoic sediments that have either escaped metamorphism (the Torridonian) or been involved in later orogenies, possibly related to ocean floor spreading processes (the Moine). Lower Proterozoic rocks are largely unknown in S. Britain, although they may be buried under the cover of upper Proterozoic sediments and volcanics that crop out as small isolated inliers in England and Wales.

Part Two

THE IAPETUS OCEAN AND
THE BUILDING OF THE CALEDONIDES

Theme

The Caledonides is the name given to the mountain chain formed during the Lower Palaeozoic Caledonian Orogeny by the deformation of a belt of late Precambrian to Lower Palaeozoic rocks that extended from Scandinavia through the British Isles into N. America. By comparison with the earlier parts of British geological history we have a detailed knowledge of the events leading up to the formation of the Caledonides, not only because the stratigraphic record is nearly continuous from late Precambrian times onwards, but also because after the appearance of shelly fauna early in the Cambrian our ability to date and correlate events is substantially improved. It was another aspect of this improved fossil evidence, the recognition of distinct Lower Palaeozoic faunal provinces, that led to the initial suggestion that an ocean basin formerly existed within the Caledonian belt and that Caledonian evolution could be considered in terms of plate tectonics. The view that a Caledonian ocean, now known as the Iapetus Ocean, opened sometime in the late Precambrian and gradually closed during the Lower Palaeozoic is now widely held. The patterns of sedimentation, volcanism, plutonism, metamorphism and deformation are compatible with this hypothesis, and recently, palaeomagnetic studies have provided a significant measure of support.

Part 2 examines the growth and decay of the hypothetical Iapetus Ocean. After discussing the geological evidence for its existence (Ch. 3), this part traces the evolution of the two opposing continental margins. The two margins were subjected to different phases of deformation which are here treated as stages of the Caledonian Orogeny. Chapters 4 and 5 consider the 'northern' margin which was subjected to the early Ordovician Grampian Orogeny, and Chapters 6 and 7 turn to the 'southern' margin which was the site of the late Precambrian Celtic (or Cadomian) Orogeny. Finally, Chapter 8 discusses the closure of the postulated ocean as reflected by the patterns of faunal migration, deformation and plutonism.

Plate 2   A view of the cliffs at Clarach Bay, near Aberystwyth. The interbedded greywackes and shales in the picture are known as the Aberystwyth Grits, a lower Silurian turbidite formation. The sands were emplaced by periodic density currents which flowed northeastwards along the floor of the contemporary Welsh Basin.

3 The Nature of the Evidence

3a Introduction

The British Isles straddle the Caledonian orogenic belt. This is a long, narrow and roughly linear belt composed of dominantly marine, late Precambrian to Lower Palaeozoic sediments and volcanics that suffered metamorphism, plutonism and deformation at various times until the end of the Lower Palaeozoic, when they were uplifted to form the Caledonian mountain chain (Fig. 3.1, 3.2).

On a pre-Permian map of the N. Atlantic region the Caledonian belt stretches from Spitsbergen, through Greenland, Scandinavia, the British Isles, Newfoundland and into the N. Appalachians (Fig. 1.2). Thickness variations across the belt (Fig. 3.1) are typical of orogens. A very thick, deformed and locally metamorphosed succession of sediments and volcanics within the belt is flanked to the northwest

and southeast by thin, undeformed and unmetamorphosed sediments resting on a stable basement or foreland (stratigraphic framework, Table 3.1). However, the Caledonian belt cannot be thought of as the product of a simple symmetrical geosyncline. Different parts of the belt have evolved in very different ways, as is illustrated by its history of deformation. There were three main episodes of Caledonian deformation: (1) the Celtic at the end of the Precambrian, (2) the Grampian during the lower Ordovician, and (3) a whole series of minor events lasting from lower Ordovician to lower Devonian times. The end-Precambrian Celtic episode affected a small area of rocks in N. Wales, SE Ireland and the Channel Isles (Fig. 3.3) and was accompanied by plutonism and low- to high-grade metamorphism. In N. Ireland and Scotland, however, sedimentation continued without inter-

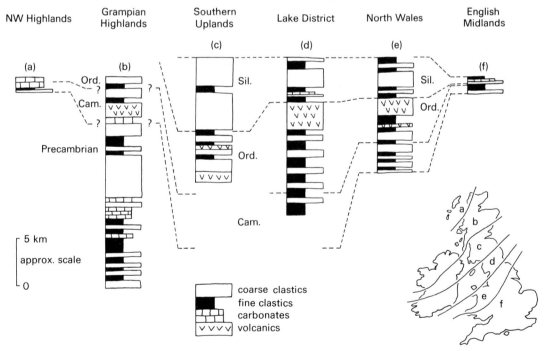

Figure 3.1 Stratigraphic logs showing the variation in thickness and stratigraphic range of the late Precambrian to Lower Palaeozoic rocks across the Caledonian belt. Areas (a) and (f) on the inset map constitute the northwestern and southeastern forelands of the Caledonian belt, which can itself be divided into the four major areas (b) to (e). Cam. = Cambrian; Ord. = Ordovician; Sil. = Silurian.

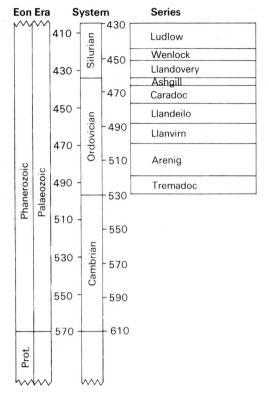

Eon	Era	System		Series
		Silurian	410 / 430	Ludlow
			450	Wenlock
			430	Llandovery
			470	Ashgill
		Ordovician		Caradoc
			450	Llandeilo
			490	Llanvirn
			470 / 510	Arenig
			490 / 530	Tremadoc
Phanerozoic	Palaeozoic	Cambrian	510 / 550, 530 / 570, 550 / 590, 570 / 610	
Prot.				

Figure 3.2 Stratigraphic terminology for the Lower Palaeozoic. The timescales are in millions of years (Ma) and are those suggested by R. St J. Lambert (1971) using Rb–Sr decay constants of 1·47 and 1·39 × 10⁻¹¹ years⁻¹ for the left and right scales respectively. More recent work suggests that the true dates of the system boundaries lie somewhere between these two scales.
Prot. = Proterozoic.

ruption during the late Precambrian and Cambrian, and in these areas the main deformation, the Grampian, is of lower Ordovician age and the metamorphism is of low- to high-grade. The lower Ordovician to lower Devonian deformation did not affect these latter areas but did affect the central and southern parts of the belt, partly overlapping with the areas affected by the end-Precambrian episode. This last episode was of uniformly low metamorphic grade. The end result of all these episodes is a markedly asymmetrical orogenic belt; low- to high-grade metamorphic rocks produced by the Grampian Orogeny are found in the northwest, but in the southeast, with the exception of the small areas exhibiting Celtic metamorphism, only very low grade rocks are found. The high-grade northwestern and low-grade southeastern parts of the orogen

are often referred to as the orthotectonic and paratectonic Caledonides respectively.

There are other major differences between different parts of the Caledonian belt. The southern part (i.e. the Lake District, Wales and southern Ireland) experienced widespread volcanism during the Ordovician but has relatively few exposed plutons. Scotland and northern Ireland exhibit little Lower Palaeozoic volcanism but are riddled with major intrusions (Fig. 3.3). The general impression then is that the various parts of the Caledonian belt have evolved in different ways, in spite of their present close proximity.

3b Plate-tectonic models

Following an idea by J. T. Wilson (1966) that the Caledonian belt was the site of an ocean that opened and then closed, Dewey (1969a*) proposed a model involving ocean floor spreading and the formation of three subduction zones to explain Caledonian sedimentation, volcanism, deformation and plutonic activity. This model was only the first of many. To date no less than nine subduction zones in different locations with different orientations have been proposed by various authors. Clearly, although there is general agreement that plate tectonics should be applicable to the Caledonian belt, considerable differences of opinion exist regarding the details.

There are three main types of evidence relevant to the interpretation of orogens as the products of plate-tectonic mechanisms: (a) the presence of ophiolites as remnants of consumed ocean floor, (b) the presence of features related to subduction zones (e.g. trench sediments, island arc volcanics), and (c) evidence of the former separation of continental plates (e.g. faunal provinces, palaeomagnetism). Let us now examine this evidence and then construct a simple model to serve as a framework for considering the evolution of the Caledonian belt. The same types of evidence are also relevant to the discussion of any other orogen, such as the Hercynian (Ch. 12), in terms of ocean floor spreading.

3c Ophiolites

Ophiolite sequences include ultrabasic rocks (often

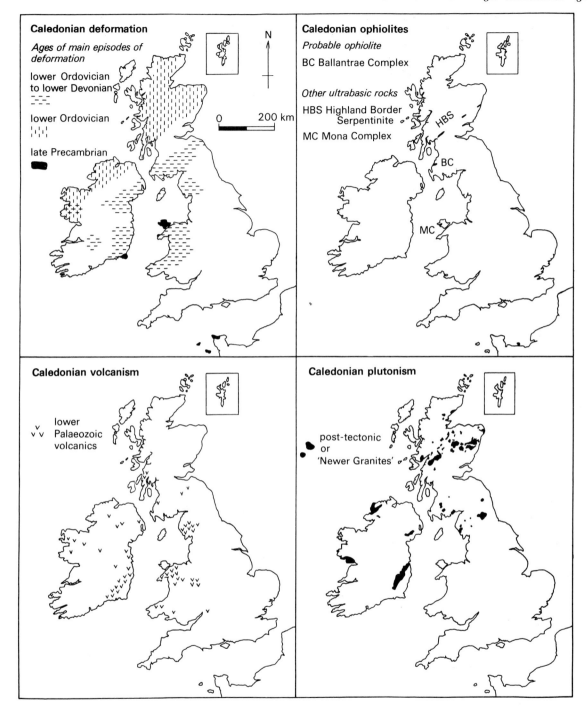

Figure 3.3 Distribution of Caledonian deformational episodes, ophiolites, volcanics and acid plutonics.

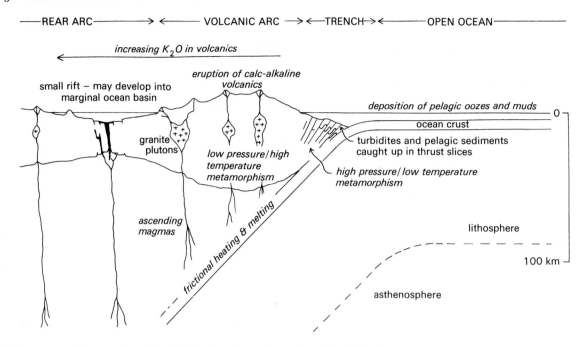

Figure 3.4   Cross-section of an idealised continental margin with subduction zone.

serpentinised), basic dyke and pillow lava complexes and cherts, commonly thought to represent slices of oceanic crust and mantle caught up within or between continental rocks during subduction or continental collision. Such sequences can mark the suture between two continental plates that have collided. However, oceanic crust can also be thrust, or obducted, on to a continental plate and emplaced some distance away from the suture, or be formed in small rifts that close without ever developing into true ocean basins (Fig. 3.4).

Several areas and linear zones of basic and ultrabasic rocks in the British Isles have been considered by various authors to be ophiolites (Fig. 3.3). In most cases good evidence for an ocean crust origin is weak or lacking. A zone of basic and ultrabasic rocks within the Moine of Sutherland and the Dalradian (?lower Cambrian), minor intrusions and volcanics which stretch from Perthshire to Mayo (Ch. 4e) are quite consistent with a continental origin. The Highland Border Serpentinite, a fault-bounded mass of serpentinised basic rock, is more difficult to interpret. Although it lies adjacent to lower Ordovician cherts and volcanics, the relationship between these rocks is unclear, and an

ocean floor origin seems unlikely (Ch. 4e). In Anglesey, the late Precambrian Mona Complex contains ultrabasic intrusions, pillow lavas and cherts. These rocks are much more like an ophiolite sequence than the previous examples. However, Maltman (1975) has shown that the ultrabasic rocks were not emplaced tectonically but as hot magmas, and again an ocean floor origin seems unlikely.

The most convincing 'ophiolites' in the British Isles are found in the Arenig (lower Ordovician) Ballantrae Complex in the Southern Uplands. This includes basic dykes, pillow lavas, tuffs and radiolarian cherts with partly serpentinised basic intrusives. Although the Ballantrae Complex may be of oceanic origin it is most unlikely that it is the exposed cap of a large mass of ocean crust concealed beneath the Southern Uplands or that it marks the position of a major suture. Rather, as Church and Gayer (1973) and Dewey (1974) have shown, either the Ballantrae sequence is an oceanic slice tectonically emplaced on to continental crust by obduction, or it represents the ocean floor of a small rear-arc (marginal) sea. If its interpretation as an ophiolite sequence is correct, it demonstrates the existence of ocean crust in the vicinity during the lower

SYSTEM	SERIES	Stage	Biozone	SYSTEM	SERIES	Stage	Biozone
ORDOVICIAN	LLANDEILO	Upper	Nemagraptus gracilis	SILURIAN	LUDLOW	post-Ludlow	facies change – no further graptolites in Britain
		Middle	Nemagraptus gracilis			Whitcliffian	M. leintwardinensis
		Lower	Glyptograptus teretiusculus			Leintwardinian	M. tumescens/incipiens
	LLANVIRN	Upper	Didymograptus murchisoni			Bringewoodian	M. scanicus
		Lower	Didymograptus bifidus			Eltonian	M. nilssoni
	ARENIG (C)	Upper	Didymograptus hirundo		WENLOCK	Homerian	M. ludensis / Gothograptus nassa / C. lundgreni
		Lower	Didymograptus extensus			Sheinwoodian	C. ellesae / C. linnarssoni / C. rigidus / M. riccartonensis / C. murchisoni / C. centrifugus
	TREMADOC (B, A)		Angelina sedgwickii / Shumardia pusilla interzone / Clonograptus tenellus / Dictyonema flabelliforme		LLANDOVERY	Telychian	M. crenulata / M. griestonensis / M. crispus
CAMBRIAN	'MERIONETH'		Acerocare / Peltura / Leptoplastus / Parabolina spinulosa interzone / Olenus / Agnostus pisiformis			Fronian	M. turriculatus / M. sedgwickii
						Idwian	M. convolutus / M. gregarius / M. cyphus
	'ST DAVIDS'		Paradoxides forchammeri / Paradoxides paradoxissimus / Paradoxides oelandicus			Rhuddanian	M. vesiculosus (= atavus) / M. acuminatus / M. persculptus
	'COMLEY'		Protolenid–Strenuellid / Olenellid / non-trilobite zone	ORDOVICIAN	ASHGILL	Hirnantian	Dicellograptus anceps
						Rawtheyan	
						Cautleyan	Dicellograptus complanatus
						Pusgillian	
					CARADOC	Onnian	Pleurograptus linearis
						Actonian	Dicranograptus clingani
						Marshbrookian	
						Longvillian	
						Soudleyan	Climacograptus wilsoni
						Harnagian	Climacograptus peltifer
						Costonian	

Table 3.1 The Lower Palaeozoic stratigraphic framework for the Caledonian belt. Correlation is generally based on the recognition of biozones (although there are now chronozones for the Wenlock; Bassett *et al.* 1975). The Cambrian biozones are based on trilobites (Cowie *et al.* 1972) and those of the Ordovician and Silurian are based on graptolites (A. Williams *et al.* 1972, Cocks *et al.* 1971). Important zonal schemes in preparation are based on conodonts and acritarchs. Note: the position of the Cambrian/Ordovician boundary is not internationally agreed (possible positions are at A, B or C; Henningsmoen 1973).
C − *Cyrtograptus*; *M* − *Monograptus*.

Ordovician but does not fix the position of the suture.

Within the Central Mobile Belt of Newfoundland there are several larger and better exposed areas of ophiolites, mostly of Ordovician age, including the ultrabasic rocks, sheeted dykes, pillow lavas, cherts and volcanic sediments of the Betts Cove Complex. To the northwest of these, on the Western Platform, a thrust sheet of ophiolites lies on Ordovician shelf sediments. This was probably obducted westwards during the middle Ordovician. The Lower Palaeozoic palaeogeography of Newfoundland is difficult to deduce because the area has been dislocated by numerous faults and thrusts. Even so there is little doubt that oceanic crust existed within the area of central Newfoundland, at least during the Ordovician.

3d Faunal provinces

Different faunas inhabit different regions of the world. These regions, termed faunal provinces, are separated by 'barriers' to the migration of the organ-isms. For terrestrial animals these 'barriers' may be wide oceans or mountain ranges, but in the case of marine fauna they are likely to be more subtle, such as changes in water temperature and salinity and the presence of ocean currents. The recognition of faunal provinces in the geological record can therefore place some restrictions on our interpretation of palaeogeography.

Some of the Lower Palaeozoic faunas in the British Isles can be divided into two provinces: the Atlantic and Pacific faunal provinces. The Atlantic faunal province includes fauna from the southern part of the British Isles together with other parts of Europe, the Avalon Platform of Newfoundland, Nova Scotia, New Brunswick and E. Massachusetts. The Pacific province includes fauna from N. Britain and most of N. America (Fig. 3.5). For example, Atlantic-province Cambrian trilobites are found in Wales, and Pacific-province trilobites occur in NW Scotland. Lower Ordovician trilobites and brachiopods fall into these provinces too. A. Williams (1969) has compared the Ordovician faunas of the two provinces and found that they were most dissimilar during the Llanvirn and Llan-

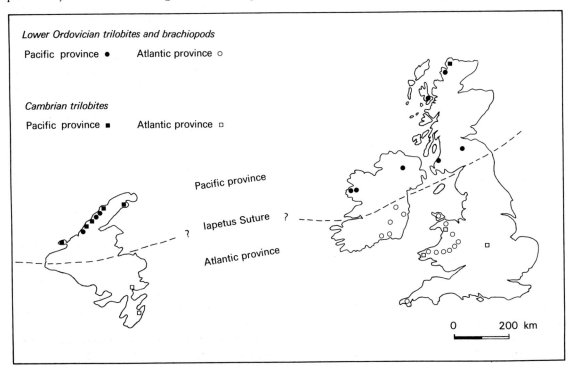

Figure 3.5 Cambrian and Ordovician faunal provinces in Newfoundland and the British Isles (after A. Williams 1969 and Palmer 1969).

deilo (Fig. 3.2); that is, the 'barrier' was most effective at this time. Although many possible barriers to the migration of benthos can be imagined, pelagic organisms or organisms with a planktonic larval stage should suffer fewer restrictions. It is therefore surprising that Lower Ordovician graptolites also define two provinces, the Atlantic faunas being found as far northwest as the Lake District and Co. Meath, and the Pacific faunas as far southeast as the Southern Uplands and Co. Mayo.

The available data (Fig. 3.5) show that there is a boundary between the two faunal provinces running roughly along the centre of the Caledonian belt. This boundary can be delimited in the Cambrian and is most clearly defined in the Ordovician, but by Silurian times the faunas had become intermingled. Although there are several possible reasons for the existence and then disappearance of the Atlantic and Pacific provinces, a change in the width of an intervening ocean offers a plausible solution. Indeed, McKerrow and Cocks (1976*) have suggested that the order in which animals appear to have migrated from one province to the other is that which would be expected if the provinces were separated by a closing ocean. Common species of pelagic graptolites appeared in both provinces first, followed by animals with planktonic larval stages, then benthos and finally freshwater fish (Fig. 3.6).

3e Sedimentary facies

There are no sediments that are diagnostic of an ocean floor environment. The fine clays, siliceous and calcareous oozes that are typical of ocean floors are also found in sediment-starved areas on continental crust. However, if these facies are found associated with basic and ultrabasic rocks of ocean floor type in orogenic belts, they provide good evidence for the former existence of oceanic crust. Similarly, the submarine fan and related facies of continental trench and slope environments are not particularly diagnostic, similar sediments being found in deep shelf basins. However, what is diagnostic of trench environments is the sequence of sediment slices produced by continuous underthrusting of the inner wall of a trench by a descending oceanic plate (Fig. 3.4). In each slice, submarine fan facies overlie abyssal plain sediments, and the age of the slices

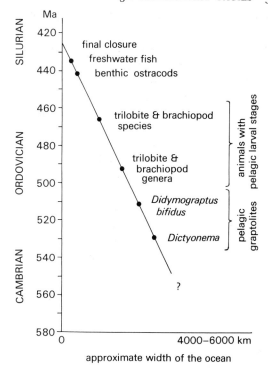

Figure 3.6 The times by which the migration of various groups of animals across the closing Iapetus Ocean produced a fauna common to both sides (after McKerrow & Cocks 1976*).

increases upwards. A good example of this type of sequence is found in the Ordovician and Silurian rocks of the Southern Uplands and the Longford Down Massif, where the facies relationships and the pattern of deformation suggest that at this time the area was underlain by a NW-dipping subduction zone (Ch. 5). Other very thick turbidite sequences, such as those found in the Dalradian Supergroup (Ch. 4e) and Lower Palaeozoic Welsh Basin (Ch. 7), lack features that can be related to subduction zones and are best interpreted as the deposits of deep shelf basins.

The presence of an ocean basin has one sedimentological effect that is recognisable on the continental shelves: namely, the deposition of sediments from strong tidal currents. Since the energy of tides is rapidly dissipated as they cross shallow shelves, strong tides are only likely to be found on shelves adjacent to an open ocean. The presence of a nearby ocean basin is therefore consistent with the record of tidal shelf sediments in the late Precambrian and Cambrian rocks of NW Britain (Chs. 4d and 4e).

3f Volcanism, plutonism, deformation and metamorphism

There is a systematic change in the chemical composition of the igneous rocks across many modern orogenic belts located above subduction zones. This has been attributed to the variations in depth of origin of the magmas (i.e. depth to the subduction zone) and may be manifested by a change from tholeiitic rocks on the oceanward side of the belt to calc-alkaline and then alkaline rocks on the landward side (Fig. 3.4). Such a change, from tholeiitic in the north to alkaline in the south, has been demonstrated by Fitton and Hughes (1970*) for the Ordovician volcanics of the Lake District and Wales. This implies the existence of a S-dipping subduction zone under these areas during the Ordovician. A similarly oriented subduction zone under SE Newfoundland has been suggested also on geochemical grounds.

Other horizons of volcanic rocks within the Caledonian belt are not developed over sufficiently large areas to show up lateral variations. However, Phillips *et al.* (1976*) have suggested that an area of acid and basic volcanics extending from Tyrone to Mayo (including the Tyrone Igneous Group) may have formed part of a lower Ordovician island arc (Fig. 3.3) that developed over a NW-dipping subduction zone.

The history of metamorphism and deformation has been outlined briefly above (Ch. 3d). One characteristic of orogenesis produced by subduction zones is the presence of paired metamorphic belts: (a) an oceanward zone of high pressure and low temperature metamorphism typified by glaucophane schists, and (b) a landward belt of low pressure and high temperature metamorphism characterised by greenschist and amphibolite facies rocks. Both the end-Precambrian Celtic and lower Ordovician Grampian episodes are dominated by the latter, although in Anglesey (in the Mona Complex) and in the Ballantrae Complex there are small patches of glaucophane schists. The presence of these schists has been cited (Dewey 1969a*) as evidence for the existence of paired metamorphic belts, and hence subduction zones, in the Caledonides. However, both occurrences are very restricted and may reflect local untypical conditions rather than the former existence of extensive glaucophane schist belts.

Many different patterns of deformation can be produced above subduction zones. It is therefore difficult to prove the presence of the latter on the evidence of the former. However, one common pattern in modern orogenic belts is that of thrust faults parallel to the subduction zone on the inner wall of the trench (Fig. 3.4). This has close parallels with the Lower Palaeozoic rocks of the Southern Uplands where the presence of thrust faults, which dipped northwestwards prior to their rotation by later phases of deformation, suggests that a NW-dipping subduction zone lay within or to the south of the area sometime during the Lower Palaeozoic (Ch. 5i, Fig. 5.13).

3g Palaeomagnetism

Many of the types of evidence discussed above can be ambiguous; they are open to non-plate-tectonic interpretations. However, palaeomagnetic pole positions can show whether or not there has been any relative motion between two areas, although it cannot prove that any motion has necessarily been due to ocean floor spreading.

To date, palaeomagnetic results from Ordovician and Silurian rocks in the Grampian Highlands, the Southern Uplands, the Lake District, Wales and N. and S. Ireland have been unable to demonstrate unequivocally the presence of a major ocean within the Caledonian belt (Fig. 3.7; Briden *et al.* 1973). However, the palaeomagnetic poles for the lower Ordovician tend to separate out into a northern and a southern group. J. D. A. Piper (1978) has argued that this result supports the thesis that there was a crustal separation (a Caledonian ocean) of *c.* 1000 km between the Southern Uplands and the northern Lake District in lower Ordovician times. It is possible that even at its maximum extension the Caledonian ocean was never very wide.

Morris (1976) has shown that, although the palaeomagnetic poles from localities throughout the Caledonian belt in Britain are similar, they do not match those from N. American rocks. This discrepancy could be due to the areas having suffered differing deformational histories. It can also be resolved by postulating major sinistral movement along a SW−NE-trending line in the northwestern part of the belt during the Devonian, possibly along

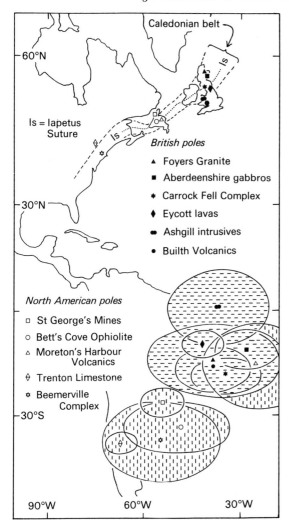

Figure 3.7 Ordovician palaeomagnetic pole positions, with 95 per cent confidence limits, for British and N. American localities on a Permian predrift reconstruction, relative to N. America. Mercator projection (after Morris 1976).

the line of the present Great Glen Fault. Palaeomagnetism therefore provides more concrete evidence of transcurrent movement than of the opening and closing of an ocean within the Caledonian belt.

3h   Summary: a simple model for the evolution of the Caledonides

The rather fragmentary evidence for a Caledonian ocean and its associated subduction zones in the British Isles is admissive rather than compelling. Nonetheless, the plate-tectonic hypothesis is attractive in that it provides a coherent framework within which to view Caledonian evolution. Evidence from other parts of the Caledonian belt, especially in Newfoundland, tends to strengthen the hypothesis of a Caledonian ocean, christened Iapetus by Harland and Gayer (1972), although the detailed evolution of the belt is still poorly understood.

If we accept a plate-tectonic hypothesis for the Caledonides, three fundamental questions are raised: (a) when did rifting and the initiation of ocean floor spreading take place, (b) when did the two continental plates finally collide, and (c) where was the suture between these two plates? There is no positive evidence bearing on the first problem. Many authors have assumed that the initial continental rifting took place sometime during the late Precambrian, presumably before the start 700–800 Ma ago, of the marine sedimentation recorded in the Dalradian. The Torridon Group clastics could then date from this rifting episode. On the other hand, rifting might have preceded the deposition of the Moine. Thus the Grenvillian events (c. 1050 Ma ago) were possibly related to subduction along the northwestern margin of the same ocean that divided Britain in the Lower Palaeozoic (Ch. 2f). Either way, the ocean must have been in existence for several hundred million years, a long time by comparison with the present Atlantic, although it was not necessarily ever very wide.

Continental collision can be more firmly dated. Upper Ordovician to lower Devonian deformation can be reasonably attributed to such a collision. The low grade metamorphism and upright folds, e.g. in the Welsh Basin and Lake District, suggest a vice-like compression across the orogen. The collision must have been fairly gentle because it did not produce the towering mountain ranges and gigantic nappe structures typical of the Alpine–Himalayan collision orogens. In fact the collision zone soon started to subside, because the Northumberland Basin, which subsequently developed over the area of the suture in N. England, was being filled with fluvial sediments by upper Old Red Sandstone times.

Where was the suture? It is not obvious (see, for example, the contrasting interpretations of Dewey 1969a*, Gunn 1973 and Phillips *et al.* 1976*). The suture can best be located by a process of elimination. Faunal evidence suggests that it lies north of the Lake District and Co. Meath and south of the Southern Uplands and Co. Mayo. Ordovician volcanism implies that it lies north of the Lake District. Geophysical data together with the presence of Lewisian-type xenoliths in a Carboniferous vent near the Southern Uplands Fault suggest that the Midland Valley and perhaps the Southern Uplands are underlain by normal continental crust, as is most of N. England (D. W. Powell 1971, Upton *et al.* 1976, Bamford *et al.* 1977). It seems that the Solway Firth area is about the only place where the suture could have been, and such an interpretation is consistent with the pattern of Lower Palaeozoic sedimentation in the area. The suture can also be traced across Ireland to the Shannon Estuary (Fig. 3.5).

It is concluded that the Iapetus Ocean was initiated in late Precambrian times, and that the two continents thus formed separated and then collided at the end of the Lower Palaeozoic along a line that can be drawn across the British Isles through the Solway Firth and Shannon Estuary. The main features of the Caledonian belt can be considered in terms of this very simple model. In the following chapters the histories of the two continental margins will be considered separately until the time of their final collision. The end-Precambrian deformation episode, peculiar to the southern margin, is used to divide its history into an early and a late phase. Similarly, the lower Ordovician Grampian orogeny is used to divide the evolution of the northern margin into two. The possibility that plate-tectonic processes, such as the formation of marginal ocean basins and subduction zones, occurred within the two continental plates, as defined above, will be discussed in the relevant chapters. The detailed history of the two continental margins is less clear than the gross evolution of the Caledonian belt as outlined in this simple model.

4 Late Precambrian to Ordovician evolution of the northern continental margin of Iapetus

4a Introduction

The late Precambrian to early Ordovician history of the northwestern continental margin is recorded in two separate well-exposed rock sequences in Scotland and Ireland. In the NW Highlands of Scotland a sequence of Cambrian and Ordovician shelf sandstones and limestones, unconformably overlying Precambrian basement, can be correlated with very similar sequences in Greenland and Newfoundland. To the southeast stretching from the Shetland Isles to Connemara, is a belt of deformed and metamorphosed late Precambrian to early Ordovician shallow- and deep-water sediments and volcanics, known as the Dalradian Supergroup (Fig. 4.1). This is correlated with the Fleur de Lys Supergroup of Newfoundland and broadly similar rocks in Greenland. The Cambro-Ordovician and Dalradian outcrops, separated by only 70 km in Scotland, probably originally formed part of two concentrically arranged subparallel belts running for over 2500 km around the margin of an ancient continent, with the shelf Cambro-Ordovician lying on the landward side of the mixed 'geosynclinal' assemblage of Dalradian lithologies. However, there is little direct evidence within these rocks for the existence of an oceanic environment to the southeast. Probably all the Dalradian and Cambro-Ordovician rocks now exposed were deposited on continental crust, and the edge of the continent must have lain southeast of their present outcrop from late Precambrian until Ordovician times.

Since the Cambrian sits unconformably on the Torridon Group in the NW Highlands, there is a gap in the record from c. 810 to c. 570 Ma ago. The base of the Dalradian is not visible in Scotland, although it may overlie Moine rocks in Ireland. Although the date of initiation of Dalradian sedimentation can only be guessed, it is unlikely to

be younger than 700 Ma and may be as old as 800 Ma ago. As there was continuous marine sedimentation throughout Dalradian times, continental break-up must have preceded this date, although we do not know exactly by how long (Ch. 3h).

4b The stable shelf: the Dalradian Grampian and Appin Groups

The oldest parts of the Dalradian Supergroup are generally found on the northwestern side of the outcrop with the younger rocks towards the southeast (Fig. 4.1). Although, in detail, the Dalradian outcrop pattern is controlled by the presence of large upright to overturned folds, further complicated by tectonic slides, this general younging southeastwards may reflect an original depositional offlap. The total thickness of the Dalradian formations is over 20 km, but it is unlikely that a vertical thickness of this order was ever deposited at one place. It is more likely that the depocentre migrated southeastwards through time.

In spite of its complex structure the stratigraphy of the Dalradian is quite well known, although many correlation problems remain. Harris and Pitcher's (1975*) review has shown that several formations can be traced from Banff to Connemara, although the correlation with Shetland is rather tenuous, and that the geological evolution of all parts of the outcrop was broadly similar. As the central part of the outcrop, in Argyll and Donegal, exhibits the lowest grade of metamorphism and deformation, the following discussion concentrates on this area. Although regional facies changes are recognised, especially in the Argyll Group where the Donegal−Argyll area is characterised by shallower-water facies than elsewhere, the geological

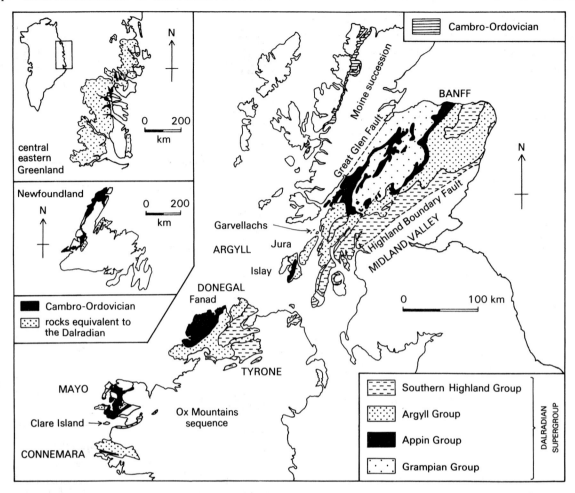

Figure 4.1 Outcrop map of Dalradian and Cambro-Ordovician rocks in NW Britain with insets showing distribution of equivalent rocks in Greenland and Newfoundland. The rocks equivalent to the Dalradian in Greenland are the Elenore Bay and Tillite Groups and in Newfoundland the Fleur de Lys Supergroup.

history of the central area can be taken as fairly typical of the whole of the Dalradian.

The Dalradian can be divided into four groups separated by distinctive formations (Fig. 4.2). The lowest group used to be named the Central Highlands Granulites and included with the Moine, because it was thought to be separated from the rocks of the Appin Group by a tectonic slide. However, this slide is now known to be present only locally, and elsewhere there is a conformable transition between the two groups. It has therefore been suggested that the term 'Dalradian' be expanded to include the Central Highlands Granulites under the term 'Grampian Group'.

The Grampian Group rocks were originally sands (quartzose to slightly feldspathic in composition) and muds, and were subsequently metamorphosed to quartzites, schists and slates. The slightly muddy sandstones have been metamorphosed to schistose quartzites which in the past were often referred to as granulites. This is an unfortunate and confusing use of the term 'granulite' because the Grampian Group is usually of greenschist or amphibolite facies.

The Appin Group is more varied, with limestones joining the clastic rocks. In Argyll, for example, there are three carbonate formations interbedded with formations dominated by quartzite and slates or schists. The quartzites are usually well bedded and often exhibit ripple marks, cross-stratification and liquefaction structures. Palaeocurrents are bimodal to polymodal (Hickman 1975*). The carbonate rocks include both limestones and dolomites. Both were originally either muddy or sandy, and they are interbedded with and pass laterally into slates, phyllites and quartzites. Stromatolites are found in the uppermost limestones, the Islay (= Lismore) Limestone, on Islay.

Most of the Grampian and Appin Group formations are thin in comparison with the younger parts of the Dalradian, being of the order of several hundred metres to a kilometre in thickness. Many formations can be traced for considerable distances, although lateral facies changes are recognised and many formational boundaries may well be diachronous. Because of the state of metamorphism and deformation of these rocks it is difficult to make detailed deductions about their depositional environment. However, the presence of shallow-water sedimentary structures, the palaeocurrent data, the blanket geometry and the persistence of many formations are consistent with a shallow shelf origin, although some of the facies changes suggest the presence of more variable marginal-marine environments.

4c Tillites and dolomites: the late Precambrian glaciation

The base of the next part of the Dalradian succession, the Argyll Group, is marked by the Port Askaig Tillite, perhaps the most interesting and useful formation within the Dalradian. It is interesting because it provides evidence of major climatic changes during the late Precambrian, and useful because it is a distinctive horizon that can be correlated with similar formations within late Precambrian sequences in other parts of the world.

The Port Askaig Tillite can be traced for nearly the whole length of the Dalradian outcrop from Connemara to Banff, but is not found in Shetland. In most places it overlies a limestone−dolomite formation which, in Argyll, contains stromatolites and flake breccias and is capped by an erosion surface. The formation contains up to forty-seven beds of tillite separated by siltstones, dolomites, con-

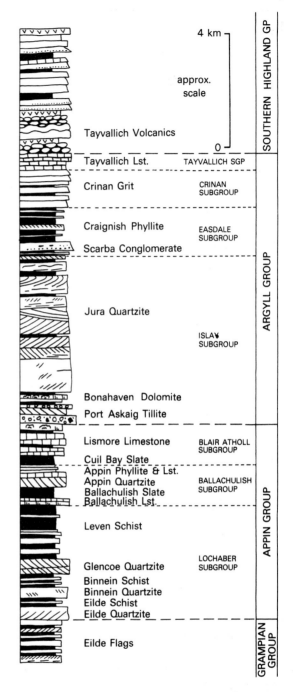

Figure 4.2 Stratigraphic log for the Argyll Dalradian.

glomerates and cross-bedded marine sandstones and quartzites. Some tillites contain lenses of bedded silt, sand and gravel, deposited by sub- or englacial streams, and have an upper surface cut by

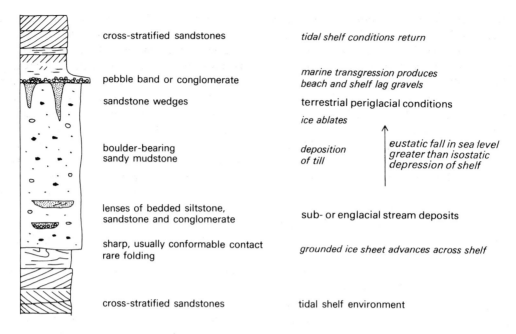

Figure 4.3   Idealised cycle for the Port Askaig Tillite. Log is of the order of metres to tens of metres thick (after Spencer 1971*).

sandstone polygons attributed to contraction cracking under periglacial conditions. Spencer (1971*) has interpreted these tillites as the deposits of grounded ice sheets advancing into a shallow sea and has attributed the alternation of marine and terrestrial features to eustatic sea-level changes (Fig. 4.3). There are also some varved beds containing dropstones, indicating the presence of floating ice. The clast content of the tillite beds varies from predominantly dolomitic in the lower parts of the formation to granitic in the upper parts. The dolomite clasts are of a very similar lithology to the underlying carbonate formations and were probably eroded locally. The granites are often albitised, have given dates of regional metamorphism of c. 1000 Ma and clearly came from a source area outside the Dalradian basin of deposition.

The Port Askaig Tillite lies within a continuous vertical succession several kilometres below rocks containing Cambrian fossils. It is therefore likely to be of late Precambrian age. A tillite horizon occurs within other sequences in the Caledonian belt, below the base of the Cambrian, in Spitsbergen, Greenland, Norway and Sweden (for review see Spencer 1975*). In N. Norway the tillite has been dated at 668 ± 7 Ma.

The individual tillite beds are from 0.5 to 65 m thick, and some can be traced from the Garvellachs to Islay and on to Fanad, a distance of 165 km. Evidently the area of deposition was virtually horizontal, and eustatic changes in sea level could easily convert it from a shelf sea to an extensive coastal plain.

The direction of the ice advance poses some palaeogeographical problems. On the southeastern side of the Caledonian belt in Norway and Sweden striated rock pavements show that the ice flowed northwestwards i.e. from the land towards the sea as one would expect. In Argyll, supposedly on the northwestern side of the Iapetus Ocean, the only evidence is the orientation of ice push folds, which suggest ice flow northwestwards. There is no obvious source for the granite clasts to the northwest of the present outcrop in Greenland, and Spencer (1975*) has suggested that they might have come from a southern extension of the Gothide complex beneath the N. European plain (Fig. 1.2). Although a southeastern source area for the Port Askaig glaciers is possible, it is not compatible with the existence of the Iapetus Ocean at this time, unless the ocean was sufficiently narrow for the glacier to float across and become grounded on its northwestern shore.

Glacial deposits of late Precambrian age which may be correlated with the Port Askaig Tillite are also found outside the Caledonian belt, e.g. in N. America, Russia, China and Africa. The glaciers seem to have been of worldwide distribution, unlike the high-latitude polar glaciations of other periods. Palaeomagnetic work on the Port Askaig Tillite, for example, suggests a palaeolatitude of only $10-15°$ at the time of deposition (Tarling 1974). The late Precambrian glaciation may be unique among the world's periodic glaciations, in both its effects and its causes. The Port Askaig Tillite is directly underlain and overlain by stromatolite-bearing dolomites indicative of warm waters. Although this may be surprising, it is consistent with the palaeomagnetic evidence for a low latitude. There is no evidence for a slow climatic deterioration preceding the deposition of the Port Askaig Tillite as is the case in Tertiary and Quaternary deposits (Ch. 17b). The late Precambrian tillites seem to have been deposited during a global glaciation that, in the case of the Caledonian belt, interrupted the deposition of warm-water deposits.

Some of the many hypotheses invoked to explain Quaternary climatic change (Ch. 17g) may also be relevant to the late Precambrian glaciation. However, this 'special' glaciation might have been due to 'special' causes. The late Precambrian glaciation slightly preceded the great expansion of metazoan life witnessed by the Ediacara fauna of Australia (Ch. 2a) and the subsequent evolution of shelly fossils in the Cambrian. Various hypotheses have attempted to relate these factors to each other and to inferred changes in atmospheric composition during the late Precambrian. A rapid increase in the mass of photosynthesising organisms at that time would have resulted in a fall in the concentration of atmospheric carbon dioxide. As this gas is very effective in absorbing the sun's radiation, such a fall could lead to a cooling of the world's climate. On the other hand, a global glaciation might have led to large scale extinctions and the vacation of many ecological niches. The subsequent return to climatic normality might then have initiated rapid evolution as the remaining organisms adapted to fill the vacant environments. There are several other possible relationships between glaciation, evolution and atmospheric chemistry, although none are sufficiently close or unique to make such hypotheses

convincing at the moment. Another specific hypothesis is that of G. E. Williams (1975), who has suggested that changes in the angle of inclination of the Earth's spin axis during the late Precambrian could have led to a global glaciation. Perhaps it is premature to speculate until we have a better understanding of the more recent glaciations. The Port Askaig Tillite and its correlatives show that the late Precambrian glaciation was unusual in extending into low latitudes, although it was similar to that of the Quaternary in the sediments deposited and in its record of repeated climatic oscillations.

4d   Shelves and basins in the Argyll Group

Following the Port Askaig Tillite, Dalradian deposition started to follow some new trends. Very thick formations are found, major lateral facies changes become important, deep-water facies are first found and the tectonic control of sedimentation becomes evident.

Overlying the Port Askaig Tillite, the dolomitic shales, siltstones and sandstones of the Bonahaven Dolomite indicate a rapid return to a warm climate. This formation includes excellent domal stromatolites which form reef-like bodies several metres across as well as extensive stratiform and rare columnar types (Spencer & Spencer 1972). Quartz–calcite nodules within the stromatolites may be pseudomorphs after anhydrite. The structures indicate shallow subtidal to supratidal conditions in a hot dry climate. The succeeding Jura Quartzite is very thick, over 5 km on Jura and often several kilometres elsewhere. It has been interpreted as a shelf deposit, where the interplay of strong tidal currents and storms resulted in the deposition of cross-stratified sands and storm layers (Anderton 1976). Palaeocurrents flowed northeastwards parallel with the present outcrop and probably with the contemporary coastline. There is little direct evidence from any of the Dalradian shelf facies to indicate whether the major landmass lay to the northwest or to the southeast. One interesting fact is that the pebble suite in the Jura Quartzite, which includes haematitic quartzites and cherts, is quite different from that of the Port Askaig Tillite, which contains exotic granite clasts, although it does show some similarities with that of the Torridon Group, which can best be matched with the Proterozoic rocks of Greenland (Ch. 2g).

Thickness and facies variations occur throughout the Dalradian rocks below the Jura Quartzite, although they are always sufficiently gradual to be explainable in terms of differential subsidence of a relatively stable continental shelf. Above the Jura Quartzite these variations become much more dramatic, can often be related to tectonic hingelines and can best be interpreted as the result of major syndepositional faults. What had previously been a stable continental margin now started to break up into a pattern of blocks and basins.

A good example of this can be seen in the rocks overlying the Jura Quartzite in Argyll (Fig. 4.2). Here, in the southwest, the Jura Quartzite passes upwards into a 500 m thick sequence of hemipelagic muds, turbidite sands and gravels, the latter forming small submarine fans derived from the northwest with a minor carbonate input from an offshore ridge to the southeast (Fig. 4.5). Two NE–SW-trending faults can be inferred here, defining the margins of a small turbidite basin. To the northeast,

however, cross faults with a NW–SE trend dropped down a much deeper basin in which about 2 km of muds accumulated, with very coarse turbidites, debris flows and slide horizons marking its faulted margins. There were syndepositional movements of several kilometres along these faults, and from mid-Argyll Group times onwards the movement of such faults was probably the major control over sedimentation. Eventually the deep basins filled up, as shown by the succeeding sediments, the Craignish Phyllite, which includes shallow marine and tidal flat facies with gypsum pseudomorphs (Anderton 1975). However, more turbidites, the Crinan Grit, overlie these shallow marine sediments, and again thickness and facies variations suggest that these accumulated in a NE–SW-trending fault-bounded basin, with palaeocurrents flowing along the basin axis towards both the southwest and northeast (Knill 1963). One point to note is that all the sands from the base of the Dalradian to near the top of the Argyll Group are petrographically mature. Both the shallow marine and turbidite sands are quartz arenites to subarkoses, although

Figure 4.4 Typical logs of Dalradian facies.
Wavy lines in shelf carbonate log indicate stromatolites; lozenges in tidal flat log indicate gypsum pseudomorphs.

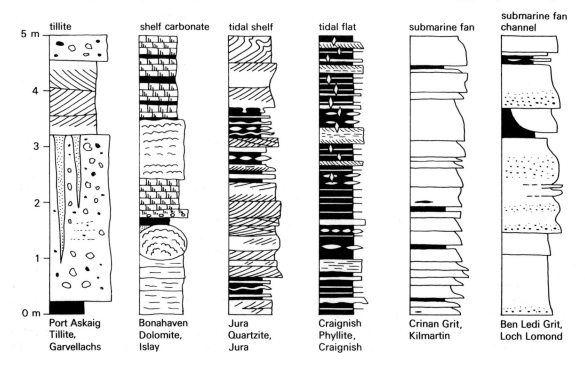

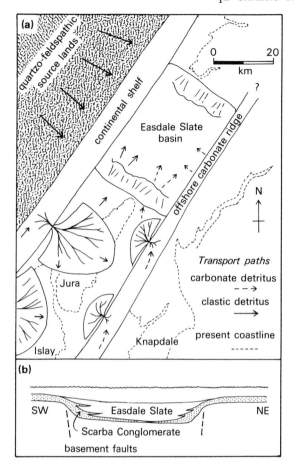

Figure 4.5 (a) Suggested palaeogeography for the Argyll seaboard area (see Fig. 4.1) during Scarba Conglomerate to Easdale Slate times, showing location of submarine fans and directions of sediment transport. (b) Sketch section showing facies relationships across the Easdale Slate basin.

they may be quite muddy. They all seem to have been derived from well-weathered quartzofeldspathic source rocks.

4e Cambro-Ordovician shelves, turbidites and volcanics

The correlation of the Port Askaig Tillite with other late Precambrian tillites, including one dated at *c.* 668 Ma, implies a Precambrian age for the Dalradian up to the lower part of the Argyll Group. Records of Palaeozoic body fossils — including the Lower Cambrian trilobite *Pagetides* from the Leny

Limestone at Callander, and the middle Cambrian sponge *Protospongia hicksi* from Clare Island, Ireland — are restricted to the Southern Highland Group. Palynological evidence confirms that the base of the Cambrian lies within the Argyll Group, somewhere below the Tayvallich Limestone (Downie *et al.* 1971).

There was a trend towards increasing instability upwards throughout the Dalradian, and the evidence for syndepositional faulting in the Argyll Group is followed by evidence of volcanism. Although the first evidence of volcanism occurs lower down within the Argyll Group, the first major outburst is at the top of the group where lavas and tuffs are interbedded with shallow water limestones (the Tayvallich Limestone). Above this, marking the base of the Southern Highland Group, is a thick sequence of basaltic lavas and tuffs with minor interbeds of marine limestones, grits and slates, the Tayvallich Volcanics. These lavas show excellent pillow structure, and in places the lava seems to have been intruded into wet sediment beneath the sea floor. Another lava formation is found above the Tayvallich Volcanics in Argyll, and sporadic tuffs and resedimented volcanic material, often known as green-beds, are found throughout the Southern Highland Group. Volcanics are not so common outside Argyll, where they are mostly restricted to the Southern Highland Group. In Argyll the lavas have suffered greenschist facies metamorphism and are now composed mostly of chlorite, calcite and quartz with some epidote and actinolite. An extensive suite of sills and occasional dykes which have a similar mineralogy and chemistry are found in the formations underlying the Tayvallich Volcanics in Argyll. They were probably contemporaneous with these lavas, and C. M. Graham (1976*) has suggested that some of the chemical variation in the lavas is due to fractional crystallisation of magma in thick sills prior to its eruption at the surface. The lavas and intrusives are of tholeiitic affinity. The parent magmas, probably having formed by shallow melting in the mantle, migrated upwards along fractures and fissures in the crust. These rocks suggest a tensional tectonic regime for the northwestern continental margin at this time and are consistent with the evidence for faulting.

Underlying the Tayvallich Volcanics in Argyll is the Tayvallich Limestone, a shallow-water facies containing small spherical grains which, although often referred to as oölites, are probably of algal origin. To the southeast a NE—SW-trending zone of limestone breccias, possibly debris flow deposits, may mark the northwestern margin of a fault-bounded basin. At the same horizon and further to the southeast a thinner succession of graded limestones, called the Loch Tay Limestone, may be turbidites deposited in the basin centre.

The stratigraphy of the rocks above the Tayvallich Volcanics is difficult to unravel because of the structural complexities of the Tay Nappe (Ch. 4f). However, in most places there is a very thick sequence of muds, sands and gravels now metamorphosed to the greenschist and amphibolite facies. The finer sediments, now transformed into slates, phyllites and schists, do not reveal much about their depositional environment. In contrast the sands and gravels often show the characteristics of submarine fan facies, with thick, often only poorly graded, laterally impersistent beds containing mud clasts and having channelled bases (Fig. 4.4). Except when interbedded with volcanics these sediments are still petrographically mature, with grains of quartz, quartzite and feldspar predominating. The Southern Highland Group may therefore have been largely deposited as submarine fan and hemipelagic sediments in fault-bounded basins fed by erosion of upfaulted ridges of quartzo-feldspathic basement. There is no evidence for the existence of any underlying oceanic crust.

Assemblages containing beds of chert and black slate are found along the Highland Boundary Fault and its western continuation in Ireland, the Leck Fault. These are of various ages, middle Cambrian on Clare Island and lower Ordovician in Scotland, which suggests the periodic existence of a sediment-starved ridge area during Cambro-Ordovician times. The Highland Boundary Fault may follow a syndepositional fault line, movement along which could juxtapose ridge and basin facies. Tuffs, spilitic lavas and intrusions of serpentinised gabbro, now found in faulted wedges, might have been erupted from and intruded along the syndepositional precurser of this fault in lower Ordovician times (Fig. 5.9). As is the case with many 'geosynclinal' sequences, there is little evidence within the Dalradian from which to construct the regional palaeogeography. Although the area of Dalradian

deposition was a NE—SW-trending shallow shelf which later gave way to a series of similarly trending ridges and basins, it is not obvious on which side of the area the major landmass lay. Shallow marine currents tended to parallel the shoreline, while turbidity currents either flowed along the long axis of the basins or entered them from all sides. It is the palaeocurrent evidence from fluvial facies that is most useful in outlining the positions of land and sea, and such facies are usually absent from 'geosynclinal' sequences such as the Dalradian.

To deduce regional palaeogeography it is necessary to look at regional facies variations. For example, we know that a predominantly shallow marine environment existed in the Dalradian area during the late Precambrian and that to the northwest there was probably a land area undergoing erosion. To the southeast there are no late Precambrian rocks until the Mona Complex of Anglesey is reached (Ch. 6c). During the Cambrian the picture is clearer. Turbidites were filling deep basins in what are now the Grampians and in NW Ireland. Further northwest, in the NW Highlands, shallow shelf sandstones and limestones were being deposited during the Cambrian. This situation continued into the lower Ordovician (Arenig), the time of the development of the Arenig Ballantrae Complex of possible ocean floor affinity on the southern side of the Midland Valley (Ch. 5b, Fig. 5.9). Thus the major landmass lay to the northwest during the whole of Dalradian times, with the open ocean, if any, lying to the southeast. Further southeast is a similar situation. No major Cambrian outcrops are found until the turbidite troughs of the Leinster, Manx and Welsh Basins are reached (Ch. 7). These are flanked to the southeast by the shallow shelf sequence of the English Midlands and S. Wales. There were broad similarities, then, between NW and SE Britain during the Cambrian. The two areas were roughly mirror images of each other. Both consisted of a shallow shelf area passing laterally into a fault-bounded turbidite basin (or basins) deriving its detritus both from a major landmass and from offshore ridges. In both cases the turbidites were deposited on continental crust. An ocean basin, if it existed, must have lain between these areas, i.e. northwest of the Lake District and southeast of the Midland Valley. As there are no late Precambrian or Cambrian rocks in this area, there is no record of any ocean floor, trench or continental slope events during this time.

The Cambro-Ordovician sequence of the NW Highlands must, as discussed above, be broadly contemporaneous with the muds and turbidites of the Southern Highland Group. In terms of facies, however, the former are not unlike the shallow shelf limestones, dolomites and quartzites of the lower part of the Dalradian Argyll Group. Indeed, we may say that the stable shallow-shelf conditions that prevailed in the southeast during early Argyll Group times must have transgressed northwestwards to reach the NW Highlands by the Cambrian.

The Cambro-Ordovician sequence in the NW Highlands consists of a lower clastic part and upper carbonate part (Fig. 4.6). A thin discontinuous basal conglomerate of local quartzites and gneisses overlies a flat erosional surface cut as the Cambrian sea transgressed northwestwards. The Eriboll Sandstone comprises the Lower Member with polymodal cross-bedding orientations, interpreted as a tidal shelf deposit, and the Pipe Rock Member with abundant vertical burrows (*Skolithus* and *Monocraterion*). There was a general decrease in clastic sediment input after this. The Fucoid Beds are dolomites, dolomitic shales and siltstones which, after a brief incursion of 10 m of sandstones with coarse well-rounded grains (the Serpulite Grit), are followed by over 1200 m of limestones and dolomites (the Durness Formation). Intertidal to shallow shelf conditions are indicated throughout, with well-developed stromatolites, oölites and flake breccias in the latter. Body fossils indicate a lower Cambrian age for the lower units and an upper Cambrian to lower Ordovician (Arenig or ?Llanvirn) age for the top of the Durness Formation. There is no obvious break between the two, but a non-sequence may be present.

The Cambro-Ordovician shelf sequences of Newfoundland and Greenland are remarkably similar to the NW Highlands succession (Swett & Smitt 1972*). In all these areas the basal clastics pass upwards into oölitic and stromatolitic carbonates of lower Cambrian to lower Ordovician age. Clearly, while these carbonates accumulated there was virtually no clastic input from the northwestern landmass due to its low relief and/or an arid climate. The Dalradian turbidites of this age to the southeast

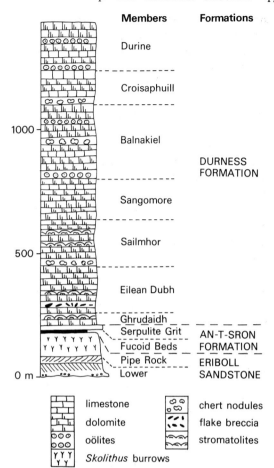

Figure 4.6 Stratigraphic log for the Cambro-Ordovician of the NW Highlands.

could not have been derived from the erosion of the northwestern landmass. They must have been derived from areas of uplifted basement lying adjacent to the turbidite basins and southeast of the carbonate shelf. The Ox Mountain sequence in Ireland is a sample of this basement, which shed detritus northwards during the Cambrian (Phillips 1973); and the large volume of the Southern Highland Group turbidites in Scotland might have been derived from the area of the present Midland Valley, uplifted along a precursor of the Highland Boundary Fault.

4f The Grampian Orogeny

By the lower Ordovician, NW Britain consisted of a

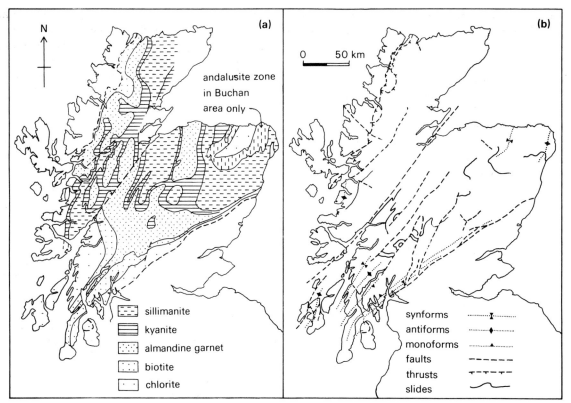

Figure 4.7 (a) Metamorphic isograd map (after Winchester 1974*). (b) Major structures of the Grampian and NW Highlands.

shallow carbonate shelf passing southeastwards into an area of turbidite basins separated by uplifted blocks undergoing erosion. The Grampian Orogeny is the name given to the period of metamorphism, deformation and igneous intrusion that, during the lower Ordovician, transformed this area of sedimentation into the mountain chain that ever since has served as a source of detritus for the surrounding areas.

The Grampian Orogeny was one of the series of orogenic events, each one localised in time and place, that together constitute the Caledonian Orogeny; and it was the only phase of the Caledonian Orogeny in the British Isles that produced a large area of metamorphic rocks, the orthotectonic Caledonides.

The metamorphism and deformation of the Dalradian was entirely due to the Grampian Orogeny. The southwestern part of the Moinian, already affected by Precambrian events (Ch. 2f), was deformed and metamorphosed again during the Grampian, although the northeastern Moines in

Sutherland may owe all their deformational history to the latter event. The major thrusts (e.g. the Moine Thrust) that cut the rocks of the NW Highlands, including the Cambro-Ordovician shelf sediments, also date from the Grampian event.

The youngest rocks affected by the Grampian Orogeny, i.e. the top of the Dalradian and Cambro-Ordovician sequences, are of lower Ordovician (Arenig or ?Llanvirn) age. The Dalradian is unconformably overlain by lower Old Red Sandstone in Scotland and upper Llandovery (Silurian) sediments in Ireland. Rb−Sr and U−Pb dates on syntectonic intrusions are in the range 480−514 Ma. These facts are broadly consistent with a lower Ordovician age for the orogeny, although there are inconsistencies between Irish and Scottish results and it is possible that the deformation was diachronous.

During the Grampian Orogeny the Dalradian rocks were deformed into large, NE−SW-trending, often tight to isoclinal folds with amplitudes of up to tens of kilometres. These folds and their associated axial-plane cleavage tend to 'fan' across the out-crop, from NW-facing in the northwest to SE-facing in the southeast where, in Scotland, most of the outcrop lies in the upside-down limb of the Tay Nappe, the Loch Tay inversion (Fig. 4.8). Several phases of deformation can be recognised, but these can be simplified into a primary phase, which produced the major folds and cleavage, and a secondary phase which refolded the major folds, also along NE–SW axes. A series of later and less important phases produced several cleavages, minor folds and kink bands (J. L. Roberts 1974). Measurements of strain (on deformed vesicles in lavas, sedimentary dykes, etc.) indicate shortening perpendicular to the main cleavage by up to 70 per cent, even in the only moderately deformed areas of Argyll. The nature of the deformation is dependent on structural level and lithology. It is less intense and less complicated at higher structural levels (Harris *et al.* 1976), and thick grits and quartzites are only slightly deformed by comparison with slates and siltstones, which are often so deformed that all sedimentary detail has been obliterated.

Metamorphism was roughly contemporaneous with deformation, although in detail several metamorphic peaks were interlaced with deformational phases. The Grampian Highlands were the location of the early classic studies of metamorphic isograds begun by Barrow (1893), and a modern version of the isograd map is shown in Figure 4.7. Metamorphic grade varies from the lowest greenschist facies (chlorite grade) up into the amphibolite facies. The index mineral sequence in the Buchan area of NE Scotland is unusual and may reflect a higher thermal gradient because of its high structural level, Northwest of the Great Glen, Morarian events were overprinted by Grampian metamorphism, and where the latter was of lower grade an area of retrogressive metamorphism resulted (Winchester 1974*).

Both basic and acid magmas were intruded during the Grampian deformation. In NE Scotland and Connemara acid and basic intrusions dated at *c.* 500 Ma postdate the main deformation but predate the secondary phases. In both the Moine and Dalradian there is a locally extensive migmatisation of a similar age to the granites. The total area of the syntectonic or 'Older Granites', such as the Ben Vuirich Granite of Perthshire, is very small by com-parison with the major suite of post-tectonic acid plutonics known as the 'Newer Granites'. These give mostly upper Silurian to lower Devonian ages of *c.* 415 Ma, and as they are also found in the Lower Palaeozoic rocks of the Southern Uplands they are clearly not related to Grampian events. They are further discussed in Chapter 8.

How can the main features of the Grampian Orogeny — the broadly contemporaneous deformation, metamorphism and plutonism — be related? The cleavage orientation and strain measurements show that there was considerable NW–SW compression across the orogen, although the major

Figure 4.8 Structural section across the Dalradian rocks of the southwestern part of the Grampian Highlands with suggested deformational history (after J. L. Roberts 1974).
AGp = Argyll Group; ApGp = Appin Group; SHGp = Southern Highland Group; AA = Ardrishaig–Aberfoyle Anticline; BLM = Ben Ledi Monoform; IA = Islay Anticline; LAS = Loch Awe Syncline; LST = Loch Skerrols Thrust; TM = Tarbert Monoform.

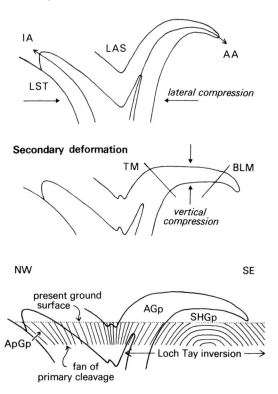

Primary deformation

Secondary deformation

structure of the Dalradian (Fig. 4.8) suggests that the rocks were transported outwards from the orogen's centre. Explanations for this range from gravitational sliding off an uplifted central block to crustal underthrusting. Metamorphism and plutonism are obviously related to heat, but how and where was this heat generated and how did it enter the orogen? There is no geochemical evidence for relating the plutonism to a subduction zone, and Richardson and Powell (1976) have shown that the metamorphic grade in the SE Highlands is no greater than would be expected from the normal conduction of mantle heat into a deeply buried sedimentary pile. It must be admitted that as yet we do not understand the causes of the Grampian Orogeny, even though there have been several recent speculative attempts to relate it to subduction zones (e.g. Wright 1976, Lambert & McKerrow 1976).

4g Summary

During the late Precambrian to early Ordovician a thick sequence of varied sediments with minor volcanics, the Dalradian Supergroup, was deposited on the continental margin northwest of the supposed Iapetus Ocean. Although its base is not seen, this sequence is thought to lie everywhere on continental crust. Initially the continental margin was stable, although gently subsiding, and clastic and carbonate shallow-marine sediments accumulated. In late Precambrian times, after the deposition of a tillite horizon widely recognised in the N. Atlantic region, the shelf area started to break up into a series of blocks and basins, and by the Cambrian a marine transgression had established a new shelf area to the northwest. The deposition of shallow marine carbonates dominated this new shelf, forming the Cambro-Ordovician succession of the NW Highlands, while turbidites and volcanics accumulated in fault-bounded basins in the unstable Dalradian area to the southeast. During the lower Ordovician, the Grampian Orogeny deformed and metamorphosed the Dalradian sediments on the southeastern edge of the margin together with older Moine and Lewisian rocks to the northwest. Although the processes of ocean floor spreading and subduction might well have been operating in the area at this time, the relationships (if any) between these processes and the Grampian Orogeny are not yet clear.

5 Late Ordovician to Silurian evolution of the northern continental margin of Iapetus

5a Introduction

The intense deformation and metamorphism associated with the Grampian Orogeny in NW Scotland and N. Ireland brought to an end a long period of marine sedimentation. In Scotland, the folded and metamorphosed rocks were raised to form mountains which shed detritus southwards towards the postulated continental margin and the Iapetus Ocean. During the Ordovician and Silurian these sediments gradually developed into the sequences now exposed in the southern part of the Midland Valley and in the Southern Uplands (Fig. 5.1). The earliest sediments recorded are mudstones and cherts of Arenig age, with associated pillow lavas, serpentinites and gabbros now situated near Ballantrae. At Girvan, a few kilometres north, the post-Arenig sequence comprises coarse marine sediments including some turbidites, but in the Rhinns of Galloway to the south the succession is largely composed of turbidites. During the Ordovician the zone of turbidite deposition gradually encroached southwards into areas formerly accreting ashy graptolitic muds. This, however, was not a simple prograding of the turbidite fans. McKerrow *et al.* (1977*) have suggested that the sediments accumulated on a continental slope which was being actively underthrust by the N-shifting oceanic plate.

In Ireland a similar situation existed in Longford Down, but in central Mayo there was a further major event as massive subsidence along a NE−SW-aligned tract caused deposition of marine sediments early in Arenig times. During lower to middle Ordovician times this area, the South Mayo Trough, was bordered to the southeast by a volcanic arc which extended northeastwards into Tyrone. By the end of the Llandeilo the trough had accumulated over 10 km of sediment.

Figure 5.1 (a) Location map of the Ordovician and Silurian rocks recording the evolution of the northern continental margin after the Grampian Orogeny. (b) Ballantrae Complex. (c) Submarine fans of mafic detritus. (d) Trench sediments: the lithic sands of the inner slope. (e) Trench sediments: the hemipelagic mud zone. (f) S. Mayo Trough and the volcanic arc. (g) Silurian transgression in W. Ireland. (h) Silurian regression in the Midland Valley.

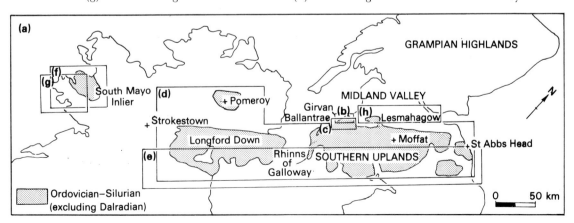

5b The Ballantrae Complex: an early Ordovician ophiolite

The Ballantrae Complex is important because it provides vital clues to the events of early and middle Ordovician times — events contemporaneous with the deformation and metamorphism of the Dalradian to the north. It also seems to represent a fragment of the terrain that supplied detritus southwards later in the Ordovician.

The complex (Fig. 5.2) consists of two main rock associations. The first comprises spilitic pillow lavas overlain by agglomerates, tuffs and red, green and grey radiolarian cherts. Black shale horizons within

Figure 5.2 (a) Map of the Ballantrae ophiolite complex (for location see Fig. 5.1). (b) Cross-section.

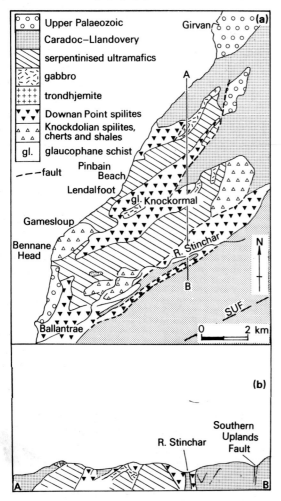

the tuffs yield the small inarticulate brachiopods *Acrotetra* and *Lingula* together with the planktonic graptoloids *Tetragraptus* and *Didymograptus*. The thinly bedded cherts display evidence of early deformation. The second association consists of serpentinised ultramafics, schists and amphibolites. The former include altered harzburgites and dunites (olivine, enstatite, chrome spinel), cumulate picrites (olivine, titanium-rich augite, plagioclase, analcite), foliated and massive gabbros, and dolerites. A subsidiary complex near Girvan comprises gabbro, diorite and trondhjemite (a granite poor in potassium). Bloxam and Allen (1960) have traced a transition from mafic volcanics through spilites bearing pumpellyite and the garnets andradite and grossularite, to greenschists, and finally to blueschists bearing glaucophane.

In recent years several authors have sought to demonstrate a trench or oceanic affinity for the complex. Dewey (1969a*) first suggested that the serpentinite was intruded into the trench sediments and volcanics; but in a revised model (Dewey 1971* and 1974) he later proposed that the complex represents the obducted (overthrust) remnants of oceanic crust from a marginal sea. In contrast, Church and Gayer (1973) favoured an hypothesis involving the folding and faulting of an obducted oceanic crust derived from the open ocean. More recently, Lambert and McKerrow (1976) have developed a model invoking the partial subduction of open ocean crust.

Wilkinson and Cann (1974) have attempted to shed light on the environmental affinities of the pillow lavas, dolerites, glaucophane schists and amphibolites by analysing their trace-element chemistry. Earlier, Pearce and Cann (1973) demonstrated that the relative proportions of titanium, zirconium, yttrium and niobium can be used to distinguish modern ocean-floor basalts, potassium-low tholeiites of island arcs, calc-alkali basalts and basalts produced within plates. The 1974 results (Fig. 5.3) show that the dolerites plot inside the field representing intrusions emplaced within plates. The lavas plot inside the fields representing intra-plate volcanism and potassium-low tholeiites of island arcs. Only the (rather weathered) glaucophane schists are comparable in trace-element content to the basalts of the modern ocean floor.

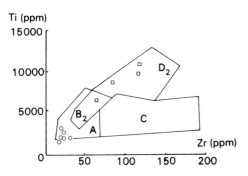

Figure 5.3 Trace element (zirconium, titanium, yttrium) composition of samples from the Ballantrae ophiolite complex. Results plotted on to fields representing different environments of occurrence established by Pearce and Cann (1973) from studies of Recent volcanics (after Wilkinson & Cann 1974).
Field A = potassium-low tholeiites of island arcs; B_1 = ocean floor basalts and island arc volcanics; B_2 = ocean floor basalts and potassium-low tholeiites of island arcs; C = calc-alkaline basalts and andesites; D_1 = within-plate or hotspot basalts; D_2 = ocean floor basalts.
Triangular plot: ○ pillow lavas from Bennane Head; ● pillow lavas from Downan Point; ▲ pillow lavas from Pinbain Beach; △ dolerites from Lendalfoot; □ glaucophane schists and amphibolites from Knockormal and Knocklaugh.
Titanium versus zirconium plot: ○ pillow lavas from near Gamesloup; □ glaucophane schists and amphibolites from Knockormal and Knocklaugh.

The results appear to deepen the puzzle. Only the glaucophane schists and the gabbro−diorite−trondhjemite complex fit an ocean floor origin. The lavas seem to give conflicting indications. As Wilkinson and Cann (1974) have suggested, it is possible that the lavas were erupted independently of the other rocks and then emplaced tectonically.

5c Submarine fans of mafic detritus

During the upper Ordovician the mafic plutons, pillow lavas, cherts and shales that form the Ballantrae Complex were smothered beneath a thick pile of coarse, neritic and deeper marine sediments (Figs 5.4 and 5.5). North of the Stinchar Valley the succession of conglomerates, sandstones, mudstones and occasional carbonates contains an assemblage of brachiopods, trilobites and graptolites; broadly, progressing northwards, successively younger strata overlap on to the Ballantrae Complex. South of the Stinchar Valley the equivalent succession comprises lithic wackes, shales, mudstones, occasional conglomerates and spilites; here the only fossils are graptolites.

The thick coarse-grained Caradocian sequence which lies north of the Stinchar Valley shows a gradual upward decline in the amount of cobbles and pebbles (areas 1 and 2 in Fig. 5.4). The basal Barr Group is dominated by two rudaceous formations: the Kirkland and Benan Conglomerates. Both are composed mainly of coarse mafic debris derived from rocks comparable with the Ballantrae Complex. The Benan Conglomerate comprises boulder beds, pebbly sandstones, weakly stratified sandstones, and thin siltstones and mudstones. The large clasts are set in a matrix of sand grade spilite, serpentinite and quartz. Crinoid columnals and the brachiopod *Leptellina* are the only fossils recorded. Separating the two conglomerates is a thin sequence (c. 50 m) of sandstones, shales and mudstones. In common with the matrix of the conglomerates, quartz and serpentinite grains abound, but in addition there are particles of feldspar and carbonate. The carbonate is composed of comminuted calcareous algae and brachiopod shells. The calcareous algae, occasionally found encrusting shells, have been identified as *Girvanella*, a form whose modern equivalents have recently been discovered growing in the tropical Aldabra Atoll in the Indian Ocean (Riding 1977). In the vicinity of the Stinchar Valley, dark calcilutites bearing brachiopods, *Girvanella* and the foraminiferan *Saccaminopsis* make up the Stinchar Limestone. Although several shallow-water indicators, occur it is not certain that the environment of deposition was inshore. The texture of the associated conglomerates suggests rapid deposition, possibly on deep-sea fans.

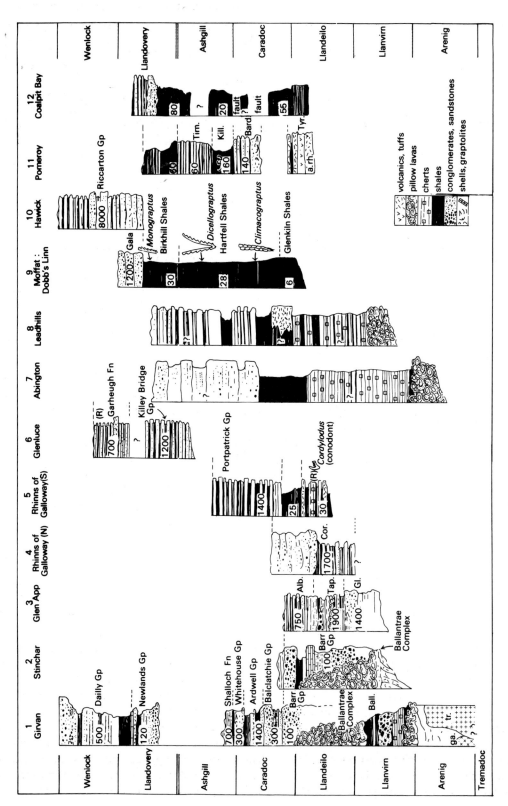

Figure 5.4a Ordovician–Silurian successions of the Southern Uplands and Longford Down. Logs in metres.
Alb. = Albany division; Bard. = Bardahessiagh Group; Cor. = Corsewall Formation; Gl. = Glen App division; Tap. = Tappins division; Tyr. =
Tyrone Igneous Group.
a. = andesitic; ga. = gabbro; (R) = red-beds; rh. = rhyolitic; tr. = trondhjemite.

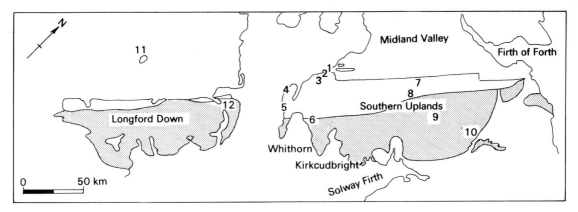

Figure 5.4b Location of successions in Figure 5.4a.

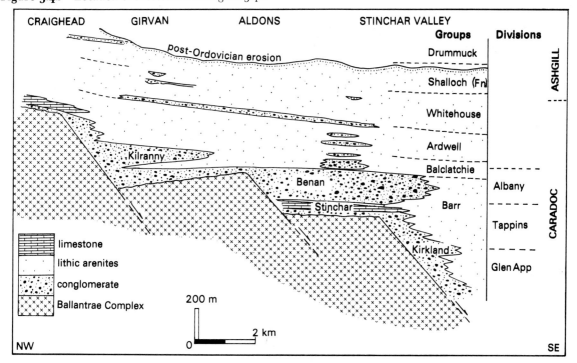

Figure 5.5 Ordovician sediments of the Girvan area (for location see Fig. 5.1) (after A. Williams 1962 and Whittington 1972).

The mudstones and conglomerates of the Balclatchie Group continued the pattern of sedimentation established during the deposition of the Barr Group. Clasts of serpentinite, gabbro and spilite are set in a calcareous matrix which is more fossiliferous than the matrices of the earlier conglomerates. Textural evidence suggests that these conglomerates were emplaced by powerful currents which entrained much of the sediment as a bedload layer. Clasts in the Kilranny Conglomerate, for example,

display a strong preferred imbrication towards the northeast. In addition, lenses of sandstone within the conglomerate are cross-stratified or horizontally laminated. As sedimentation continued there was no further influx of mafic clasts. At first sandy or silty laminated muds were intercalated with thin conglomerates, thin sandstones and thin black shales. Hubert (1969) has noted that the conglomerates of the Ardwell Group are composed of calcite-cemented imbricated intraformational debris. He has further observed that some intraformational conglomerates pass laterally into sediments displaying penecontemporaneous folds with axes orientated ENE—WSW. The sandstones display flute casts and climbing ripples indicative of SE- and SSE-flowing currents. Henderson (1935) invoked earthquakes and tsunamis to explain the folds; but Hubert, noting the perpendicular relationship between fold axes and current vectors, has postulated the action of powerful neritic bottom currents.

The succeeding sediments of the Whitehouse Group record a substantial influx of bioclastic carbonate. The skeletal debris accumulated in beds up to 1·5 m thick, interbedded with sands and muds. From flute marks and cross-stratification, Hubert (1969) has shown that the fragmented crinoids, brachiopods, corals, bryozoans and trilobites were derived from the west or northwest. However, the alternating thin sandstones and shales of the Shalloch Formation which completes this upper Ordovician sequence show that the source of carbonate soon declined. The sandstones, which are not generally graded, display flute marks, current ripples and convolute lamination. The flutes and ripples indicate the passage of flows southwestwards, and the axes of the convolute lamination are almost perpendicular to this flow path. It is interesting to note that these vectors represent a clockwise rotation of approximately 90° in comparison with the values obtained from the Ardwell Group.

South of the Stinchar Valley the upper Ordovician succession is thicker (4·8 km compared with 2·6 km) and more monotonous than the varied sequence to the north. Grey mudstones, coarse lithic wackes and conglomerates predominate, but at some levels there are red mudstones, cherts, ashes and spilites. At one horizon there is a spilitic agglomerate. The graptolites *Climacograptus*, *Dicellograptus* and *Dicranograptus* affirm the (?Llandeilo) Caradoc age.

In his monograph, A. Williams (1962) has suggested that upper Ordovician sedimentation in the Girvan area was controlled by three faults, each with a downthrow towards the southeast (Fig. 5.5). The hypothesis of fault-controlled sedimentation has several points of merit: (a) it explains the substantial thickness of the sedimentary pile, as fault movement provides a mechanism for rapid subsi-

dence; (b) it accounts for the general northward overlap of the rudaceous units on to the Ballantrae Complex; and (c) contemporaneous fault movement provides an explanation for the liquefaction of the sediments deformed by the passage of marine currents during the deposition of the Ardwell Group.

5d   Trench sediments: lithic sands of the inner slope

From Girvan the evidence is clear that during Llandeilo—Ashgill times a northern zone of deep sea fans accumulated coarse mafic debris. From the northern belt of the Southern Uplands we learn that this zone of lithic sand deposition gradually extended southwards during the Ordovician and Silurian. This has been demonstrated by Kelling (1961, 1962*) in the Rhinns of Galloway. In the northern extremity of the Rhinns, conglomeratic lithic wackes were deposited in Llandeilo times (area 4 in Fig. 5.4); 25 km to the south, deposition of lithic wackes did not commence until Caradoc times (area 5). More recently, McKerrow *et al.* (1977*) have traced the southward extension of the zone of lithic wacke deposition in the central Southern Uplands. At Coulter—Noblehouse coarse lithic sands were deposited during the late Llandeilo, but at Craigmichen, 26 km to the south, there was no influx of coarse detritus until upper Llandovery times some 40 Ma later. As Lapworth recognised long ago, the first influx of sands at Moffat in the central Southern Uplands occurred early in Silurian times.

Through his detailed sedimentological studies, Kelling (1962*) has recognised four groups of lithic wackes, distinguished by their respective compositions, in the Rhinns of Galloway. The Corsewall Group contains boulders of weathered hornblende—albite granodiorite, granophyre, quartz porphyry, andesite, rhyolite, dolerite and spilite. In addition there are fragments of serpentinite, phyllite and scarce glaucophane schist. Much of this debris compares with the rocks of the Ballantrae Complex. The Kirkcolm Group is distinguished from the Corsewall because in addition it contains garnet—muscovite schist, garnet—talc schist, graphite schist, andalusite schist, hornfels and garnet-bearing gneiss. The diversity of schist types is accompanied by a paucity of ferromagnesian minerals. The schists were presumably derived from the regionally metamorphosed Dalradian to the northeast. Compared with the Kirkcolm, the Galdenoch Group has a higher content of ferromagnesian minerals but a smaller range of schist types. Lithic wackes of the Portpatrick Group are so rich in altered plagioclase, fresh pyroxene, amphiboles, epidote and fragments of andesite that they resemble tuffs.

There is a link between the above petrographic data and palaeocurrent results. The Corsewall Group, with its abundant volcanic debris, displays structures which suggest derivation from the north-northwest. The Galdenoch and Portpatrick Groups also yield volcanic debris but seem to have been derived from the west and south. The finer-grained lithic wackes of the Kirkcolm Group, with the high content of schistose grains, were emplaced by currents flowing from the northeast and southwest.

Thus the picture of southeastward progradation is not simple. During the Llandeilo and lower Caradoc, an exposed volcanic terrain of Arenig age, lying to the north, supplied the very coarse detritus of the Corsewall Group. Later, in the upper Caradoc, there was a general supply of regionally metamorphosed rock, presumably derived from the uplifted Dalradian pile; this sediment formed the Kirkcolm Group. In the Caradoc and Ashgill, new sources of volcanics fed detritus from the west and south to form the Galdenoch and Portpatrick Groups.

5e Trench sediments: the hemipelagic mud zone

It is now important to consider the nature of the sedimentation southeast of the prograding zone of lithic wacke deposition. Generally there was a slow accumulation of mud, but Weir (1973) has pointed out that the Ordovician−Silurian sequences tend to have a cyclical aspect. His ideal sequence displays a passage from dark fissile pyritic shales with thin radiolarian cherts, to thickly bedded grey mudstones overlain by brown mudstones bearing trilobites. Two cycles have been identified: one late Ordovician (Llandeilo−Ashgill), the other early

Silurian (Rhuddanian−Fronian). The first is recognised in the Southern Uplands (Moffat) and in Longford Down (Coalpit Bay), but the second is satisfactorily demonstrated only in the Southern Uplands (Dobb's Linn, near Moffat) (Fig. 5.4).

Weir (1973) has postulated that each cyclic sequence reflects increasing current activity. Thus the pyritic shales and radiolarian cherts are believed to represent quiet anaerobic sea-floor conditions. Periods marked by slow settling and the accumulation of clay minerals alternated with periods characterised by the steady 'rain' of opaline radiolarian tests. Modern cores of ocean sediment show clearly that the final chert may bear little evidence of the formerly abundant radiolaria. With burial and time, the opaline silica of radiolaria dissolves, migrates and reprecipitates as microspherular clusters (lebispheres) of crystals composed of the open framework form of silica, cristobalite. Eventually the cristobalite inverts to the stable cryptocrystalline quartz (chert).

In contrast, the lighter grey mudstones of the upper part of the cycles are indicative of more aerobic sea-floor conditions. The source of the muds is uncertain, but it is possible that the fall-out of terrigenous pelagic mud was supplemented by the deposition of silts and clays resuspended from shelf-sea floors by storm waves, and transported oceanwards on to the continental slope by thick, slowly moving bodies of water. In recent oceans, such water masses, characterised by slightly higher concentrations of suspended mud particles than the ambient water, are known as nepheloid layers.

5f The South Mayo Trough and volcanic arc

Southwest of the Ballantrae−Girvan area, in W. Ireland, there is further evidence of Ordovician volcanic activity and instability. During the Arenig a NE−SW-aligned tract started to subside. The base of the Ordovician sequence that filled this South Mayo Trough is not seen, but the exposed succession commences with Arenig spilites, rhyolites, tuffs and associated red and green cherts (Fig. 5.6). These are overlain by, and pass northwards into, a thick (5 km) series of tuffaceous turbidites of Arenig−Llanvirn age. Llanvirn slates are overlain by siltstones and ignimbrites possibly of Llandeilo

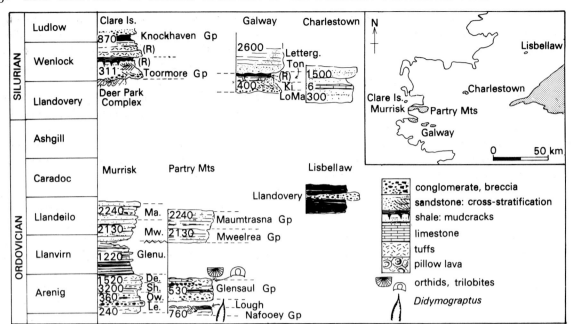

Figure 5.6 Ordovician—Silurian successions of W. Ireland. Logs in metres.
De. = Derrylea Group; Glenu. = Glenummera Group; Ki. = Kilbride Formation; Le. = Letterbrock Group; Letterg. = Lettergesh Formation; Lo. Ma. = Lough Mask Formation; Ma. = Maumtrasna Group; Mw. = Mweelrea Group; Ow. = Owenmore Group; Sh. = Sheefy Group; Ton. = Tonalee Formation.
(R) = red-beds.

age. Flute marks indicate that many of the tuffaceous turbidites were derived from the south, where a volcanic arc comprising mafic and silicic eruptives extended northeastwards to Tyrone.

5g The Silurian transgression in W. Ireland

During late Ordovician times the thick sequence of sediments and volcanics in the South Mayo Trough was subjected to deformation and uplift, but by upper Llandovery times the area was once more invaded by the sea. In NW Galway, basal cross-stratified feldspathic sandstones with S-facing foresets pass up into mudstones, followed by quartz arenites with *Skolithos*, the athyrid brachiopod *Eocoelia* and occasional channelled pebble—granule conglomerates (Fig. 5.6). The sandstones pass gradationally up into red mudstones bearing *Clorinda*. D. J. W. Piper (1972) has suggested that the basal feldspathic sandstones are fluvial and that the overlying association represents a lagoon—barrier complex. Eventually the sea deepened and sedimenta-

tion was reduced to the slow accretion of offshore shelf muds.

In late Llandovery—Wenlock times there was a major change in the type of sedimentation. Submarine canyons, incised into the shelf, guided sands and gravels entrained in turbidity currents to deep marine basins. There, massive bedded sands with thin, occasionally graptolitic, muds accumulated to a thickness of 1500 m (Lettergesh Formation). The debris included both Dalradian and Ordovician fragments. To the north in Clare Island, SW Mayo, Phillips (1974) has described a thick (1700 m) Silurian sequence of marginal marine and fluviatile sediments resting on the Dalradian and the Deer Park Complex. From the absence of Ordovician rocks beneath this sequence and the lack of included Ordovician debris, Phillips has argued that the sediments are of middle—upper Wenlock age — that during the lower Wenlock SW Mayo was the *source* of Ordovician detritus. The middle—upper Wenlock date is supported by a K—Ar isotopic age determination of 412 + 10 Ma for potassic feldspars in a welded tuff. Phillips has explained the great thickness of marginal marine deposits by invoking the contemporaneous development of a monoclinal scarp downwarping to the south.

5h The Silurian regression in the Midland Valley

In the Midland Valley there is no trace of the widespread Llandovery transgression seen in W. Ireland. The numerous Silurian inliers of the Midland Valley (Fig. 5.7) clearly record a general regression continuing through the Llandovery into the Wenlock. Marine turbidite sequences pass up into shallow marine, brackish or fluvial sediments.

A most informative succession has been documented at Lesmahagow by Jennings (in Walton 1965). The lowest sediments comprise graded lithic wackes and laminated siltstones interbedded with grey mudstones and carbonaceous siltstones. The basal laminae of solemarked lithic wackes yield a diverse allochthonous shelly assemblage including brachiopods, bivalves, trilobites, bryozoans, orthoceratids and crinoids. In contrast, the laminated siltstones contain fish scales (*Thelodus*), eurypterid remains (*Ceratiocaris*) and gastropods (*Platyschisma*) suggestive of a brackish marine source. The overlying olive-grey mudstones with carbonaceous silty mudstones are characterised by a prolific fauna of eurypterids. The cross-stratified sandstones that follow contain pebbles of silicic igneous rock and are interbedded with mudstones featuring carbonaceous silt laminae which bear fragments of fish. The top of the sequence comprises a very coarse vein-quartz/jasper/acid-igneous conglomerate.

To the south, the Silurian sediments of the Hagshaw Hills similarly record the regressive event, but above the turbidite sequence there is a 30 m unit composed of pebbles and cobbles of spilite, granite and microgranite. Northeast of the Hagshaw Hills, the Llandovery sediments of the Pentland Hills consist of sandstones with thin limestone horizons crowded with the corals *Palaeocyclus* and *Pleurodictyum*, tentaculitids, *Atrypa*, strophomenids, monograptids and crinoid columnals. The sandstones also include the distinctive and complex dendroid graptolites *Retiolites*, *Koremagraptus* and *Acanthograptus*. As at Lesmahagow, the Hagshaw Hills expose a conglomerate of igneous pebbles between a sequence of marine sediments and an overlying sequence of sandstones yielding freshwater fish remains.

In the southwestern extremity of the Midland Valley at Craighead and Girvan the Silurian successions are entirely marine. The Llandovery

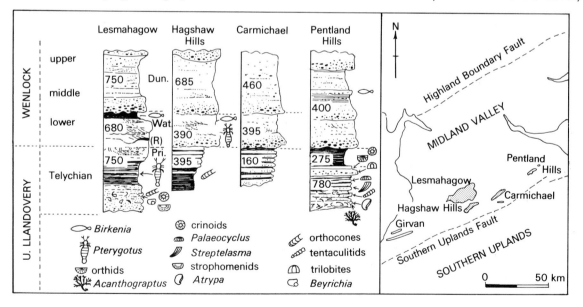

Figure 5.7 Silurian successions of the Midland Valley. Logs in metres. Dun. = Dungavel Group; Wat. = Waterhead Group; Pri. = Priesthill Group. (R) = red sediments.

sequence displays sandstones overlain by a stratified boulder conglomerate including fragments of Arenig mafic volcanics, soda granite, microgranite, chert and quartz — an assemblage altogether more silicic than the Caradocian conglomerates beneath. An overlying sequence of sandstones, grey, green or red mudstones and shales is variably fossiliferous. The fauna include tabulate corals, orthids, pentamerids and phacopids in arenaceous beds, with diplograptids and monograptids in argillaceous beds. Despite the abundance of shallow marine fossils there is little evidence to suggest that the environment of deposition was shallow. Indeed the water may have been deep. The Craigskelly Conglomerate at Girvan displays massive (3−4 m) fining-upwards units of highly rounded cobbles and pebbles, suggestive of a resedimentation mechanism of emplacement.

5i   Synthesis

We now turn to the problem of how the northern continental margin evolved during the closure of the Iapetus Ocean. Hypotheses have been advanced by Dewey, Gunn, Church and Gayer, Mitchell and McKerrow, Phillips *et al.*, Lambert and McKerrow, and McKerrow *et al.* (Fig. 5.8). Initially Dewey (1969a*) proposed that the Ordovician−Silurian shales and cherts of the Southern Uplands rest directly on ocean crust, considering the Ballantrae volcanics to have erupted along the inner slope of the trench. But later Dewey (1971*, 1974) revised his model and invoked a rear arc and interarc origin for the Newfoundland and Ballantrae ophiolites. Gunn (1973) believed that the suture zone separating the northern and southern continents was sited in the region of the Midland Valley, the Southern Uplands being considered part of the southern continental margin. Church and Gayer (1973) developed an hypothesis in general agreement with Dewey's revised model. Mitchell and McKerrow (1975) (and later Lambert & McKerrow 1976) postulated that the Iapetus Ocean crust of Arenig times extended as far north as the Highland Boundary Fault on the northern boundary of the Midland Valley. According to this model, uncoupling occurred in the vicinity of the fault in Arenig times, and the oceanic lithosphere was subducted northwest-

wards beneath the rising mountains of the Grampian region. A second zone of uncoupling developed in the region of the Southern Uplands in Caradoc times and, as subduction commenced along this new zone, so the first subduction zone became extinct. According to Phillips *et al.* (1976*), however, there was only one NW-dipping subduction zone, positioned beneath the Southern Uplands in agreement with Mitchell and McKerrow's (1975) second zone, but originating in the Arenig and not the Caradoc. The NW-dipping subduction continued until Wenlock or early Ludlow times (McKerrow *et al.* 1977*).

Recent geophysical investigations of the crustal structure of Scotland have had an important, but not yet a decisive, role in the above arguments. From Mitchell and McKerrow's (1975) model one would predict that the Midland Valley is underlain by crust of oceanic affinities. However, this does not appear to be the case. Using seismic refraction methods, Bamford *et al.* (1977) have studied a deep (7−20 km) crustal layer characterised by its capacity to transmit seismic waves at high velocities (at least 6·4 km/s). The layer, which has been interpreted as granulitic, can be traced from beneath the Grampians to beneath the northern margins of the Southern Uplands. This geophysical evidence is strongly supported by the discovery (Upton *et al.* 1976) of xenoliths of garnet-bearing gneiss in Carboniferous volcanic necks at Parton Crag on the northern boundary of the Southern Uplands block (Max 1976a). However, the positive gravity and magnetic anomalies recorded over the Midland Valley indicate that mafic rocks may exist at depth. Sedimentary evidence from the upper Dalradian is also inconsistent with the suggestion that the ocean crust of Iapetus extended as far as the Highland Boundary Fault. Regional facies variations point to a southern source in the vicinity of the Midland Valley, and petrographic studies show that this source was composed of continental crust. Thus the balance of geological evidence just seems to tip the scales in favour of the hypothesis of Phillips *et al.* (1976*).

If then we interpret the geophysical evidence as indicating that in pre-Arenig times the edge of the northern continental margin was situated in the central−southern region of the Southern Uplands, it follows that the Ballantrae ophiolite complex,

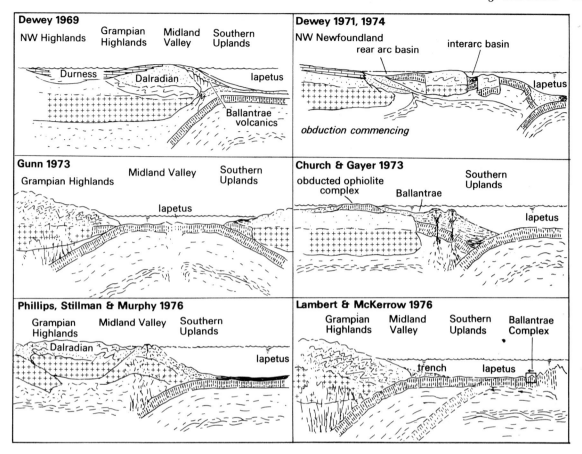

Figure 5.8   Rival hypotheses for the evolution of the northern continental margin in Ordovician (late Arenig) times.

which appears to have both oceanic and continental margin affinities, was either of rear-arc origin or was obducted from an open ocean site of origin. The widespread occurrence of the Arenig ophiolites, from NW Newfoundland and Scotland to Norway, favours Dewey's (1974) view that there was a number of small rear-arc (marginal) seas. Thus as uncoupling of oceanic and continental crust occurred in late Cambrian times, the distended crust of the continental margin was subjected to rifting and minor ocean-floor spreading. As subduction of the main oceanic lithosphere continued, however, tensional stresses within the continental margin changed to compressional stresses. Spreading processes ceased by late Arenig times, and blocks of ocean crust (the marginal sea floors) were fractured by a series of high-angle thrusts (Fig. 5.9). Elevation and thrusting of this young marginal ocean-crust

may have been facilitated by its heat and its slender thickness. Contemporary movements of the oceanic crust of the Ballantrae Complex are recorded by the bedded cherts exhibiting plastic deformation. During Llanvirn times the complex was uplifted and eroded, but during the Llandeilo and Caradoc the area was subjected to subsidence. Mafic debris spread southwards in large submarine fans, some reaching the Rhinns of Galloway in the Southern Uplands (Fig. 5.10). As subsidence continued, debris of the fan apex gradually overstepped northwestwards on to the degrading complex, which was presumably more extensive than the remnants now exposed. In the Girvan area sedimentation continued into Ashgill times, but there was extensive erosion of Ashgill sediment before the deposition of the Llandovery. In W. Ireland subsidence of the South Mayo Trough probably ceased in Llandeilo

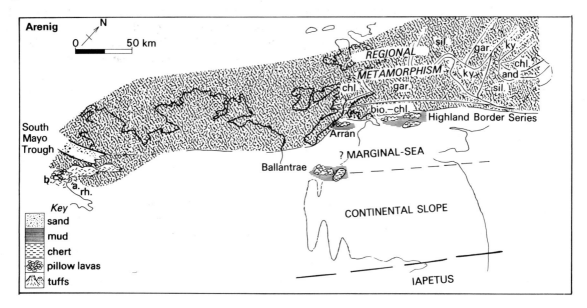

Figure 5.9   Late Arenig palaeogeography of the northern continental margin (metamorphic zones in the Grampian sourcelands after Winchester 1974*). Note: the Recent coastline of S. Scotland is palinspastic.
a. = andesite; b. = basalt; bio. = biotite; chl. = chlorite; gar. = garnet; ky. = kyanite; rh. = rhyolite; sil. = sillimanite.

Figure 5.10   Caradoc palaeogeography of the northern continental margin. Note: the Recent coastline of S. Scotland is palinspastic.
Co. = Corsewall Formation; Ki. = Kirkcolm Formation; Po. = Portpatrick Formation;
a. = andesite; ap. = apatite; rh. = rhyolite.

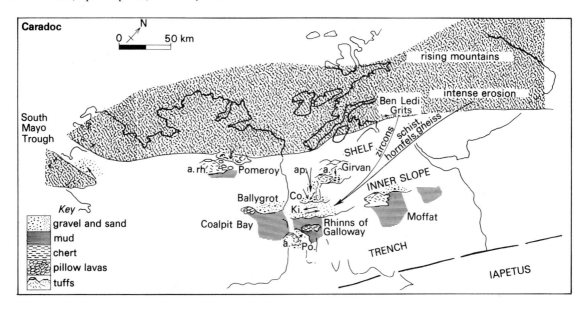

times. During the Llanvirn and Llandeilo the trough was filled by deltaic sediments and ignimbrites derived from the southeast. As at Girvan there was uplift and erosion before the deposition of marine sediments recording the Llandovery transgression.

During Ordovician–Silurian times, northwestward subduction of ocean crust beneath the northern continental margin was active. Perhaps the strongest evidence in favour of this view is structural. Craig and Walton (1959) have reinterpreted the structure of the SW Southern Uplands as comprising a series of compound monoclines descending northwestwards and cut by N-dipping reverse faults (high-angle thrusts). Successive large-scale blocks of strata bounded by major reverse faults are found to be younger towards the southeast, but *within* each block the stratigraphic sequence ascends northwestwards (Fig. 5.11). McKerrow *et al.*

(1977*) have recently compared this structure to an idealised trench slope model. Close to the trench margin, underthrusting of the oceanic lithosphere is thought to produce blocks of sediment bounded by low-angle thrusts directed oceanwards and containing folds facing oceanwards (Seely *et al.* 1974*; Fig. 5.12). The youngest blocks occur closest to the trench, but within each block the sedimentary sequence ascends in a landward direction. As underthrusting continues, the thrust blocks are rotated through 40–50° and become bounded by steeply dipping high-angle thrusts (reverse faults). This is broadly the structure that appears to be preserved in the Southern Uplands.

The trench model has further implications for the formation of cleavage and the phases of deformation in the Southern Uplands. In Wigtownshire, in the SW Southern Uplands (Rust 1965), the first phase of deformation (D1) developed NW-facing mono-

Figure 5.11   The structure of the Ordovician–Silurian succession of the Southern Uplands (after Dewey 1969b*).

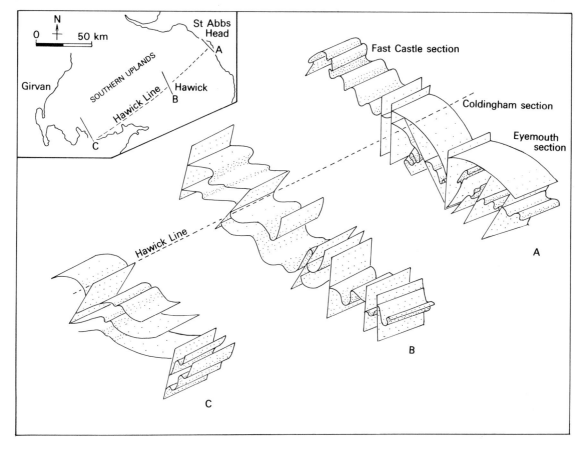

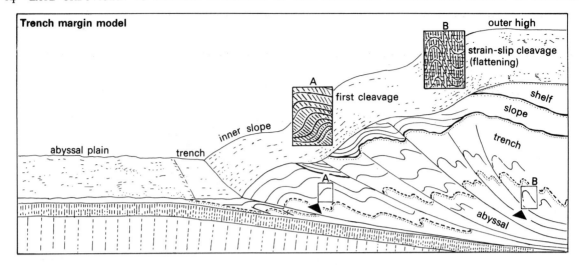

Figure 5.12 Trench margin model (after Seely *et al.* 1974*). Note the progressive rotation of the folds and the planes of underthrusting. Inset A — formation of a first cleavage associated with the development of folds; inset B — formation of strain-slip cleavage as further underthrusting causes flattening of the rotated folds.

clines. Some cleavage development and steepening of fold plunges took place towards the end of this first phase. During the second phase (D2) S-facing monoclines were superimposed on to the D1 monoclines, resulting in conjugate box folds. A further phase (D3) of NW—SE shortening produced a strike-slip (wrench) fault regime. However, the fourth phase (D4) was marked by vertical shortening. This caused the development of flat strain-slip and crenulation cleavage. The final phase of deformation (D5) resulted in kink bands. Northeast of Wigtownshire in the central and NE Southern Uplands there is some variation in the style and facing direction of the folds. At Hawick asymmetrical folds verge towards a synclinal axis. Further northeast, along the Berwickshire coast, the D1 folds at East Castle are upright, although those at Coldingham are recumbent and downward-facing towards the southeast.

Following the trench slope model of Seely *et al.* (1974*), the monoclinal folds (D1 and D2) may represent rotated folds which formerly faced the closing Iapetus Ocean (Fig. 5.13). The strike-slip faults represent a phase of axial extension. As underthrusting continued, squeezing of the rotated thrust blocks, containing the folds resulting from the early phases of deformation, might have caused flattening and induced the formation of strain-slip cleavage (D4). The final phase of deformation might

have been associated with the time of collision.

The trench margin model predicts the oceanward transportation of sediment over the growing pile of imbricate thrust blocks, and the possession by the inner slope of a moderately undulating topography aligned parallel to the continental margin. The lithic sands of the Kirkcolm Group were emplaced by turbidity currents (or geostrophic flows) which flowed along a depression possibly situated at the foot of the contemporary continental slope. Muds and siliceous oozes (radiolaria) accumulated on an adjacent rise to the south. Volcanism, which had been important during the Arenig and Llanvirn, gradually declined in the Caradoc and Ashgill. In modern oceans the continued underthrusting of sediment blocks at the continental margin may eventually cause shoaling (or even emergence) of the pile in the vicinity of the shelf edge. This is exemplified by the present-day shelf of S. Oregon near 44°N (Kulm & Fowler 1974). Such an event appears to be recorded in the Silurian of the Southern Uplands and the Midland Valley (McKerrow *et al.* 1977*). The turbidites of the Hagshaw Hills in the Midland Valley indicate that in late Llandovery times lithic wacke and shale detritus was emplaced by currents flowing from the *south*. Clearly the former Midland Valley block had started to subside, resulting in the invasion of the sea. It is generally thought that the northern part of the imbricate pile of

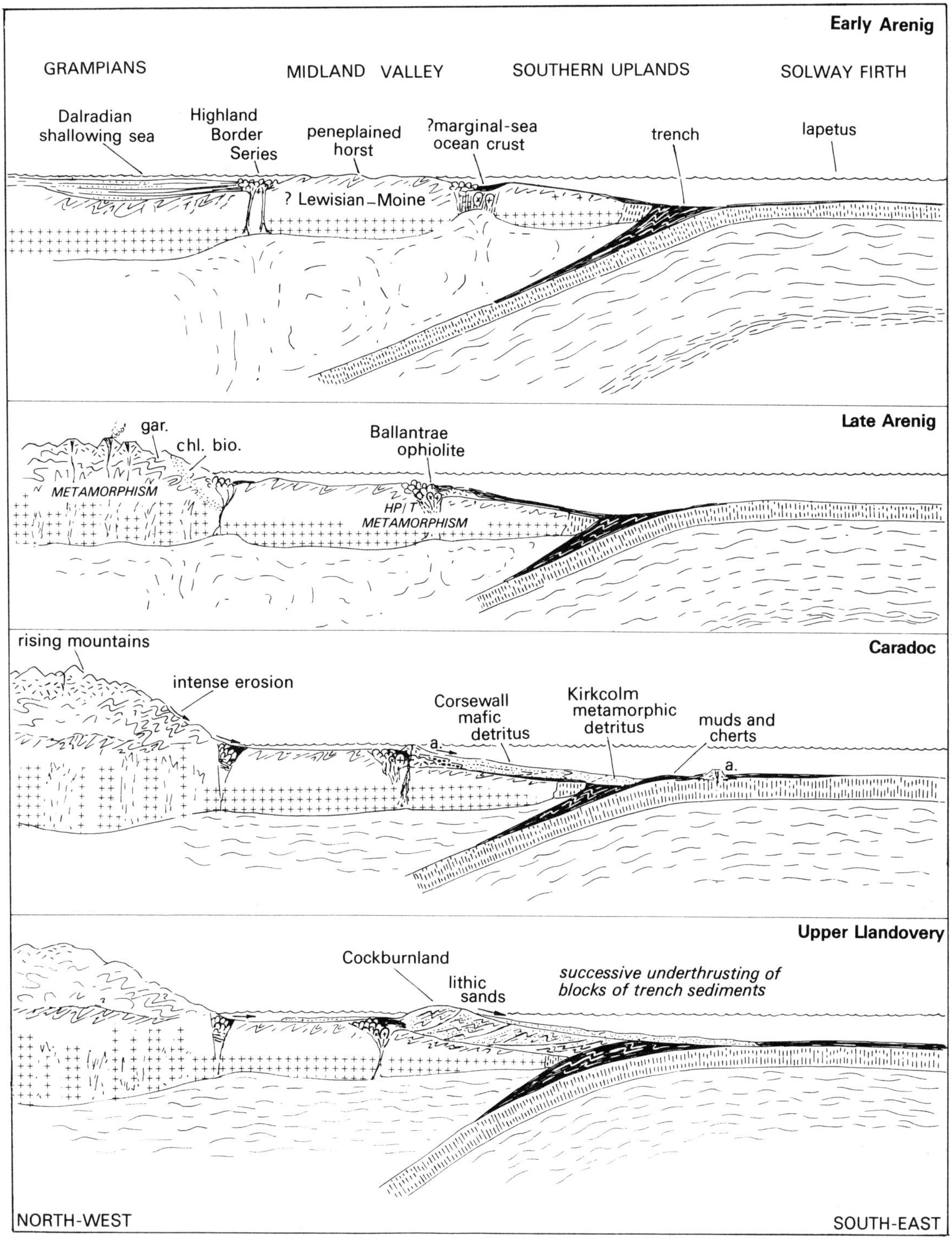

Figure 5.13 Hypothetical evolution of the northern continental margin in Ordovician–Silurian times.
a. = andesite; bio. = biotite; chl. = chlorite; gar. = garnet.

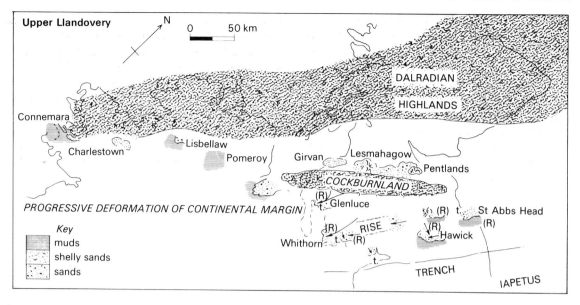

Figure 5.14   Upper Llandovery palaeogeography of the northern continental margin.
(R) = red-beds; t. = turbidites.

slope—trench sediments, just to the south of the Midland Valley Graben, was uplifted to the point of emergence in upper Llandovery times (Dewey 1971*, Mitchell & McKerrow 1975). But when Walton (1963) first invoked this hypothetical landmass, which he named Cockburnland (Fig. 5.14), he envisaged an earlier, Caradoc, origin.

The rise of Cockburnland was accompanied by a distinctive change in sedimentation; there was widespread deposition of reddened sediment. Ziegler and McKerrow (1975) have proposed that the oxidised sediment was derived from andesitic volcanoes situated in the northwestern part of the Midland Valley. However, as the newly emergent terrain of Cockburnland became weathered, it too might have supplied reddened mud. As the uplift of Cockburnland continued and the once narrow landmass widened, sands were shed northwards into the Midland Valley Graben and southwards down the continental slope. In the north the influx of sands caused shoaling and the establishment of coastal, and eventually alluvial, conditions. To the south, lithic sands spread generally southwards over the Moffat area, which may have been a contemporary rise. A thick upper Llandovery sequence of coarse sandstones and shales at Kirkcudbright, in the southernmost part of the Southern Uplands, possibly indicates that there were other

contemporary sources of sediment (Clarkson *et al.* 1975). Essentially, then, the rise of Cockburnland caused the spread of molasse facies northwards and the further extension of turbidite facies southwards over an undulating continental slope.

5j   Summary

This chapter has traced the evolution of the continental margin during the Ordovician—Silurian closing phase of the Iapetus Ocean. As deformation and metamorphism of the Dalradian pile in Ireland and Scotland commenced in basal Ordovician times, the continental margin experienced a tensional regime. Ophiolites in NW Newfoundland and at Ballantrae have been interpreted as remnants of the thin oceanic crust of a marginal sea, which were tectonically emplaced as the principal stresses became compressional. However, trace-element studies show that some of the Ballantrae lavas have continental margin affinities. Southwest of Ballantrae a tract of continental crust known as the South Mayo Trough started to subside and accumulate sediments.

During the Llandeilo and Caradoc, mafic debris was transported southwards to the northern belt of the Southern Uplands where the faulted continental margin was subsiding rapidly. Schistose detritus eroded from the rising mountains produced by the Grampian Orogeny was swept southwestwards to the foot of the contemporary continental slope. Still further south, muds steadily accumulated on an adjacent rise.

There is an impressive similarity between the structure of the Southern Uplands and the model structure of modern continental slope–trench sequences. It is therefore likely that the Ordovician–Silurian sedimentary sequences of the Southern Uplands have been telescoped by the underthrusting of the N-moving plate. By late Llandovery times underthrusting caused part of the imbricate sedimentary pile to be uplifted to the point of emergence. Sediments were shed northwards into the developing Midland Valley Graben, and southwards over the contemporary continental slope. It therefore seems that some of the deformation recorded in the Southern Uplands occurred before continental collision in Wenlock–Devonian times.

6 Precambrian evolution of the southern continental margin of Iapetus

6a Introduction

Like its northern counterpart, the southern margin of the postulated Iapetus Ocean subsided considerably during upper Precambrian times and accumulated a thick pile of marine sediments. However, unlike on the northern margin, the deposition of sediments was interrupted by lava flows and augmented by the extensive fall-out of pyroclastics during long periods of upper Precambrian volcanic activity. Upper Precambrian sediments and volcanics recording the early evolution of the southern margin of the Iapetus Ocean are located in the Maritime Provinces of Canada, S. Britain and Brittany (Fig. 6.1). They rest on a basement consisting of folded schists, quartzites and limestones, gneisses and plutons (Ch. 2d). The sequences are generally considered to range in age between 900 and 590 Ma, but such is the difficulty of dating rocks that correlation of the sequences discussed below remains speculative.

6b Maritime Provinces

Thick sequences (up to 5 km) of volcanics rest, or are inferred to rest, on the basement in New Brunswick, Cape Breton and SE Newfoundland. In Newfoundland the acidic and basic lavas and ashes of the Love Cove and Harbour Main Groups are overlain by a thick (up to 2·5 km) sequence of turbidites (Fig. 6.1, sequence 3), but elsewhere they are overlain by late Precambrian to early Cambrian molasse (sequences 1 and 2). The conglomerates and red sandstones of the molasse pass up into quartz arenites, the lithology which is almost ubiquitous at the base of the Cambrian. That these Canadian arenites *are* basal Cambrian is supported by the presence of the late lower Cambrian tilobite *Protolenus* in the

overlying shales and interbedded sandstones. In New Brunswick a late Precambrian dyke swarm intrudes the basement and forms a sill within the overlying volcanics. The sill is associated with plutons ranging from quartz diorite to adamellite, none of which intrude the Cambrian. The Holyrood Granite in SE Newfoundland, which intrudes the late Precambrian volcanics, has given Rb−Sr isochron ages of 574 ± 11 Ma 609 ± 11 Ma. There is little doubt that these plutons in the Maritime Provinces provide evidence of an important late Precambrian event.

6c Anglesey and SE Ireland

In Rosslare, SE Ireland, upper Precambrian sediments of the Cullenstown Group are thought to rest on a high-grade metamorphic basement which is probably more than 1600 Ma old (Max 1976b). But across the waters in Anglesey the situation is less clear. Baker (1969) has claimed that gneisses in Lleyn belong to the basement rather than to the overlying metamorphosed upper Precambrian Monian sediments (Fig. 6.1, sequence 5) because they show evidence of retrograde metamorphism. However, R. M. Shackleton (1975*) has regarded the gneisses as highly metamorphosed Monian, because pelitic gneisses tend to crop out close to pelitic schists and hornblende gneisses are associated with hornblende schists.

The oldest exposed beds of the Monian, the New Harbour Group, include coarse-grained greywackes displaying bottom structures, graded bedding and some cross-bedding. R. M. Shackleton (1969) has interpreted these as turbidites. Curiously the greywackes are associated with massive quartz arenites which Shackleton, noting a paucity of cross-bedding, has tentatively suggested represent

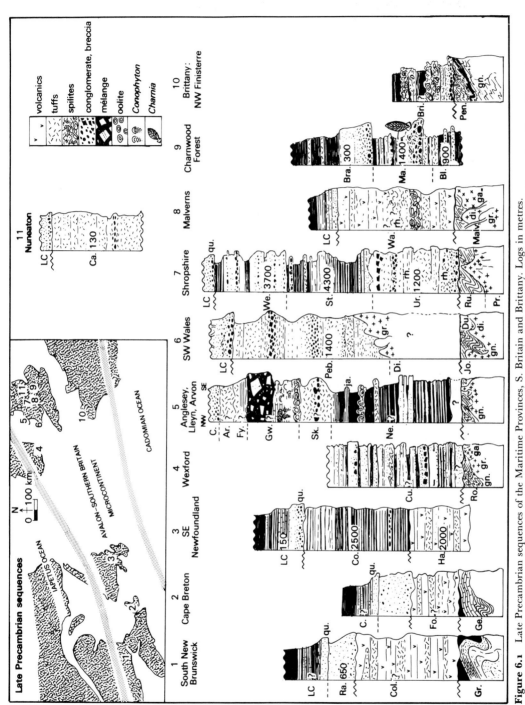

Figure 6.1 Late Precambrian sequences of the Maritime Provinces, S. Britain and Brittany. Logs in metres.

Ar. = Arvonian; Bl. = Blackbrook Formation; Bra. = Brand Group; Bri. = Brioverian; Ca. = Caldecote Volcanic Formation; Co. = Conception Group; Col. = Coldbrook Group; Cu. = Cullenstown Formation; Di. = Dimetian Complex; Du. = Dutch Gin Schists; Fo. = Forchu Group; Fy. = Fydlyn Felsitic Formation; Ge. = George Group; Gr. = Greenhead Group; Gw. = Gwna Group; Ha. = Harbour Main Group; Ho. = Holyhead Quartzite Formation; Jo. = Johnston Complex; LC = lower Cambrian; Ma. = Maplewell Group; Malv. = Malvernian; Ne. = New Harbour Group; Peb. = Pebidian; Pen. = Pentevrian; Ra. = Ratcliffe Brook Formation; Ro. = Rosslare Complex; Ru. = Rushton Schists; Sk. = Skerries Group; St. = Stretton Group; Ur. = Uriconian; Wa. = Warren House Group; We. = Wentnor Group.
di. = diorite; ga. = gabbro; gn. = gneiss; gr. = granite; ja. = jasper; qu. = quartz arenite; rh. = rhyolite; sl. = slump.
In Anglesey, Ne., Sk., Gw., Fy and Ar. = Monian; in Shropshire, St. and We. = Longmyndian; in Charnwood Forest, Bl., Ma. and Bra. = Charnian.

subaqueous sandflows. Wood (1974) has discovered large flutes on the soles of the quartz arenites and supported this interpretation.

The argillaceous sediments that follow are characterised by pillow lavas and beds of jasper. These pass up into coarser tuffaceous beds with boulders of felsite and granodiorite and pebbles of green and purple mudstone, white quartzite and jasper. The succeeding Gwna Group shows continuing volcanic activity and also provides valuable clues as to the contemporary palaeogeography. In NW Anglesey the Gwna Group features quartz arenites with oölitic and stromatolitic limestones, but to the southeast these lithologies thin and are replaced by pillow lavas and jasper. R. M. Shackleton (1969) has postulated a SE-facing palaeoslope. The outstanding member in the sequence is the thick (over 900 m) Gwna mélange, which consists of blocks, cobbles and pebbles of limestone, quartzite, grit and pillow lava, enveloped in a matrix of mud and sand. The mélange crops out intermittently across the strike over 80 km and includes some blocks several kilometres in size. There is, however, some doubt as to the age of the mélange. Despite its Precambrian context it has yielded some late Cambrian to early Ordovician acritarchs. There are no striations on the blocks to suggest a glacial origin, and the absence of a tuffaceous matrix is incompatible with a volcanic mudflow or lahar. R. M. Shackleton has interpreted the mélange as a subaqueous slide breccia or olistostrome. Modern olistostromes are indicators of gross tectonic instability. Moore *et al.* (1976) have reported examples from the foot of the narrow southwestern shelf bordering the Bay of Bengal, where an oceanic lithospheric plate is being subducted beneath a continental plate to the east. However, olistostromes are also known from 'passive' continental margins. Above the Gwna Group, the top of the Monian succession is marked by a sequence of sodic volcanics.

As in the Maritime Provinces there was a late Precambrian thermal event. The Monian succession was folded and metamorphosed. The metamorphism locally attained high grades. The sediments described above pass into marbles, mica and graphite schists, hornblende and biotite gneisses and migmatites. In Lleyn, gradational contacts between sediments and metamorphosed strata indicate that at least some gneisses are metamorphosed Monian sediments rather than basement. Outcrops of glaucophane schist in S. Anglesey have aroused special interest. Lawsonite and pumpellyite indicate that the schist, which is associated with greenschist and local gneiss, was formed under conditions of high pressure but low temperature (2−6 Kb, 100−300 °C).

Acidic and basic plutons were also emplaced. The earliest granite was the potassium-rich Coedana Granite, which is comparable in age with the Holyrood Granite in Newfoundland. Dates from the Coedana Granite and its hornfels suggest that it was emplaced 580−600 Ma ago. The basic plutons consist of gabbros and serpentinites. According to Thorpe (1974*), their composition is similar to equivalent rock types on modern ocean floors.

6d The Welsh Borderland and the Midlands

The upper Precambrian successions of this region bear a general resemblance to those of the Maritime Provinces and Anglesey. Volcanic activity was intense and there is further evidence for a late Precambrian thermal event. In Shropshire, as in Anglesey, there is controversy as to the status of small inliers of highly metamorphosed rock. Baker (1973*) has interpreted the outcrops of garnetiferous Primrose Hill Gneiss and Rushton Schist as part of a regionally metamorphosed basement. But according to Wright (1969), the exposures of sheared tuff and granophyre (Primrose Hill 'Gneiss') and epidote−quartz−mica schist (Rushton Schist) are dynamically metamorphosed zones of Uriconian volcanics and Longmyndian sediments respectively (Fig. 6.1, sequence 7). Thorpe (1974*), in support of Baker, has described the gneiss as comprising irregular bands of green tremolite alternating with leucocratic bands of quartz, plagioclase and alkali feldspars which are clear and untwinned or show microcline twinning. Compared with intermediate calc-alkaline igneous rocks the gneiss is high in MgO and K_2O but low in Al_2O_3 and Na_2O. This unusual composition has suggested to Thorpe an origin from hybrid igneous rocks, representatives of which are unknown in the Uriconian.

The Uriconian at the Wrekin comprises some 1500 m of rhyolitic lavas, pyroclastic breccias, agglomerates and tuffs. The lavas are spherulitic, banded or porphyritic, and the tuffs are locally welded. The breccias, with fragments of andesite, oligoclase andesite and dacite, record the presence of more intermediate volcanic-rock types. The Uriconian was intruded by granophyre and a swarm of dolerite dykes. The overlying Longmyndian consists of two series. The lower, the Strettonian Series, is dominantly argillaceous. Grey, green and purple shales are intercalated with thin fine sandstones and occasional tuffs. The sediments occasionally exhibit sole structures, grading and convolute lamination, but generally it is hard to discern sedimentary structures in the field. The Wentnorian Series above is dominantly rudaceous and arenaceous but includes grey, green and purple shales. Wentnorian pebbles and granules include soda granite, rhyolite, felsite, andesite, silicified tuff, jasper, quartzite, quartz, oligoclase and garnet. Broadly, the succession has been interpreted as a regressive sequence passing from marginal marine to fluvial, but more detailed sedimentological investigations are required. The source of fragments has been traced to the Uriconian and the Monian.

The age of the Uriconian and Longmyndian is problematical. It was formerly thought that these rocks were more than 700 Ma old, but Fitch *et al.* (1969) have obtained K−Ar ages of 677 ± 72 Ma and 632 ± 32 Ma for rhyolites from the Uriconian at the Wrekin. A recent proposal for the age of the Longmyndian sediments is consistent with these figures. Bath (1974) has conducted Rb−Sr isotopic analyses of the illitic shales of the Strettonian Series and obtained isochrons of 452 ± 31 Ma and 529 ± 6 Ma, which he has interpreted as dating the final movements of ions by pore waters. Extrapolation of the $^{87}Sr/^{86}Sr$ ratios back to reasonable (assumed) initial values indicates that the time of deposition was about 600 Ma ago. Bath has proposed that the deformation of the late Precambrian Longmyndian pile took place very soon after deposition. The Cambrian isochrons may indicate subsequent mild metamorphic events.

To the south, in the Malvern Hills, the Malvernian represents basement composed of schists, sheared diorites and tonalites associated with foliated granite, riddled by syenite and granite pegmatite veins (Fig. 6.1, sequence 8). But the K−Ar mineral and Rb−Sr biotite isotopic dates are late

Precambrian, indicating an important event about 590 ± 20 Ma ago (Lambert & Rex 1966). R. St J. Lambert (1969) has since quoted an Rb−Sr whole rock isochron at 635 Ma. The basement is associated with the Warren House Volcanics, a sequence consisting of rhyolites, soda trachyte, spilites and tuffs, but the age of these volcanics is uncertain. To the east, in the Midlands, upper Precambrian Charnian volcanics and sediments exceed 2·6 km in thickness (Fig. 6.1, sequence 9). Tuffs − lithic and crystal, coarse and fine, sometimes banded or graded − seem to predominate in the lower part of the Charnian sequence. Lapilli include clasts of porphyritic andesite, rhyolite, vein quartz, well-foliated quartzite and aggregates of epidote, chlorite and secondary feldspar. Evans (1968) has suggested that certain slump horizons indicate a N-facing palaeoslope. One horizon, known as 'slate agglomerate', has many of the characteristics of a lahar. Contorted fragments of fine tuff, and dust tuff, together with lapilli of rhyolite and andesite, are set in a coarse tuffaceous matrix. The contorted fragments of tuff in this unit suggest a westward flow. Of special interest are the tuffaceous silts resting on the 'slate agglomerate'. These have yielded blocks featuring frond-like impressions with basal discs, which have been described and demonstrated by Ford (1968) to represent late Precambrian metaphytes or metazoans. The affinities of these organisms *Charnia* and *Charnodiscus* are uncertain; they have been compared with siphonalean algae and modern sea pens. The upper part of the Charnian sequence comprises conglomerates, sandstones and purple and grey−green slates which are comparable in lithology with the Wentnorian. The succession of pyroclastic sediments was intruded by porphyritic dacitic and rhyodacitic volcanics, some of which might have been vent plugs. These have given K−Ar whole-rock dates of 684 ± 29 Ma and 595 ± 26 Ma (Meneisy & Miller 1963) which confirm the late Precambrian age. During late Precambrian and Cambrian times the Charnian sequence was intruded by a number of granophyric diorites. The intrusions divide into two groups: the Charnwood diorites, and the S. Leicestershire diorites. The Charnwood diorites are further divided into a northern and southern group. The southern diorites give a Rb−Sr isochron of 552 ± 58 Ma, but the northern diorites are Upper Palaeozoic. Using

Rb−Sr dating Cribb (1975) has shown the S. Leicestershire diorites to be of a similar age (546 ± 22 Ma) to the S. Charnwood plutons. Cribb has further demonstrated that the $^{87}Sr/^{86}Sr$ ratios of the Precambrian−Cambrian intrusions are low (0·7056−0·7061) compared with the average for Phanerozoic granites of N. America (0·707).

6e S. Wales

Tightly folded quartzose metasediments, the Dutch Gin Schist, appear to represent the basement rocks in Dyfed (Pembrokeshire), SW Wales. Wright (1969) has suggested that these represent formerly high-grade garnetiferous schists that have been subjected to retrogressive greenschist metamorphism and degraded to their present state of chlorite−quartz schist. Closely associated with the schists are the acidic and basic gneisses of the Johnstone Complex. The latter also includes acidic and intermediate granites, tonalites, quartz diorites and hybrids cut by pegmatite veins. The schists, gneisses and plutons of S. Dyfed are accompanied by a poorly exposed sequence of spherulitic rhyolites, tuffs and pyroclastic breccias known as the Benton Volcanics. The latter have given K−Ar dates of 613 and 625 Ma (R. M. Shackleton 1975*), but their age relationship to the above rocks is uncertain.

In N. Dyfed tuffaceous rocks dominate the Precambrian. A thick succession of rhyolites, quartz keratophyres, feldspathic and banded silicic tuffs (halleflinta), altogether known as the Pebidian, is intruded by a plutonic group called the Dimetian. The intrusives are probably contemporaneous in part with the Pebidian extrusives and include quartz porphyries, granophyres and soda granites. Towards the end of Precambrian times the volcanics were folded, the plutons were uncovered by erosion, dolerite dykes were intruded and the terrain was deeply weathered (R. M. Shackleton 1975*).

6f Brittany and Normandy

In Brittany and Normandy the upper Precambrian succession, the Brioverian, rests on Pentevrian basement. Brioverian sequences appear to vary both W−E and N−S. In NW Brittany the succession commences with shales, spilitic lavas and associated siliceous shales. These coarsen upwards into interbedded greywackes and shales. To the east, in N. Cotentin, the marine sediments and volcanics rest on thin conglomerates and feldspathic sandstones carpeting a weathered Pentevrian basement. But to the south, in S. Brittany, the sequence is ophiolitic and a major E−W fault zone is associated with blueschists and eclogites. The age of these blueschists and eclogites is, however, uncertain; they may be Upper Palaeozoic.

Towards the end of Precambrian time the Brioverian succession was subjected to deformation, plutonism and metamorphism. In NW Brittany the 'Gneiss de Brest' Granodiorite was emplaced about 690 ± 40 Ma ago, after the main phase of folding but before the metamorphism. In Guernsey the foliated l'Eree Adamellite (660 ± 25 Ma) was intruded during the phase of compression (Bishop et al. 1975). The igneous activity continued well into Cambrian times. The Cobo Adamellite on Guernsey is dated at 570 ± 15 Ma and the NW and SE Jersey granites are 520 ± 4 and 490 ± 15 Ma respectively. Clearly plutonism continued as Brioverian sediments were eroded and the Cambrian seas transgressed.

6g Evolution of the late Precambrian margin: rival hypotheses

Our knowledge of the late Precambrian evolution of the southern continental margin is far from satisfactory. This stems from the nature of the record discussed above. Rock sequences have been contorted and dislocated; sedimentary structures and textures have been masked by the effects of diagenesis and metamorphism; and isotope ratios (so precious for correlation) have suffered from the effects of subsequent thermal events. This low level in the quality of the information on the original nature of the deposits has resulted in the development of several distinctly different hypotheses. Most authors accord with the view that crustal tension, rifting and distension took place in the Precambrian. Certainly during Cambrian times the Iapetus Ocean was sufficiently wide to cause provincialism of the trilobite fauna (Ch. 3e). The general problem is to produce an hypothesis that satisfactorily explains the late

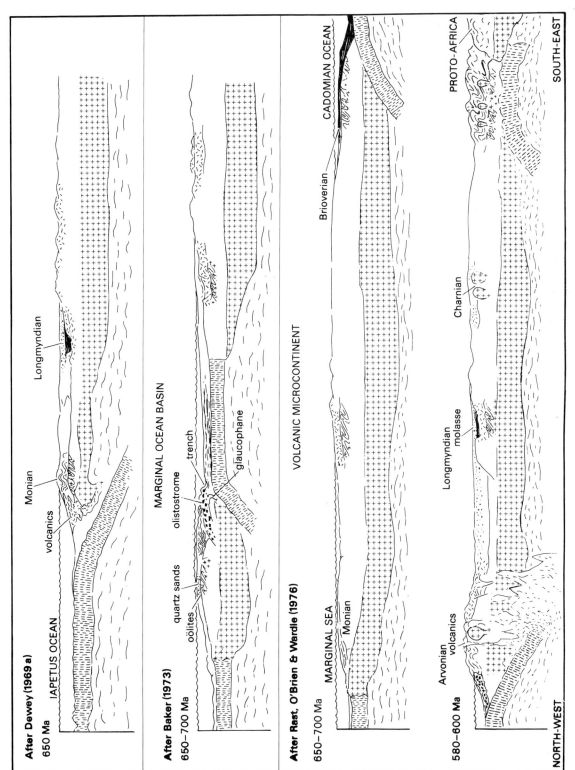

Figure 6.2 Rival hypotheses for the late Precambrian evolution of the southern continental margin.

Precambrian Celtic deformation, metamorphism and plutonism in SE Newfoundland, S. Britain and Brittany.

Hypothesis 1

J. F. Dewey (1969a*) developed Wilson's thesis that in Palaeozoic times a proto-Atlantic ocean, Iapetus, had existed in the general vicinity of the modern Atlantic (Ch. 3b). He proposed that the Longmyndian and Monian constituted respectively the fluviatile–coastal and offshore marine sediments on the southern margin of the newly opened ocean (Fig. 6.2). By late Precambrian times (c. 650 Ma ago) a thick pile of sediments tapered northwestwards on to the mafic ocean crust. Dewey suggested that towards the end of Precambrian times the formerly passive continental margin transformed into an active margin as southeastward subduction of the ocean crust commenced. A trench was envisaged just northwest of Anglesey. Frictional heating in the Benioff zone was considered responsible for the eruption of volcanics, the emplacement of plutons and the thermal metamorphism.

Hypothesis 2

In 1973* J. W. Baker proposed major modifications to Dewey's hypothesis. He argued that Dewey had not taken into account the evidence for shallower water towards NW Anglesey during the deposition of the Monian. This evidence supported the proposition that an area of deeper water lay between a shallow water bank in the vicinity of Anglesey and the main shoreline to the southwest. Baker also noted that the style of the folds in Anglesey, upright to S-facing, was not compatible with deformation resulting from a SE-dipping subduction zone. Baker therefore proposed that as the Iapetus Ocean rifted and subsequently widened, ocean crust was emplaced between SE Ireland, Anglesey and the main part of the continental margin to the southwest (Fig. 6.2). An analogy was drawn between this late Precambrian marginal basin and the Rockall Trough situated on the eastern margin of the modern N. Atlantic. The Rockall Trough, now filled with nearly 2 km of sediments, is underlain by ocean crust which appears to date from the earliest phase of Atlantic floor spreading (D. G. Roberts 1974).

According to Baker, the late Precambrian compressional regime may have been the result of the development of a subduction zone dipping northwestwards beneath Anglesey. As the ocean crust was consumed, so the basin contracted. Slices of ocean crust were serpentinised and thrust upwards into the sediments above. The high pressure and low temperature conditions associated with the higher levels of the subduction zone were considered responsible for the glaucophane schist. Close by, the underthrusting of ocean crust generated the olistostrome which incorporated blocks of sediment and ophiolite.

Thorpe, in his 1974* analysis, also envisaged a late Precambrian basin floored by ocean crust, but he differed over the nature of the subduction. The postulated NW-dipping subduction zone did not account for the isotopic results obtained from the Precambrian rocks of the Welsh Borderland, nor for the emplacement of late Precambrian plutons in the Midlands. Thorpe presented two solutions, each invoking subduction along two zones. In one proposal a SE-dipping subduction zone was sited beneath the Midlands as a counterpart to the zone envisaged by Baker. In the second, Thorpe invoked a pair of SE-dipping subduction zones: one descending from a trench positioned northwest of Anglesey, the other descending from a trench in the vicinity of the Welsh Borderland.

Hypothesis 3

In 1976* Rast et al. pointed out that the successful hypothesis must consider the evidence from the Maritime Provinces, S. Britain and Brittany. Baker's (1973*) envisaged subduction zone dipping northwestwards beneath Anglesey accounted for the deformation and metamorphism of the Monian, but it did not explain the late Precambrian granites and volcanics of Avalon in SE Newfoundland; it was positioned too far north. In their proposal Rast et al. centred attention on the orogenic belt of central and S. Brittany, a site that they regarded as on the suture of a Cadomian ocean (Fig. 6.2). Closure of this ocean culminated in the collision of a 'European–N.American' plate and an 'African' plate. Rast et al. tentatively suggested that in its early stages the opening Iapetus Ocean was genetically a back-arc marginal sea floored with ocean

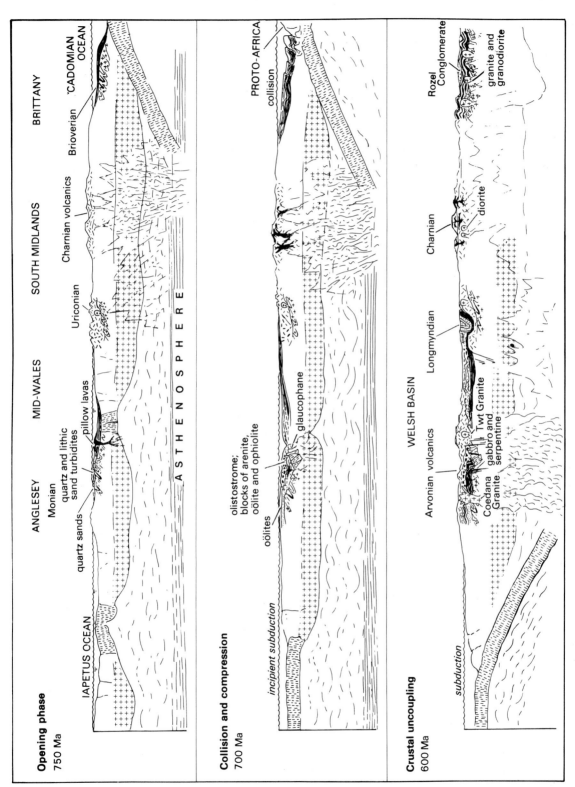

Figure 6.3 Hypothetical evolution of the southern continental margin in late Precambrian times.

crust. Hence the Avalon—S. Britain zone represented a volcanic microcontinent separated from the 'European—N.American' plate to the north. S. Brittany, with its late Precambrian(?) blueschists, ophiolites and eclogites, was considered the site of a trench associated with a major zone of subduction which passed northwestwards beneath the volcanic microcontinent. Compression of the Monian sequence was regarded as tectonism at the edge of the back-arc basin. Rast *et al.* incorporated Wood's (1974) suggestion that the granites of Anglesey and Arvon are the plutonic equivalents of the Padarn —Arvonian rhyolitic volcanics.

Early in Cambrian times, as ocean floor spreading continued in the back-arc marginal sea, the associated tensional regime caused rifting and subsidence of the Welsh and Nova Scotian Basins. By middle to upper Cambrian times the former marginal sea had become the Iapetus Ocean.

6h Synthesis and summary

All three hypotheses postulate the emplacement of ocean crust about or before 700 Ma ago. M. J. Kennedy (1976) has suggested that the Harbour Main and Love Cove Volcanics of SE Newfoundland represent this phase of distension, but there is a paucity of evidence. As Rast *et al.* (1976*) have pointed out, this early ocean crust might have represented a marginal sea associated with a major contracting Cadomian ocean lying to the south. If this is correct, then for much of late Precambrian times a zone of subduction would have dipped northwestwards beneath S. Britain (Fig. 6.3). It is a matter of speculation as to whether the Charnian volcanics (*c.* 700—600 Ma) represent activity of this zone. Cribb (1975) has argued that the low $^{87}Sr/^{86}Sr$ ratios indicate either partial melting of the basal crust or mixing of basaltic and silicic material during partial melting of an oceanic plate along a subduction zone.

R. M. Shackleton's (1969) evidence for the presence of a SE-facing palaeoslope during the deposi-

tion of the Monian in NW Anglesey clearly indicates an element of continental margin topography. The Monian may represent the sediment infill of a shelf basin partially floored by a narrow segment of oceanic crust. The late Precambrian was probably the time of initiation of the NE—SW-trending faults of the Anglesey region — faults that were later to act as pathways for rising magma and to delimit the northwestern margin of the Palaeozoic Welsh Basin. One fault at least, the Dinorwic (Fig. 8.3), remains active today (Wood 1974).

Around 700 Ma ago the state of crustal tension transformed to one of crustal compression and heating. The time of collision between the 'African' plate and the Avalon—S. Britain microcontinent of the 'European—N. American' plate is clearly implicated. This collision not only caused deformation and plutonism (the Celtic Orogeny) at the site of impact, but was probably also responsible for the disruption of the Monian sedimentary pile. The collapse of resultant fault scarps and steep slopes perhaps provided the blocks that formed the olistostrome. Subduction of the postulated narrow ocean crust might have produced the S-facing folds noted by Baker (1973*) and caused the formation of the high-pressure and low-temperature minerals of the glaucophane schists. The serpentinites may be supposed to represent sheared slithers of ocean crust thrust into the succession by the process of subduction, but Maltman (1975) has found no evidence for forceful emplacement in either the serpentinite bodies or the overlying sediments. Maltman has postulated the ascent of the mafic and ultramafic magmas up deep vertical fractures.

Subduction of ocean crust northwestwards directly beneath Anglesey (Fig. 6.2) does not seem an adequate explanation of the granitic plutons and volcanics of the very late Precambrian. The site of the trench would be improbably close to the site of magma genesis and intrusion. Yet there was clearly a major source of heat. Frictional heating resulting from the partial underthrusting of the 'African' plate beneath the Avalon—S. Britain microcontinent might have been responsible (Fig. 6.3).

7 Lower Palaeozoic evolution of the southern continental margin of Iapetus

7a Introduction

At the dawn of the Phanerozoic *c.* 590 Ma ago, the areas now comprising Scandinavia, England, Wales, S. Ireland and the southeastern parts of the Maritime Provinces probably formed part of the NW-facing continental margin of the Iapetus Ocean (Fig. 7.1). During the late Precambrian the margin had been subjected to considerable subsidence, volcanism, plutonism, metasomatism and deformation (Ch. 6), but early in Cambrian times this activity declined as a major transgression flooded the continental margin. In Scandinavia the sea spilled eastwards from late Precambrian basinal areas. In England and Wales the sea submerged areas of subsidence (e.g. the Welsh Basin) and relative stability (e.g. the Midland Platform) alike.

However, the lower Cambrian sea did not transgress the entire land surface. Probable Cambrian fluviatile sediments are recorded from the Channel Isles, and in parts of central Sweden it is the middle rather than the lower Cambrian that rests on the Precambrian.

The lower Palaeozoic continental margin consisted of two parts: (a) a northwestern (oceanward) zone characterised by structural basins accumulating thick sequences of shallow- and deep-water sediments; and (b) a southeastern (landward) platform zone comprising areas of relative stability accumulating thin sequences of shallow water origin. This chapter will trace the evolution of the shelf basins and platforms during the Lower Palaeozoic. The development of one shelf basin, the Welsh Basin, will be followed in detail.

Figure 7.1 Late lower Cambrian palaeogeography of the southern continental margin. FB = Finnmark Basin; LB = Leinster Basin; WB = Welsh Basin. t. = turbidites.

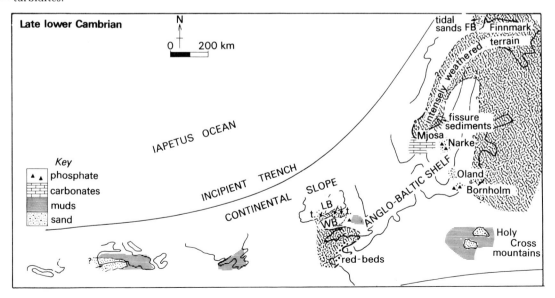

7b The Cambrian intrashelf Welsh Basin

The Welsh Basin, an area of approximately 200 × 100 km, was flanked to the northwest by a periodically emergent platform situated in the region now extending from Anglesey across to SE Ireland. To the southeast lay the platform area of the Welsh Borderland and English Midlands. Subsidence commenced in late Precambrian or lower Cambrian times and by Ludlow times, when subsidence gave way to uplift, the basin had locally accumulated several kilometres of sediments, ashes and lavas. Indeed by the end of the Cambrian period alone (520 Ma ago), the northern parts of the basin had subsided 1−2 km.

At St Davids in SW Wales, on the southern margin of the Welsh Basin, lower Cambrian marine gravels were deposited over a weathered Precam-brian plutonic and volcanic terrain; but in Arvon, on the northern margin of the Welsh Basin, undated terrestrial rhyolitic lavas with ignimbrites pass up gradationally into conglomerates and sandstones assumed to be of early Cambrian age. A borehole recently sunk in the Harlech Dome, in northern −central Wales, penetrated volcanics at the base of the lower Cambrian Dolwen Grits (Fig. 7.2).

The basal Cambrian conglomerates of SW Wales (Fig. 7.3) display large-scale planar cross-stratification and channel structures. The foresets suggest that palaeocurrents directed southwards (? shorewards) dominated. The finding of epsilon cross-stratification, suggestive of point-bar deposition, accompanied by the vertical cylindrical burrows of *Skolithos*, has led Crimes (1970a*) to invoke an intertidal origin. The basal conglomeratic units pass up into coarse green feldspathic sandstones.

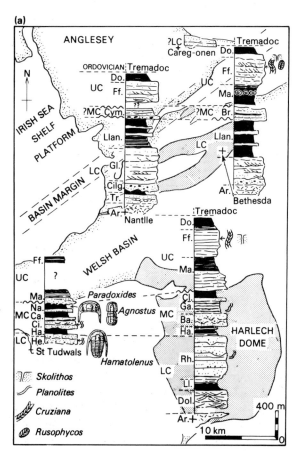

Figure 7.2a Cambrian successions in the northern part of the Welsh Basin.
Ba. = Barmouth Grits; Br. = Bronllwyd Grits; Ca. = Caered Mudstones; Ci. = Cilan Grits; Cilg. = Cilgwyn Grits; Cl. = Clogau Shales; Cym. = Cymffyrch Grits; Do. = Dolgelley Beds; Dol. = Dolwen Grits; Ff. = Ffestiniog Beds; Ga. = Gamlan Shales; Gl. = Glog Grit; Ha. = Hafotty Formation; He. = Hell's Mouth Grits; LC = lower Cambrian; Ll. = Llanbedr Slates; Llan. = Llanberis Slates; Ma. = Maentwrog Beds; MC = middle Cambrian; Na. = Nant-pig Mudstones; PaV = Padarn Volcanics; Prec. = Precambrian; Rh. = Rhinog Grits; UC = upper Cambrian.

Figure 7.2b Location of Cambrian outcrops in the Welsh Basin and Midland Shelf Platform.

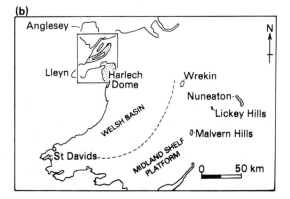

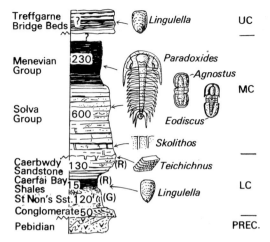

Figure 7.3 The Cambrian succession at St Davids in the southern part of the Welsh Basin (for location see Fig. 7.2b). Log in metres.
Prec. = Precambrian; LC = lower Cambrian; MC = middle Cambrian; UC = upper Cambrian.
(G) = green sediments; (R) = red sediments.

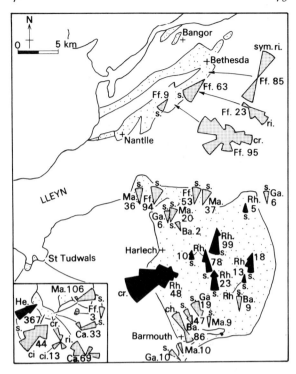

Figure 7.4 Cambrian palaeocurrent evidence from the northern part of the Welsh Basin (for location see Fig. 7.2a). Figures indicate number of readings (data from Crimes 1970a*).
Vectors in solid black = lower Cambrian; vectors in stipple = middle and upper Cambrian.
ch. = channels; cr. = cross-stratification; ri. = ripples; s. = solemarks; sym. ri. = symmetrical ripples.
For key to formations see Figure 7.2a.

The latter are not structureless, as at first they appear; they are in fact intensely bioturbated. Since shelly fauna is lacking, the evidence indicates the former presence of a flourishing soft-bodied infauna. The sandstones grade upwards into red shales characterised by thin graded tuffs. Sporadic burrows and a limited body fauna comprising the inarticulate brachiopods *Lingulella* and *Discina*, and the bradoriid crustacean *Indiana*, have been recorded. The shales in turn pass up by intercalation into laminated and cross-laminated red sandstones. These lack body fossils but display the deposit-feeding trace *Teichichnus*. Overall the sequence is compatible with a shelf origin; but Crimes (1970a*), noting the lack of symmetrical wave-formed ripples in the non-burrowed horizons, has favoured deposition below the effective wave base (100−200 m). The sequence of sandstones is terminated by a marked erosion surface with the deep and narrow crevices first recorded by Jones (1940).

The lower Cambrian of N. Wales similarly records deepening of the sea. Basal gravels passing up into large-scale cross-stratified unfossiliferous sands possibly reflect shallow water deposition, but there are no shells or animal tracemarks recorded. The overlying shales, although purple and green in colour, contain graded and cross-laminated siltstones and therefore compare with the equivalent red shales of SW Wales. There are, however, two important differences in that the N. Wales shales (a) lack both body and trace fossils, and (b) include coarse, graded and laminated sandstones of probable turbidity-current origin. Both differences point to a substantial deepening of the water in the northern part of the Welsh Basin during lower Cambrian times. This deepening is confirmed by the widespread occurrence of turbidite and mass flow deposits of late lower Cambrian age in N. Wales. The Rhinog Grits (*c.* 700 m) exposed in the Harlech Dome correlate with the Hell's Mouth Grits, a turbidite formation to the west in Lleyn (Fig. 7.2). Solemarks indicate, clearly and consistently, palaeocurrent flow towards the south and southwest (Fig. 7.4). These sands probably repre-

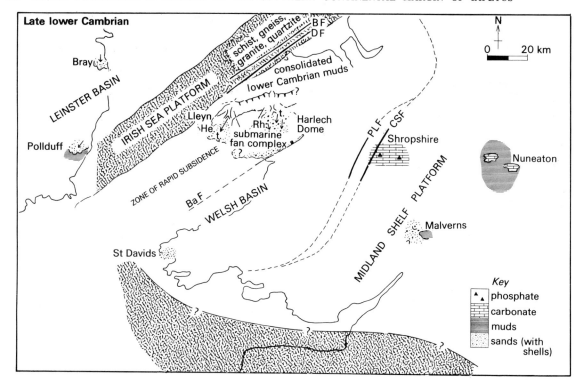

Figure 7.5   Late lower Cambrian palaeogeography of the Welsh Basin.
He. = Hell's Mouth Grits; Rh. = Rhinog Grits.
t. = turbidites.
BF = Berw Fault; BaF = Bala Fault; DF = Dinorwic Fault; PLF = Pontesford–Linley Fault; CSF = Church Stretton
Fault.

sent part of a late lower Cambrian submarine-fan complex (Figs 7.5 and 7.6).

Many of the pebbles and cobbles in the lower Cambrian of SW Wales match with the silicic tuffs of the Pebidian and the granophyres of the Dimetian (Ch. 6e). Fragments of jasper and schist, however, were probably derived from rocks no longer exposed. In N. Wales pebbles of quartzite, albite, granite, jasper and greenschist compare with rocks exposed to the northwest in Anglesey. Blue quartz grains probably come from older high-grade metamorphic rocks. Such formations occur in the Rosslare Complex of SE Ireland and possibly cropped out in the Irish Sea area in Cambrian times.

The earliest middle Cambrian sediments of N. Wales show evidence of a pause in sedimentation, but there is no indication that the sea shallowed. Turbidite deposition ceased, but muds substantially enriched in manganese continued to accumulate. The proportion of manganese in the shales (1·2%)

is of an order of magnitude greater than in the lower Cambrian Llanbedr Shales (0·12%). At one horizon in the Harlech Dome, a finely laminated 'ore' bed shows a hundredfold enrichment (12·3% manganese) but is quite unlike the spheroidal manganese nodules located on modern sea floors. Following a detailed study of the manganese-rich shales, Mohr (1964) has postulated intense weathering of a source terrain comprising spilitic and keratophyric lavas. He has further suggested that the sea was isolated under conditions of strong evaporation. The manganese was precipitated with silica in a carbonate form.

Middle Cambrian times next witnessed the initiation of a new phase of sedimentation in the foundering Welsh Basin. To the southwest sedimentation resumed. In addition to the dwelling burrow *Skolithos*, the basal sandstones of the Solva Series display large-scale planar cross-stratification, channels and wave-formed ripples, all signs of shallow water deposition. Crimes (1970a*) has

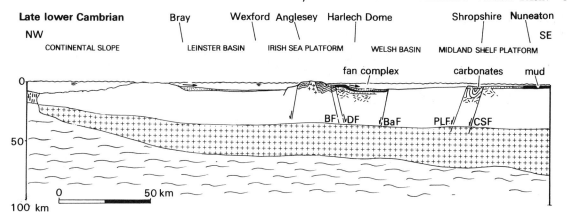

Figure 7.6 Hypothetical crustal section of the southern continental margin in late lower Cambrian times. For key to faults see Figure 7.5.

described reddened sandstones interbedded with green shales containing the horizontal burrows *Planolites* and *Sinusites*. In N. Wales the rate of sedimentation intensified as manganiferous muds gave way to coarse sands introduced by turbidity currents. But compared with the lower Cambrian turbidites the middle Cambrian sands show less consistency in derivation. Whereas in Lleyn they record SW-flowing palaeocurrents, those in the Harlech Dome indicate transport towards the northwest. Thus in the Harlech Dome there was a reversal of palaeocurrent trend, a switch from a northern to a southern source (Fig. 7.4).

During later middle Cambrian times the influx of coarse sands into the northern part of the Welsh Basin became spasmodic. Graded sands and cross-laminated silts were separated by sequences of mud. Current-formed sedimentary structures indicate that northward sediment transport now operated in Lleyn as well as to the east, but the vectors show wide variation. Gradually the supply of sand ceased. Throughout much of the Welsh Basin grey or pyritous grey—black muds accumulated above the purple and green sediments of earlier times. Although feeding burrows and resting traces are common in the sediments of lower and middle Cambrian age, the remains of body fossils are most characteristic of the late middle Cambrian. The fauna is dominated by the large spinose trilobite *Paradoxides* and the small eyeless trilobite *Agnostus*, but it also includes early orthid brachiopods.

Towards the end of middle Cambrian times mud deposition ceased in the northern part of the basin.

Meanwhile, to the south there was a conspicuous coarsening of the incoming detritus. At Lleyn the upper Cambrian commences with a thin calcirudite, bearing abraded limestone pebbles containing fragments of middle Cambrian trilobites. The interbedded sandstones and grey shales that rest upon the calcirudite are deep water in character, with solemarks and internal structures including parallel, ripple and convolute lamination. These features, together with the rarity of shells and the finding of the deep-water trace fossil ?*Nerietes*, give strong evidence for a turbidite origin for the sandstones. Just as this turbidite sequence at Lleyn grades upwards from thick coarse beds to thinner finer beds, so too does the contemporaneous but thicker sequence in the Harlech Dome. Crimes (1970a*) has used this evidence to invoke a late middle Cambrian regression on the adjacent shelf platform. As the upper Cambrian sea advanced, the basinward supply of sands over the edge of the shelf platform diminished.

In the later upper Cambrian times the trend towards water deepening reversed yet again. Distal turbidites pass up into sediments of shallow marine origin. The latter, the Ffestiniog Beds, display abundant evidence of their shoal water origin. The foraging trails, *Cruziana*, and resting traces, *Rusophycos*, of trilobites are associated with the burrows *Planolites* and *Skolithos*. Body fossils are limited to the inarticulate brachiopod *Lingulella*, the orthid *Orusia* and the trilobites *Parabolina*, *Peltura*, *Olenus* and *Agnostus*. These shoal-water sediments provide two clues as to the location of the shoreline: (a) the sands of the Ffestiniog Beds coarsen from the Harlech Dome towards the northwest; and (b) there is a preferred NE−SW crestline orientation of symmetrical ripples (Fig. 7.4). Presumably the shoreline lay just to the northwest of the Arvon outcrop, close to the structural margin of the Welsh Basin.

7c The intrashelf Leinster and Finnmark Basins

Northwest of the stable fault-bounded Irish Sea Platform lay the less well-known Leinster Basin. The rarity of body fossils has made dating of the sediments uncertain, but Crimes and Crossley (1968) and others have suggested that the Cambrian succession, with no exposed base, is at least 2 km thick. Southernmost exposures display shales with coarse turbidites, slumps and pebbly mudstones. These sediments pass northwards into finer-grained turbidites not associated with slumps. Grooves and cross-lamination confirm that turbidity currents transported sediment generally northwards. Only the radiate grazing traces *Oldhamia*, possibly of annelid origin, indicate the presence of life. Further north, near Bray, the sediments display slumps and structures indicative of derivation from the northwest. It is possible that there was a second horst forming the northwestern margin of the Leinster Basin. Far to the northeast of the Leinster Basin,

in Finnmark, Cambrian sedimentation continued in an established late Precambrian area of subsidence. However, sedimentation kept pace with subsidence (*c.* 2 km), and the entire Cambrian succession is of shallow-water origin. Banks (1973) has ascribed the formation of later lower Cambrian large-scale cross-stratified quartz arenites to powerful tidal currents.

7d The Cambrian shelf platform

The southeastern continental margin featured extensive areas of relative stability. The Midland Platform accreted just 500−900 m of shallow marine sediments (Fig. 7.7). To the northeast, in Sweden, the Cambrian shelf accumulated 100−200 m of sediments, considerably less than the 1−2 km pile in the Welsh Basin. As it transgressed the Midland Platform, the lower Cambrian sea reworked rhyolitic and granitic detritus. Basal conglomerates are succeeded by quartz arenites. In Shropshire the latter display lenses of pebbles and bedding planes with small symmetrical ripples featuring rounded crests which are possibly indicative of tidal reworking. Body fossils are not recorded, but the U-shaped trace *Arenicolites*, described by Callaway (1878), confirms the intertidal sandflat to nearshore subtidal interpretation. Upwards through the sequence an increasing content of authigenic glauconite is

Figure 7.7 Cambrian successions of the Midland Shelf Platform. Logs in metres.
Ab. = Abbey Shales; Be. = Bentleyford Shales; Ca. = Caldecote Volcanics; CoGp = upper Comley Group; CoLst = lower Comley Limestone; CoSst = lower Comley Sandstone; HaQ = Hartshill Quartzite; Ho. = Hollybush Sandstone; Ma. = Malvernian; MaQ = Malvern Quartzite; Man. = Mancetter Grits; Mo. = Moor Wood Flags; Mon. = Monks Park Shales; Ou. = Outwood Shales; Pu. = Purley Shales; Sh. = Shoot Rough Road Shales; Ur. = Uriconian; Wh. = White-Leaved-Oak Shales; WrQ = Wrekin Quartzite.
gl. = glauconite; ▲ = phosphatic limestone.

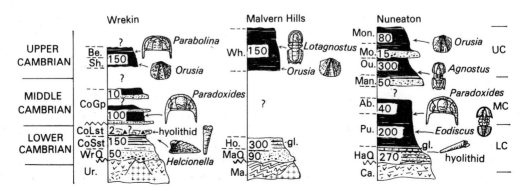

accompanied by thin interbeds of shale which occasionally yield the trilobite *Callavia*, the inarticulate brachiopods *Obolella* and *Paterina*, and fragments of calcareous conical and pyramidal shells belonging to an extinct class of benthic molluscs known as hyolithids. Porrenga (1966) has pointed out that authigenic glauconite forms on the Recent sea floor in regions where the rate of sedimentation is low and the water is cool (10–15°C).

Towards the end of lower Cambrian times, low sediment influx allowed the localised deposition of carbonates (Fig. 7.5). In Shropshire, close to the margin of the Midland Platform, the Comley Limestone comprises 1·8 m of thin nodular phosphatic and glauconitic limestone yielding the trilobites *Callavia* and *Strenuella*, inarticulate brachiopods, hyolithids and the gastropod *Helcionella*. Only thin carbonate bands are located in the contemporaneous glauconitic sandstones exposed to the south in the Malverns and in the shales that crop out to the east at Nuneaton. This phase of slow carbonate deposition on the platform margin appears to correlate with the late lower Cambrian pause in sedimentation in S. Wales. Modern glauconitic and phosphatic carbonates have been discovered on the S. African continental shelf at depths of 100–200 m. R. J. Parker (1975) has suggested that, during the late Tertiary to early Quaternary period of lower sea level, estuarine waters with phosphorous-rich organic matter percolated down through fine-grained carbonates and converted them to calcium phosphate. However, on the Peruvian shelf this diagenetic phosphatisation operates today at 250–500 m, where the cool percolating ocean waters are rich in phosphorous but low in dissolved oxygen. The origin of the recurrent Cambrian phosphatised carbonates remains uncertain.

Cobbold (1927) has presented evidence for mild deformation of the lower Cambrian platform-margin sequence in Shropshire in early middle Cambrian times. As sedimentation resumed, thin conglomerates and muds, bearing the remains of the spinose trilobite *Paradoxides* and the small elliptical agnostids *Peronopsis* and *Eodiscus*, flanked an eroded fold. The time of uplift correlates with the early middle Cambrian sedimentation of the manganese-enriched muds in the Welsh Basin. It is interesting to note that as basinal sedimentation resumed in middle Cambrian times, the main sources of detritus had switched from the north to the south. Shoaling probably occurred once more late in middle Cambrian times. Breaks in the shelf platform successions (Fig. 7.7) match with pauses recorded in the Welsh Basin sequences (Fig. 7.2).

The upper Cambrian silty shales of Shropshire thicken eastwards into interbedded burrowed mudstones and laminated shales. Clearly there was, even after the Cambrian transgression, an important source of mud, but its location is uncertain. Traced eastwards to Sweden, the platform succession thins markedly. The early Cambrian sea inundated an irregular and fissured platform terrain comprising gneiss and pegmatites. Cambrian sediments infilled the fissures and, as in England and Wales, a thin conglomeratic 'carpet' with occasional wind-faceted dreikanter was widely developed. The conglomerates fine upwards into bioturbated sandstones and shales containing *Cruziana* and *Diplocraterion*. The middle Cambrian of Sweden also compares with the Midland Platform. Phosphatic, glauconitic conglomerates and thin trilobite-rich carbonate horizons intercalated with shales are common. In late middle and upper Cambrian times sedimentation became dominantly argillaceous. The shales, rich in illite but poor in chlorite, have a high potassium content (alum shales). Björlykke (1974) has proposed that the muds were derived from the granitic Precambrian shield of low relief which was weathering in a warm climate.

7e The late Cambrian to early Ordovician mud blanket

The nature of late Cambrian to early Ordovician (530–510 Ma) sedimentation on the southern continental margin is decidedly curious. There was widespread deposition of mud (Fig. 7.8), but the source of the mud is unknown. During Tremadoc (lowest Ordovician) times, parts of the formerly stable western margin of the Midland Shelf Platform subsided greatly. This was the first of several indications that the margins of the Lower Palaeozoic Welsh Basin varied in position with time. While 300 m of argillaceous Tremadoc sediments accreted at Portmadoc in the northern part of the Welsh

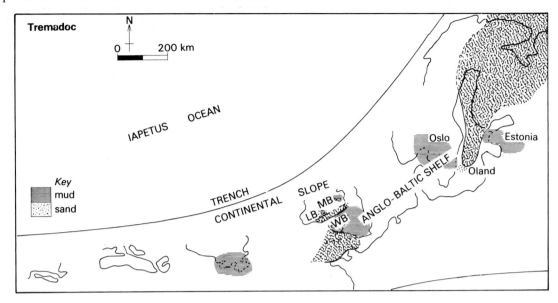

Figure 7.8   Tremadoc palaeogeography of the southern continental margin.
LB = Leinster Basin; MB = Manx Basin; WB = Welsh Basin.

Basin, approximately 1000 m of muds accumulated on the shelf platform in Shropshire. At Tortworth, 50 km to the south, the thickness of the time-equivalent sediments probably exceeds that in Shropshire. However, the Tremadoc sequence thins eastwards from the shelf platform margin. At Nuneaton in the Midlands, 100 m of sediments are preserved; but far to the northeast in Norway, equivalent potassium-rich illitic muds and overlying muds amount to just 20 m.

Following the deposition of the rippled and burrowed sands marking the upper Cambrian shallowing, there was a prolonged period of mud supply to the northern-central, southern and eastern parts of the Welsh Basin. Topmost Cambrian muds with thin, closely spaced, silt laminae are overlain by basal Tremadoc muds. The latter coarsen upwards into laminated sandstones, which are in turn overlain by late Tremadoc muds with isolated ripples of sand associated with burrows. The upper Cambrian fauna includes the small spinose trilobites *Parabolina* and *Peltura* together with the orthid *Orusia* and *Lingulella*. The basal Ordovician Tremadoc sediments yield the wide-bordered *Asaphellus*, the small blind *Shumardia* and the planktonic dendroid graptolite *Dictyonema*.

A similar but substantially thicker sequence developed on the shelf platform margin in Shrop-

shire. Muds gradually coarsen upwards into rippled and burrowed sands, just as they do in the northern-central part of the Welsh Basin. The fauna is similar too. Remains of *Shumardia* and *Asaphellus* are associated with inarticulate brachiopods and dendroid graptolites. The muds are also notable for the abundance of the microscopic spinose cysts known as acritarchs. The suspicion that the platform margin sequence was deposited in shallow water is confirmed by evidence presented by Curtis (1968), who has investigated the sequence developed at Tortworth. Curtis has described thin toolmarked sandstones interbedded with shales. The sandstone tops are marked by wave-formed oscillation ripples and rill marks. These signs of near-emergence are supported by the occurrence of mud cracks. Annelid burrows are associated with the feeding trails of olenid trilobites. North and west of the Welsh Basin, closer to the contemporary continental margin, there was deposition of mud in the Leinster and Manx Basins, and in the vicinity of the Lake District.

7f The Arenig transgression of the Welsh Basin

During the Cambrian period there had been strong

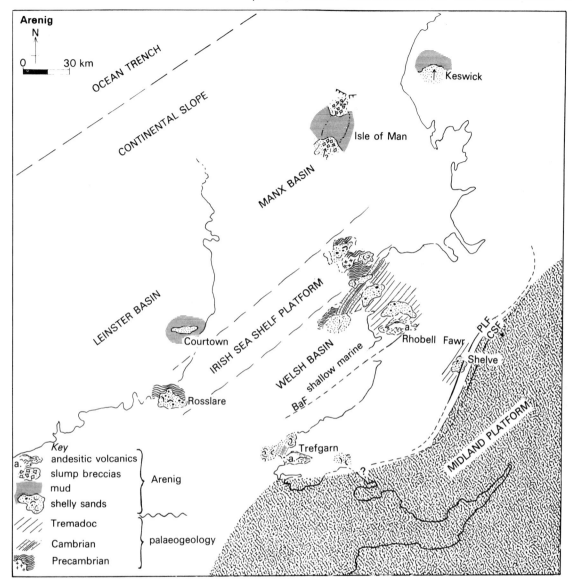

Figure 7.9 Lower Arenig palaeogeography of S. Britain. For key to faults see Figure 7.5.

hints of basin margin or intrabasinal fault movement, reflected, for example, by the middle Cambrian switch in sediment source from the north to the south. Further indications of fault movement in earliest Ordovician times are even more clear. In NW and SW Wales consolidated Cambrian muds and sands were tilted and eroded during the Tremadoc. Thus as the early Arenig sea transgressed northwestwards on to the Irish Sea Platform, it overlapped the sediments of the Tremadoc and the Cambrian and finally inundated the irregular topography of the Precambrian schist, gneiss and granite (Fig. 7.9). Similarly in SW Wales the Arenig sea overlapped westwards on to tilted and eroded upper Cambrian sediments. In Shropshire, the eastern margin of the Welsh Basin, the sea transgressed over an eroded surface of the Tremadoc muds.

A basal conglomeratic sand was widely developed at the basin margin, but it varied greatly

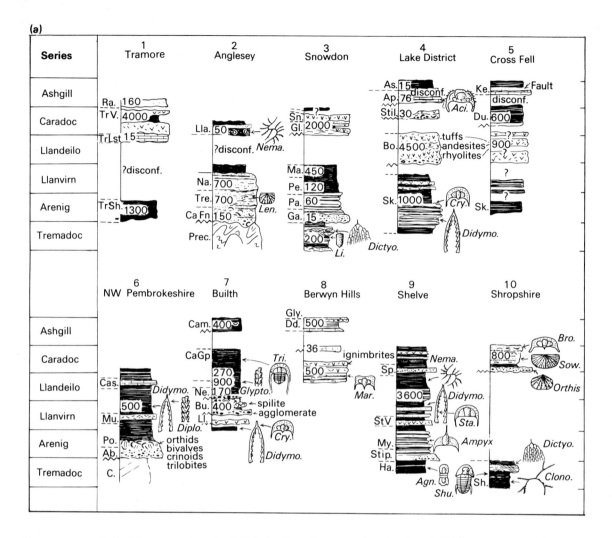

Figure 7.10a Ordovician successions in S. Britain. Logs in metres (partly after A. Williams *et. al.* 1972).
Ab = Abercastle Beds; Ap. = Applethwaite Beds; As. = Ashgill Shales; Bo. = Borrowdale Volcanics; Bu. = Builth
Volcanics; C. = Cambrian; CaFn = Carmel Formation; CaGp = Carmel Group; Cam. = Camlo Hill Group; Cas. =
Castell Limestone; Dd. = Dholhîr Beds and Limestone; Du. = Dufton Shales; Ga. = Garth Flagstones and Grits; Gl. =
Glanrafon Beds; Gly. = Glyn Grit; Ha. = Habberly Shales; Ke. = Keisley Limestone; Ll = Llandrindod Volcanics; Lla.
= Llanbabo Formation; Ma. = Maesgwm Slates; Mu. = Murchisoni Ash; My. = Mytton Flags; Na. = Nantannog
Formation; Ne. = Newmead Volcanics; Pa. = Pant-y-wrach Beds; Pe. = Pennant Slates and Quartzites; Po. = Porth
Gain Beds; Prec. = Precambrian; Ra. = Raheen Group; Sh. = Sheinton Shales; Sk. = Skiddaw Slates; Sn. = Snowdon
Volcanic Group; Sp. = Spy Wood Grit; Stil. = Stile End Beds; Stip. = Stiperstone Quartzite; StV = Stapely Volcanics;
TrLst = Tramore Limestone; TrSh = Tramore Shale; TrV = Tramore Volcanics; Tre. = Treiorworth Formation.
disconf = disconformity.
*Aci. = Acidaspis; Agn. = Agnostus; Bro. = Broeggerolithus; Clono. = Clonograptus; Cry. = Cryptolithus; Dictyo. = Dictyonema;
Didymo. = Didymographtus; Diplo. = Diplograptus; Glypto. = Glyptograptus; Len. = Lenorthis; Mar. = Marrolithus; Nema. =
Nemagraptus; Shu. = Shumardia; Sow. = Sowerbyella; Sta. = Stapleyella; Tri. = Trinucleus.*

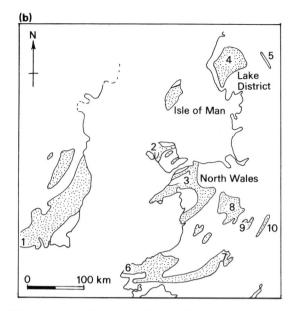

(b)

N

4

5

Lake
District

Isle of Man

2

North Wales

3

8

9

10

1

0   100 km

6

Figure 7.10b Location of Ordovician outcrops in S. Britain.

in thickness. In N. Wales the sand varies from 10–30 m on the eastern flanks of the Harlech Dome to *c.* 1000 m in N. Anglesey. In the northern part of the Harlech Dome where it rests on Tremadoc, the clasts include slate, rhyolite, quartzite, vein quartz and andesite set in a matrix of slate and quartz granules. Lewis (1926) has noted that many of the rounded pebbles are encrusted by black, cherty, phosphatic overgrowths, '*Bolopora undosa*'. Some of these abiogenic structures are mammillated; others are tubular with transverse laminae. The Arenig sequence in SW Wales gradually fines upwards into blue-black shales. The sands and sandy muds provided a substrate for orthids, palaeotaxodont bivalves, gastropods, trilobites and the early crinoid *Ramseyocrinus*; but the overlying muds only entombed the remains of benthic or planktonic dendroid graptolites and the nektonic *Trinucleus* (Fig. 7.10).

The Arenig sediments of the northwestern margin of the Welsh Basin also supported a rich benthic fauna. Above a thin basal conglomerate in the Lleyn Peninsula a sequence of silty mudstones, rippled sandstones, thin cherts and a single horizon of ferruginous oöliths is frequently burrowed. Crimes (1970b) has noted *Cruziana* and *Skolithos* from the sandstones and recorded the feeding trace *Teichichnus* in silty mudstones. The planktonic graptoloids

Azygograptus and *Didymograptus* are located in the mudstones. Further north in Anglesey, the time-equivalent sandstones are notable for a varied fauna of brachiopods. The transgression of the eastern margin of the Welsh Basin is also recorded by a basal conglomeratic sandstone known as the Stiperstone Quartzite. This quartz arenite contains burrows, but the overlying interbedded mudstones and sandstones yield both brachiopods and trilobites. As in SW Wales the sequence gradually fines upwards into shales.

7g Ordovician—Silurian volcanism

The late Tremadoc movements and the Arenig transgression in the marginal regions of the Welsh Basin coincided with another event of very great geological importance. This was the commencement of volcanism on a large scale. First in the Welsh Basin, later in the Leinster Basin and in the region of the Lake District, the volcanicity was to be a conspicuous feature of unstable shelf areas from late Tremadoc to Wenlock times, a period of about 100 Ma.

Dewey (1969a*), later supported by the chemical evidence of Fitton and Hughes (1970*), has related the volcanism to the development of a SE-dipping subduction zone. By analogy with modern island arcs it is supposed that consumption of the 'cold' oceanic lithospheric slab caused frictional heating and melting, and that upward-rising mafic diapirs penetrated the continental crust. Independently, Rast (1969*) has invoked the upward movement of mafic magmas into the continental crust of the Welsh Basin during Ordovician times. These magmas are considered responsible for partially melting siliceous crustal material which rose as magmatic blisters above the mafic magmas. In this way Rast has explained two conspicuous, but not immediately compatible, features: (a) the widespread distribution of rhyolitic and andesitic volcanics in Wales; and (b) the existence of high residual-gravity anomalies close to the main centres. From their studies of the volcanics in the Lake District and Wales, Fitton and Hughes (1970*) have observed that, as in modern island arcs, the lavas closest to the inferred position of the ocean trench were tholeiitic whereas the contemporary lavas to the

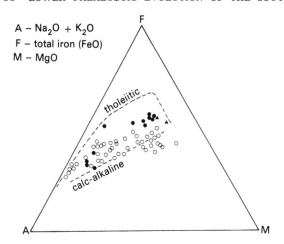

A – Na$_2$O + K$_2$O
F – total iron (FeO)
M – MgO

Figure 7.11 A–F–M (weight %) plot of analysed samples from the Borrowdale Volcanics (after Fitton & Hughes 1970*).
○ = lavas and tuffs from the southern outcrop (Scafell area); ● = acid rocks and aphyric basic lavas from the northern outcrop (Binsey–Eycott region); ▲ = mechanically separated groundmasses from porphyritic (Eycott-type) lavas from the northern outcrop.
Dashed lines = typical calc-alkaline (Cascades) and tholeiitic (Thingmuli) trends.

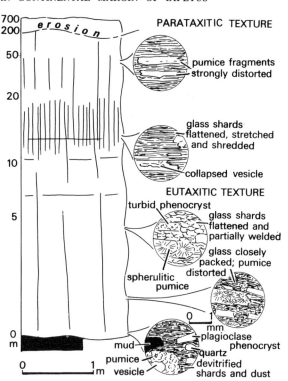

Figure 7.12 Diagramatic section through the Llwyd Mawr Ignimbrite from N. Wales (data from B. Roberts 1969).

south were more calc-alkaline (Fig. 7.11). More recently, Stillman *et al.* (1974*) have documented a similar situation in SE Ireland. Caradocian volcanics north and west of Dublin were mainly tholeiitic andesites and basaltic andesites; but to the southeast, in the Wicklow–Waterford area, the contemporaneous volcanics were calc-alkaline.

Petrologically, the volcanics range from silicic to mafic, but the silicic–intermediate type dominate. The subvolcanic intrusive bodies are mainly flow-banded and flow-brecciated rhyolites, rhyodacites, keratophyres, microgranites and granophyres; but there are also alkali dolerites. There is growing evidence that the extrusive volcanics are dominantly pyroclastic. Rhyolitic, andesitic and spilitic lavas occur, but Oliver (1954), Rast (1969*) and others have drawn attention to much former misidentification. It was formerly thought that rhyolites were widespread and abundant, but now it appears that many supposed lavas are the welded portions of pyroclastic flow deposits known as ignimbrites (Fig. 7.12). This discovery has led to a substantial reinterpretation of the Caradocian palaeogeography in Snowdonia. It has been widely argued that the

ignimbrites were emplaced by nuées ardentes, i.e. dense flows of hot gas and suspended dust, vitric shards and tuffaceous clasts resulting from subaerial eruptions. Thus the silicic volcanics of Snowdonia have come to be regarded as being generally subaerial in origin. Howells *et al.* (1973), however, have noted that some welded tuffs are closely associated with marine sediments and that some units of welded tuff pass eastwards into non-welded tuffs. They have argued that some flows, at least, are products of submarine emissions of an intensely turbulent mixture of hot gas and ash which maintained its state, as it flowed over the sea floor, by virtue of its high density. Thus much of the volcanic terrain has been 'lowered' once again beneath the Caradocian waves. The matter remains an issue of controversy. Spilitic rocks have also been misidentified. Hyaloclastites (pillow-lava breccias), formed by *in situ* pseudobrecciation of pillow lavas in contact with water, have been incorrectly described as spilitic tuffs and agglomerates.

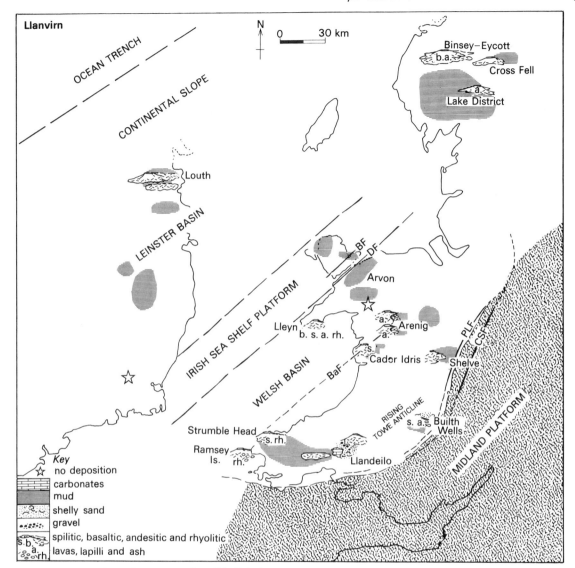

Figure 7.13 Llanvirn palaeogeography of S. Britain. For key to faults see Figure 7.5.

In addition to the above types there are pyroclastic fall deposits, notably rhyolitic and andesitic but occasionally basaltic. Some exhibit cross-stratification and pass laterally into tuffaceous sediments containing shells. Others display graded bedding, load casts, slumps and accretionary lapilli. Such marine deposits possibly flanked emergent volcanoes. Lahars or volcanic mudflow breccias have also been recognised. These thick units consist of a chaotic assortment of lava, sandstone and shale blocks set in a completely unsorted tuffaceous muddy matrix.

The evolution of the Ordovician—Silurian volcanic province may be envisaged as a sequence of four phases.

Phase 1: Late Tremadoc—Llanvirn (Fig. 7.13)

Volcanism commenced in the Welsh Basin during late Tremadoc times. Centres of andesitic volcanism were located at Rhobell Fawr in northern—central Wales and at Trefgarn in SW Wales. By Llanvirn times the volcanicity had spread to closely

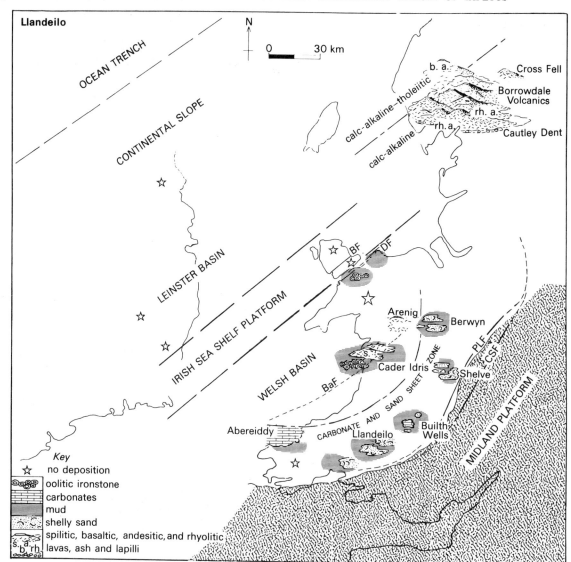

Figure 7.14 Llandeilo palaeogeography of S. Britain. For key to faults see Figure 7.5.

adjacent areas and diversified in nature to rhyolites and spilites. Rast (1969*) has noticed the NE−SW alignment of the volcanic tract and suggested that fractures associated with the Bala Fault, which passes close to Cader Idris and Rhobell Fawr, might have acted as pathways for the rising magma. By late Arenig−Llanvirn times volcanism was also developed close to the northwestern margin of the Welsh Basin in Lleyn and close to the southeastern margin at Llandeilo, Builth Wells and Shelve. North of the Welsh Basin, tholeiitic volcanoes

commenced activity in the northern part of the Lake District (the Eycott Volcanics) and in E. Ireland at Louth.

Phase 2: Llandeilo (Fig. 7.14)

The middle Ordovician period witnessed a general upsurge of volcanic activity in the north but a corresponding decline in the south. P. J. F. Jeans (1973) has attributed this change to a steepening of the SE-dipping subduction zone. Tholeiitic basalts

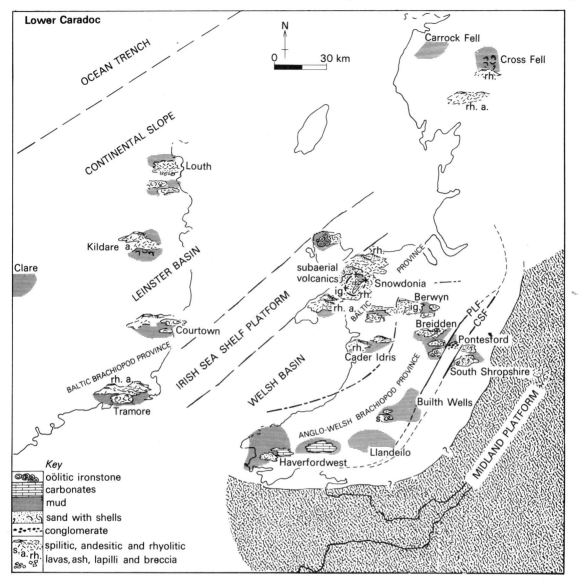

Figure 7.15 Lower Caradoc palaeogeography of S. Britain.
ig. = ignimbrites.
For key to faults see Figure 7.5.

and andesites in the Binsey region (the Eycott Volcanics) pass southwards into calc-alkaline andesites and rhyolites of the Scafell region (the Borrowdale Volcanics). But in the Welsh Basin volcanism became confined to the Cader Idris—Arenig zone where it had erupted first in late Tremadoc times.

Phase 3: Lower Caradoc (Figs 7.15 and 7.16)

During the third phase of volcanism the main zone of volcanic activity shifted southwestwards from the Lake District to the southern part of the Leinster Basin and the northern part of the Welsh Basin. In Snowdonia, partially emergent volcanoes formed part of a major complex dominated by pyroclastic flow and fall deposits. R. M. Shackleton, later supported by Rast (1969*) and A. V. Bromley (1969), has suggested that rising magma caused local updoming and the formation of a major rim syncline. Collapse of the dome resulted in a caldera.

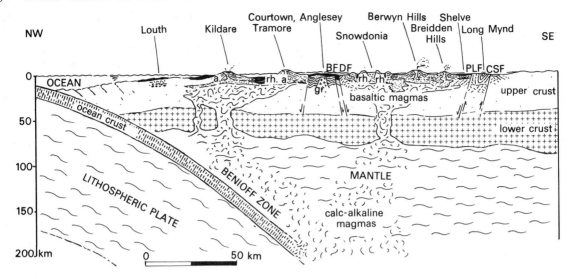

Figure 7.16 Hypothetical crustal section across the southern continental margin in lower Caradoc times.
a. = andesites; gr. = granite; rh. = rhyolites.
For key to faults see Figure 7.5.

Rhyolite and alkali dolerite intrusions later plugged the rim syncline.

Phase 4: Late Ordovician–Wenlock

The fourth and final phase of volcanism reflected a further southward and westward shift of the main zone of activity. Alkali basalts of the Skomer Volcanic Series were erupted in SW Wales in late Ordovician to lower Llandovery times. Tuffs are rare. To the east, calc-alkaline andesitic and rhyodacitic lavas were extruded in the Mendip region during early Wenlock times. Further Wenlock volcanism is recorded to the west at Dingle, where tuffs and both silicic and mafic lavas were erupted. It is now clear that during the Ordovician there was little or no truly deep water (over 200 m) sedimentation in the Leinster and Welsh Basins (Fig. 7.16). Indeed, major tracts of the basinal environments were subaerial or shallow marine. These areas were only basinal in the sense that they were unstable regions of the continental shelf. Brenchley (1969) has recorded lower Caradoc sediment thicknesses across the Welsh Basin as follows; from east to west:

E. Shropshire	(shelf platform)	200 m
Shelve area	(basin)	1000 m
Breidden Hills	(basin)	>500 m
Berwyn Hills	(basin)	1500 m
Bala	(basin)	1000 m
Snowdonia	(basin)	2000 m

Since the general environments of deposition changed little during the lower Caradoc, it may be assumed that these values provide some estimate of the relative rates of subsidence. It may be concluded that, at this time of intense volcanic activity, the basin was subsiding at a rate at least five times as great as that of the adjacent Midland Shelf Platform.

There is little evidence of sediment supply from the shelf platform regions. Preliminary petrographic results from Brenchley (1969) suggest that much of the sand may be of volcanic origin. During Llanvirn times the volcanic tract from northern-central Wales to SW Wales was surrounded by sea floor on which muds and planktonic graptolites were accumulating; but along the southeastern margin of the basin the Llandeilo–Builth–Shelve volcanoes were associated with tuffaceous sands which supported a rich benthic shelly fauna. During the Llandeilo this arcuate zone of shelly sheet sands and muds persisted despite the decline in volcanism. In the northern–central part of the Welsh Basin muds associated with ferruginous oölite horizons flanked active volcanoes. Later, in basal Caradoc times, the sea transgressed the northwestern and southeastern platform margins and caused a shift in sediment zones. Shell-rich sands were deposited on the western edge of the Midland Shelf Platform, while offshore the former zone of sheet sands and muds was overlain by less silty muds and thin carbonates (Fig. 7.15). A similar transgressive episode is recorded in the Lake District where it followed a middle Ordovician phase of folding.

A. Williams (1969a, 1976) has shown that the Ordovician brachiopod faunas belonged to two distinct provinces. In the northwestern part of the Welsh Basin and in the southern part of the Leinster Basin the brachiopods are 'Baltic' in affinity. But to the southwest they are 'Anglo–Welsh' in character. Although there was some mixing of these faunas, the provincialism, together with palaeocurrent evidence, may eventually provide some clues as to the nature of the shallow marine currents responsible for dispersing the planktonic brachiopod larvae.

7h Active faults and turbidite sedimentation in the Welsh Basin

Although the episode of volcanicity persisted until Wenlock times through the activity of basaltic and andesitic volcanoes in SW Wales and the Mendips–Tortworth region, volcanism to the north, in the Welsh Basin, effectively ceased in lower Caradoc times. With the demise of the volcanoes it seems that sediment supply to the basin, particularly sand, was very much reduced. Late Caradoc muds such as the Nod Glas in the northern part of the Welsh Basin are phosphatic and pyritous, indicating anaerobic conditions, if not at the former sediment/water interface, then at only a few millimetres depth.

Lower Caradoc sediment thicknesses, discussed above, have indicated the high rate of subsidence. Sediment starvation and continued subsidence probably led to a deepening of the water in this unstable shelf basin. James and James (1969) have demonstrated that during the ensuing Ashgill–Llandovery period sedimentation was greatly influenced by intrabasinal and basin margin faulting. Fault movements triggered slumps and turbidity currents which resulted in the progradation of small submarine fans at the foot of fault scarps. Transport was generally northwestwards (across the basin) and northeastwards (parallel to the basin margin). The position of the contemporary south-

Figure 7.17 Lower Llandovery palaeogeography of S. Britain. b. = basalts; g. = graptolites; rh. = rhyolites; t. = turbidites. For key to faults see Figure 7.5.

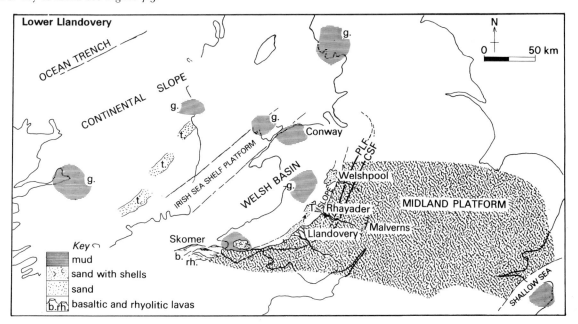

eastern basin margin is indicated by the curvilinear distribution of small submarine fans and slump horizons closely parallel to, but some 10 km west of, the Pontesford–Linley Fault (Fig. 7.17). James and James (1969) have postulated the presence of an infra-Towy–Severn Fault at depth beneath the late Ordovician to early Silurian basin margin. Thus it appears that the structural margin of the Welsh Basin shifted slightly basinwards during upper Caradoc times. It is possible that this change coincided with the tilting and open folding of the Ordovician sequence of muds, sands and ashes on the Shelve block between the Pontesford–Linley Fault and the inferred infra-Towy–Severn Fault.

During the lower and middle Llandovery, before the Llandovery transgression, there was substantial lateral supply of sediments to the basin. Woollands (1970) has shown that the lobes of a birdfoot delta prograded basinwards in the Llandovery–Garth region of the southeastern basin margin. At Rhayader, 30 km north, the massive Caban Conglomerate (Fig. 7.18) has been interpreted by Kelling and Woollands (1969) as detritus resedimented in a submarine channel. The sediments were em-

Figure 7.18 Silurian marine facies: (a), (b), (c) and (d) = Skomer Volcanic Group, Marloes, Dyfed; (e) = Wenlock Limestone, Wenlock Edge, Shropshire (after Scoffin 1971); (f) = Ludlow, Cwm Blithus, Powys (after Woodcock 1976); (g) = Lower Caban Conglomerate, Caban Coch Dam, Rhayader (after Kelling & Woollands 1969); (h) = Aberystwyth Grits, Clarach Bay, Aberystwyth.
Location of Llandovery marine communites: *Lingula* community – facies (a) and (b); *Eocoelia* community – facies (b) and (c); *Pentamerus* community – facies (c) and (d); *Stricklandia* community – facies (c) and (d); *Clorinda* community – facies (d).

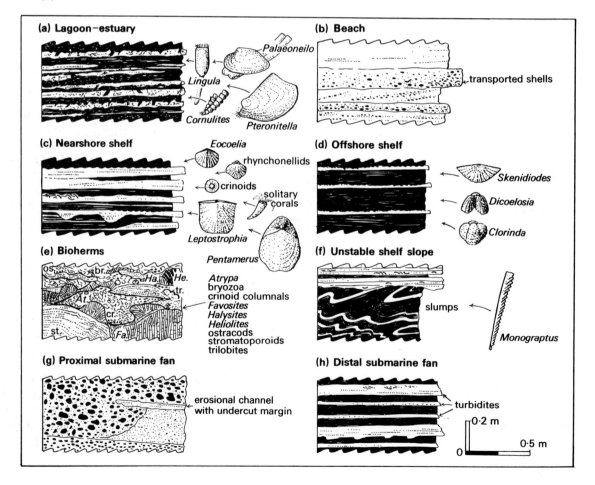

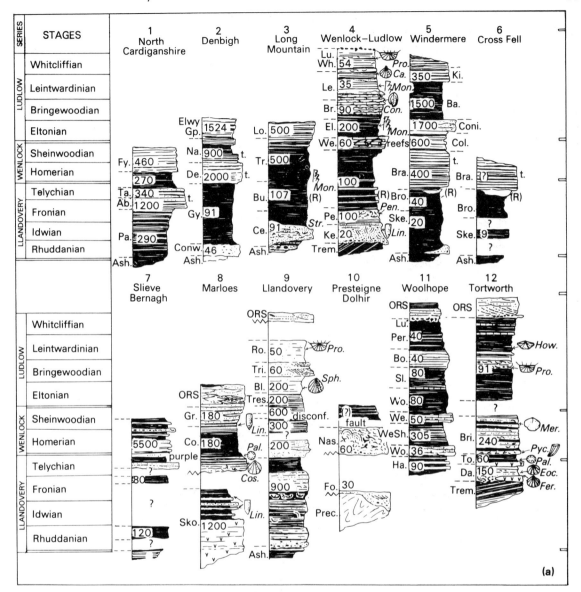

Figure 7.19a Silurian successions in S. Britain. Logs in metres.

Ab. = Aberystwyth Grits; Ash. =Ashgill; Ba. = Bannisdale Slates; Bl. = Black Cock Beds; Br. = Bringewood Beds; Bra. = Brathay Flags; Brink. = Brinkmarsh Beds; Bro. = Browgill Beds; Bu. = Buttington Formation; Ce. = Cefn Beds; Co. = Coralliferous Group; Col. = Coldwell Beds; Coni. = Coniston Grits; Conw. = Conway Castle Grits; Da. = Damery Beds; De. = Denbigh Grits Group; El. = Elton Beds; Fo. = Folly Sandstone; Fy. = Fynyddog Grits; Gr. = Gray Sandstone Group; Gy. = Gyffin Shales; Ha. = Haugh Wood Beds; Ke. = Kenley Grit; Ki. = Kirkby Moor Flags; Le. = Leintwardine Beds; Lo. = Long Mountain Siltstone Formation; Lu. = Ludlow Bone Bed; Na. = Nantglyn Flags Group; Nas. = Nash Scar Limestone; ORS = Old Red Sandstone; Pe. = *Pentamerus* Beds; Prec. = Precambrian; Ro. = Roman Camp Beds; Ske. = Skelgill Beds; Sko. = Skomer Volcanic Group; Ta. = Tallerddig Grits; To. = Tortworth Beds; Tr. = Trewern Brook Mudstone Formation; Trem. = Tremadoc; Tres. = Tresglen Beds; Tri. = Trichrûg Beds; We. = Wenlock Limestone; WeSh = Wenlock Shale; Wh. = Whitcliffe Beds; Wo. = Woolhope Limestone.
(R) = red-beds; t. = turbidites; v. = volcanics; disconf. = disconformity.
Ca. = *Camarotoechia; Con. = Conchidium; Cos. = Costistricklandia; Eoc. = Eocoelia; Fer. = Ferganella; How. = Howellella; Lin. = Lingula; Mer. = Meristina; Mon. = Monograptus; Pal. = Palaeocyclus; Pen. = Pentamerus; Pro. = Protochonetes; Pyc. = Pycnactis; Sph. = Sphaerirhynchia; Str. = Stricklandia.*

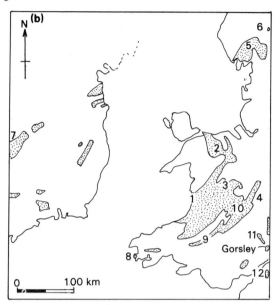

Figure 7.19b Location of Silurian outcrops in S. Britain.

placed by mass flow and turbidity current mechanisms. Curiously, the Caban channels appear to have twisted northwards at the foot of the NW-facing palaeoslope.

Longitudinal sediment transport by NE-moving turbidity currents persisted from Ashgill to lower Ludlow times, a period of 40 Ma. The lithic and feldspathic sands, now diagenetically altered to greywackes, were deposited in an elongate turbidite-facies zone (Fig. 7.18h). In chronological order these greywackes are represented by the Aberystwyth and Talerddig Grits (upper Llandovery; Plate 2), the Denbigh Grits (Wenlock) and the Nantglyn Flags and lower Ludlow Grits (Ludlow; Fig. 7.19). Both Jones (1938) and Cummins (1969) have pointed out that the eastern limit of the turbidite zone shifted eastwards with time. By lower Ludlow times the limit was 35 km east of its upper Llandovery position. In the northern part of the Welsh Basin E-advancing turbidity currents deposited sands in a separate turbidite zone during the Wenlock and lower Ludlow. Cummins (1969) has suggested that the division of the area of turbidite deposition into two zones reflected the development of basinal topography. The E–W Denbigh Trough was separated from the N–S Montgomery Trough by an area where only mud accumulated, the Der-

wen Ridge. This was the first sign of the uplift that was to continue during the Ludlow.

The structural margin of the Welsh Basin in Wenlock–Ludlow times is uncertain. Sediment thicknesses suggest considerable subsidence of the region to the west of the Church Stretton Fault. This is further evidence that, throughout the long history of the Welsh Basin, the southeastern structural margin fluctuated in position.

7i Llandovery transgression of the Midland Platform

As the lower Caradoc volcanicity in the northern and eastern parts of the Welsh Basin drew to a close, the sea retreated from the margin of the Midland Platform. Thus during the upper Caradoc to middle Llandovery interval the sea was confined to the unstable shelf basin.

During the middle and upper Llandovery, a period of 8 Ma, the sea transgressed eastwards once more across the Midland Platform. But it was not a localised event. Contemporaneous transgressions occurred in W. Ireland (Ch. 5g), N. and E. Russia, S. China, Australasia and N. America; the rise in sea level was worldwide. Berry and Boucot (1973*) have pointed out that the late Ordovician Saharan ice sheet was probably melting at this time, signifying perhaps that the transgression was eustatic.

The transgression has been traced in some detail on the Midland Shelf Platform (Welsh Borderland) where two distinct phases have been recognised (Bridges 1975). During the first phase the sea inundated a terrain of irregular topography comprising Precambrian–Ordovician volcanics and sediments (Figs 7.17 and 7.20). By the close of middle Llandovery times the sea had reached a line extending from close to the Church Stretton Fault in the north to the Malverns in the south. In the northern Welsh Borderland the Long Mynd appears to have formed a local S-facing peninsula (Fig. 7.21). To the west (seawards), fine laminated sands with a diverse open-marine shelly fauna flanked the lithified, tilted and folded Ordovician sediments of the Shelve bank. To the east (landwards), the contemporaneous coarse sands were generally poorly sorted, silty or conglomeratic and poor in fauna. The silty sandstones contain vertical burrows and the phosphatic

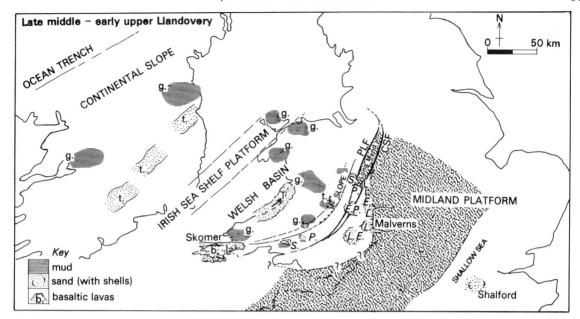

valves of the burrowing *Lingula*. They were probably estuarine in origin. The conglomeratic sands,
devoid of signs of life, may represent stream-fed
beach deposits consisting of Precambrian clastics
derived from the Long Mynd. Only at the southern
tip of the Long Mynd Peninsula did waves substantially rework the coarse sands, enhancing the compositional maturity and bringing in disarticulated
valves of the thick-shelled *Pentamerus*. As the first
transgressive phase ceased, muds began to accumulate in the wave-sheltered estuarine and offshore
environments.

During the second phase the shoreline shifted 50
km eastwards; indeed the sea may have united with
a westward incursion over SE England. As nearshore marine currents operating in the central part of
the Midland Shelf Platform reworked quartz
arenites derived from lower Cambrian (? and Precambrian) outcrops, deposition of muds took place
offshore in areas that had been formerly current
scoured. Thus south of the former Long Mynd
Peninsula, late upper Llandovery silts accumulated
above Carodoc siltstones.

On the southern margin of the Welsh Basin the
two phases of the Llandovery transgression can be
recognised in SW Wales. Here the sea inundated the
decaying late Ordovician to early Silurian volcanic
centre (Figs 7.17 and 7.20). During the first phase,

Figure 7.20 Late middle to early upper Llandovery
palaeogeography of S. Britain.
E. = *Eocoelia* community; *L.* = *Lingula* community; *P.* =
Pentamerus community; *S.* = *Stricklandia* community.
b. = basalts; g. = graptolites; t. = turbidites.
For key to faults see Figure 7.5.

Figure 7.21 Late middle to early upper Llandovery
palaeogeography of Shropshire. The bathymetry is
hypothetical. Key as in Figure 7.20.
L. = *Lingula* community; *P.* = *Pentamerus* community; *R.*
= *Rostricellula* community; *S.* = *Stricklandia* community.

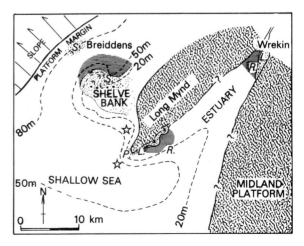

periodic local uplift and the extrusion of basalt flows interrupted the transgression. A longshore supply of non-volcanic detritus allowed the growth of barrier islands and lagoons, which retreated westwards over the basalt flows as the transgression resumed (Bridges 1976). By upper Llandovery times the volcanic centre had expired. Late in upper Llandovery times the second transgressive phase caused a substantial deepening of the sea. Bioturbated muds and thin very-fine shelly sands accumulated in offshore conditions.

Ziegler *et al.* (1968) have demonstrated that the brachiopod-dominated shelly fauna contained within this transgressive Llandovery sequence can be ascribed to five depth-related assemblages or communities. From the shoreline seawards they have recognised the *Lingula*, *Eocoelia*, *Pentamerus*, *Stricklandia* and *Clorinda* communities (Fig. 7.18a, b, c, d). During the first transgressive phase only the *Lingula*, *Pentamerus* and *Stricklandia* communities were well developed on the shelf platform. The *Lingula* community − sometimes including the rhynchonellid *Ferganella*, shallow burrowing bivalves and gastropods − inhabited wave-sheltered environments. The *Pentamerus* and *Stricklandia* communities − including the distinctive pentamerids, atrypids, strophomenids, trilobites, rugose and tabulate corals and tentaculitids − colonised mud and fine sand substrates in nearshore and offshore marine environments. Sometimes they were smothered as life clusters beneath fine laminated sands, but more often they occur as transported, disarticulated and broken valves concentrated in coquinas enveloped in sand. These shell-rich sands were probably emplaced during shelf storms. The less diverse faunas of the *Clorinda* community were located on the platform slope environment. Fürsich and Hurst (1974) have made the attractive suggestion that Silurian brachiopod distribution was largely controlled by turbulence and the concentration of nutrients. Thus turbulent nearshore environments were dominated by the rhynchonellids, which had a strong pedicle attachment and a dentate commissure designed to hinder the entry of sand particles into the mantle cavity containing the delicate filter-feeding lophophore. Offshore environments, with lower concentrations of microplankton and suspended organic particles, were dominated by pentamerids, spiriferids and specially adapted orthids, which featured plicae, sulci and wings designed to aid the separation of inhalant and exhalant currents and so to enhance the filter-feeding capacity of the lophophore.

7j Wenlock carbonates

By Wenlock times the remnant Ordovician topography had been smothered by the deposition of Llandovery sands and muds. West of the Church Stretton Fault a thick sequence (*c.* 1000 m) of muds and laminated silts, frequently containing monograptids with hook-shaped thecae, accumulated in relatively deep water (over 100 m). There is little evidence of wave reworking. East of the Church Stretton Fault the more stable shelf platform accreted muds rich in acritarchs, chitinozoans and carbonates (Fig. 7.22). The carbonates are abundant at two horizons. In the southern Welsh Borderland they are essentially basal Wenlock (Woolhope Limestone) and topmost Wenlock (Much Wenlock Limestone) in age, but in the north only the younger is developed. The lower horizon generally comprises allochthonous corals, crinoids and bryozoans, but at the outer margin of the shelf platform, at Presteigne, the calcareous alga *Solenopora* and bryozoans formed biostromes. The upper carbonate horizon is most thickly developed (30 m) at Wenlock Edge in the northern Welsh Borderland. There Scoffin (1971) has demonstrated the former existence of small oval patch-reefs 0·5−3·0 m high and 12 m in diameter on the Wenlock sea floor.

Reef growth commenced on soft muddy substrates and on mounds of crinoidal debris. Massive and branching tabulate corals, branching rugose corals, stromatoporoids and bryozoans acted as reef binders (Fig. 7.18e). Crinoid ossicles, brachiopods, ostracods, pieces of coral, stromatoporoids and bryozoans together with carbonate sand and mud trickled into the reef interstices. Unlike modern patch reefs there was a dearth of reef-boring organisms. The inter-reef sediments consisted of bioturbated crinoidal biomicrites. Scoffin (1971) has deduced from the presence of the green alga *Girvanella* that the water depth did not exceed 30 m at the time of reef formation. Reefal conditions extended eastwards from the platform margin to the

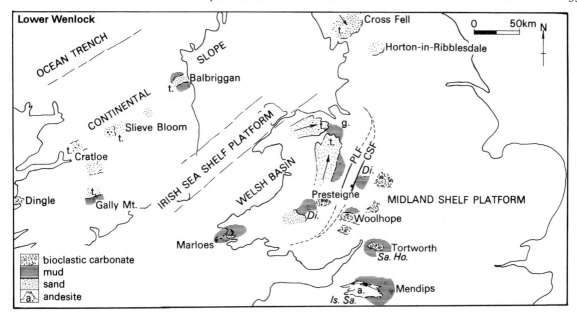

Figure 7.22 Lower Wenlock palaeogeography of S. Britain (data from Calef & Hancock 1974). Common brachiopods: *Di.* = *Dicoelosia; Ho.* = *Homoeospira; Is.* = *Isorthis; Sa.* = *Salopina.* a. = andesites; g. = graptolites; t. = turbidites. For key to faults see Figure 7.5.

central part of the Midland Shelf Platform at Dudley. Still further east, contemporaneous reefs developed in the vicinity of Gotland.

7k Uplift of the Welsh Basin: Ludlow silts and slumps

The Ludlow period witnessed the final demise of the Welsh Basin which had developed some 160 Ma earlier at the beginning of Cambrian times. Throughout the Welsh Borderland the sea floor of the shelf platform accumulated cross-stratified, laminated and rippled silts, bioturbated silts, shell-rich coquinas and silts with finely comminuted shells which now form thin nodular limestones. During one period, however, the accumulation of carbonates (the former Aymmestry Limestone) dominated over the deposition of silt. The silts of the Midland Shelf Platform accumulated to the thickness of 300–400 m, but at Gorsley in the south (Fig. 7.19b) they thinned to just 2 m. West of the line of the Church Stretton Fault the platform silts passed into a basinal sequence of shales and silts 1000–2000 m thick.

The fauna was abundant and diverse; fossils range from scolecodonts (annelid jaw elements), ostracods and conodonts to corals, stromatoporoids, brachiopods, bivalves, gastropods, trilobites, orthoconic nautiloids and graptolites. Calef and Hancock (1974) have found that the spinose flat-lying strophomenid *Protochonetes ludloviensis*, the orthid *Salopina*, the small spiriferid *Howellella* and the rhynchonellids *Sphaerirhynchia* and '*Camarotoechia*' dominate rippled siltstones. These inshore brachiopods were associated with *Pteronitella* and *Palaeopecten*, two epifaunal bivalves, and *Fuchsella*, an infaunal form. Deeper shelf substrates were colonised by brachiopods of smaller adult size, including *Protochonetes minimus* and the orthids *Dicoelosia, Skeniodiodes* and *Visbyella*. The corals and stromatoporoids generally lived in the carbonate-rich environments.

At Leintwardine in the southern Welsh Borderland, close to the western margin of the Midland Shelf Platform, two features of significance developed as the silts accumulated: large channels and slumps. The channels, approximately 0.8 km wide and up to 4 km long, sloped towards the west-northwest at 10° (Whitaker 1962). Marine currents eroded into the early Ludlow silts. The minor slumps at Leintwardine indicate the instability of local palaeoslopes. Further west, in the Welsh Basin, slumped silts were common (Fig. 7.18f). Near Builth Wells and Bishops Castle, close to the line of the Church Stretton Fault, the noses of Eltonian slumps face northwestwards, but some 30 km to the north near Newtown the slumps are overfolded southeastwards. Woodcock (1976) has described extensive slump sheets commonly 10−25 m thick. The lower contacts of the slump sheets are usually gradational, but upper contacts are commonly sharp and erosive. Some units lack folds; others display tight or isoclinal folds with axial microfold lineations and axial planar cleavage. The slumps appear to have developed on the slopes of a major NE−SW-orientated depression in the Welsh Basin. Turbidites, contemporaneous with the slumps and composed of silt and shells, display sole structures, which indicates that the flows advanced northeastwards (Bailey 1969). To the north, turbidites of the same composition bear structures that confirm the northward direction of transport. Cummins (1969) has pointed out that this depression is the same (Montgomery) trough that had originated earlier, in Wenlock times. The Ludlow turbidity currents, transporting shelf sediments, followed similar paths to those which had emplaced the lithic sands of the Wenlock Denbigh Grits and the early Ludlow Nantglyn Flags. However, in the Denbigh Trough north of the Derwen Ridge, the deposition of lithic sands swept in from the west continued.

Thus by middle Ludlow times much of the Welsh Basin was already in a state of uplift; indeed the Montgomery and Denbigh Troughs might have been all that remained of the former marine basin. Coarse sediments were transported southwards and eastwards. Near Llandovery the coarse sands of the Trichrûg Beds were laid down in deltaic conditions (Potter & Price 1965), but the sea returned before the final and widespread regressive phase at the end of Ludlow times. To the west, at Marloes, the progradation of deltas and rivers marking this regressive episode took place in late Wenlock times (Walmsley & Bassett 1976).

7l The Silurian outer shelf and slope

North of the Welsh Basin and the Irish Sea Platform

the outer shelf continued to be subjected to considerable subsidence. The deposition of mud predominated in the lower Silurian, but in the Lake District the basal Llandovery sediments were calcareous. During the Wenlock the deposition of muds continued, with the occasional influx of silt in the form of thin (3−5 mm) laminated and rippled bands interpreted as turbidites. These Brathay Flags yield graptolites and orthoceratids and compare closely with the Nantglyn Flags of the Welsh Basin. Towards the end of Wenlock times the sands coarsened. Sedimentary structures suggest that the currents mainly flowed from the northwest and southeast.

Calcareous silts marking the base of the Ludlow gave way to argillaceous silts with both monograptids and shelly fossils. These in turn coarsened upwards into a thick (1700 m) series of sands, the Coniston Grits, containing grains of quartz, orthoclase and muscovite, and fragments of spilite, soda trachyte, rhyolite and chert with radiolaria (Furness 1965). Sole structures suggest that these lower Ludlow beds were emplaced by currents flowing from the northwest and northeast (Furness et al. 1967) (Fig. 7.22). A. M. Ziegler (1970) has postulated that the zone of turbidite deposition extended towards the west-southwest into Ireland, because in the Devilsbit Mountains area sands were similarly emplaced by currents from the northwest and northeast. Mud deposition resumed in the Lake District later in lower Ludlow times, but in upper Ludlow times there was once again an influx of sands. Generally the sequences yield graptolites with hooked and simple thecae, but shelly faunas also occur. Weir (1975) has documented a rich assemblage of Wenlock brachiopods and trilobites enveloped in thick coarse-grained turbidites in the Cratloe Hills. In W. Wicklow and Slievenamon turbidites, transporting volcanic detritus derived from the southeast, constructed fans which gradually prograded northwestwards (Brück 1972).

7m Summary

This chapter has traced the evolution of the southern continental margin during the Lower Palaeozoic contractional phase of the Iapetus Ocean, a time span of 160−170 Ma. The ancient margin consisted of unstable shelf basins separated by relatively stable shelf platforms. The origin of the basins is uncertain, but there is a link between shelf instability and Ordovician−Silurian volcanicity.

The crustal development of the continental margin has been illustrated by tracing the development of the Welsh Basin and the Midland Platform. Although rapid subsidence of the basin broadly contrasted with slow subsidence of the platform, the detailed picture is complex. First, subsidence was not uniform throughout the basin. During late Cambrian to early middle Cambrian times greatest subsidence was located in the north where fans developed at the foot of submarine slopes. Later, during the Ordovician, 'blisters' of silicic magma caused local updoming which temporarily countered the effects of subsidence. Secondly, the southeastern margin of the basin shifted restlessly with time. During the lower−middle Ordovician period the Pontesford−Linley Fault was the critical fracture, but by upper Ordovician times the structural margin had shifted westwards (basinwards) to the line of the inferred infra-Towy−Severn Fault. By the Wenlock the margin edge had returned eastwards to the position of the Church Stretton Fault. Mild deformation, tilting and erosion accompanied the late Ordovician period of change.

The Ordovician volcanism has provided some of the strongest evidence for ocean crust subduction. Tholeiitic volcanoes in the northern Lake District passed southwards into calc-alkali volcanoes. This gradation appears to reflect an increase in the depth of origin of the magma. Rhyolitic and andesitic pyroclastics were predominant. Many volcanoes were probably emergent, but their flanks were submerged and colonised by shelly faunas. As volcanism declined in late Ordovician times the supply of volcanic sand diminished. Continued subsidence led to a deepening of the water; however, the subsidence was uneven, and intrabasinal faults produced local scarps and associated turbidites.

The Lower Palaeozoic evolution of the southern continental margin was punctuated by transgressive events in the lower Cambrian, Arenig, Caradoc and Llandovery. The lower Cambrian and Llandovery transgressions were worldwide, but those of the Arenig and Caradoc appear to have been more localised. The climax of the worldwide transgressions was the development of carbonates. Continental collision in upper Ordovician−Silurian times (Ch. 8) finally destroyed the pattern of shelf basins and platforms developed during the contractional phase of the Iapetus Ocean.

8   Closure, collision and deformation

8a   The pattern of closure

Part 2 has traced the geological events that accompanied the growth and demise of the Iapetus Ocean. That closure of this ocean had commenced by Tremadoc times is almost certain; the intense deformation and metamorphism of the Grampian Orogeny (Arenig—Caradoc), coupled with the early Ordovician extrusion of ocean crust within the margin of the northwestern continent, attest to this. However, it is Wright's (1976) thesis that Iapetus began to close very much earlier, in Precambrian times. Wright has proposed that the series of orogenies recorded in the British Isles and adjacent areas between 1050 Ma and 450 Ma ago reflects successive phases of subduction activity. He has further noted that the orogenies alternately affected the northwestern and southeastern continents. Thus the Morarian Orogeny (1050—730 Ma) is regarded as marking the first phase of NW-directed subduction. This was the first deformation of the Moine. Later the subduction activity turned southeastwards. The Celtic Orogeny (700—600 Ma) affected the sediments of the Monian and Brioverian. Early in Ordovician times the Grampian Orogeny (530—465 Ma) reflected a return to NW-directed subduction. But c. 500 Ma subduction became active along both margins and continued until continental collision early in the Wenlock.

Wright has invoked a neat but speculative explanation for the switching pattern. Subduction at a given margin would tend to cease as the hot light mid-ocean ridge approached the trench. As subduction of the main ocean lithosphere arrested, so the spreading activity of (any) back-arc marginal ocean basin would cease. Indeed, the crust of the marginal basin would be *subducted* until the volcanic back arc made contact with the mainland continent. At this point subduction would commence at the opposing continental margin on the far side of the ocean.

8b   The rate of closure

Two independent lines of evidence have been emp-
loyed to estimate the rate of closure: (a) the progressive migration of fauna, and (b) the times and sites of magmatic activity. McKerrow and Cocks (1976*) (Ch. 3e) have clearly demonstrated that there was progressive migration of faunal groups across Iapetus. First to establish representatives in shelf seas of both margins were colonies of the pelagic dendroid *Dictyonema flabelliforme*. Later, species of benthic ostracods and freshwater fish became common to both sides of the former ocean.

Using knowledge of the longevity of modern pelagic larvae, McKerrow and Cocks have suggested that Ashgill (late Ordovician) brachiopod and trilobite larvae might have survived seven to fourteen weeks before reaching suitable substrates for development. From this they have estimated that the contemporary ocean, which was then being traversed by such species for the first time, was approximately 2000—4000 km wide. Obviously much depended on the pattern of the surface ocean currents. If final ocean closure occurred in the late Wenlock, the average rate of closure between Ashgill (460 Ma) and Wenlock (420 Ma) times was about 7.5 cm a^{-1}.

In their analysis of the times and sites of magmatic activity Phillips *et al.* (1976*) have derived rather more conservative values. Allowing for 50% strain incurred by post-collision deformation of the Lower Palaeozoic of SE Ireland, it appears that the upper Ordovician volcanic arc of SE Ireland was aligned at 14° to the line of the Iapetus Suture. Along the length of this volcanic arc the ending of volcanism occurred at progressively later dates towards the southwest (Fig. 8.1). Phillips *et al.* have attributed this to the SW-migrating point of collision at a rate of 1.7 cm a^{-1}. From this they have evaluated the closure as 0.4 cm a^{-1}. Using this figure as an average rate operating between the times of the commencement and cessation of subduction, Phillips *et al.* have calculated the amount of closure attributable to consumption on the southeastern side during the Ordovician to be about 280 km. A similar figure has been rather tenuously derived from consideration of

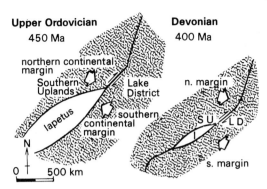

Figure 8.1 Hypothetical diagram showing the final closure of the Iapetus Ocean. Note the continental misfit and the dextral movement along the Iapetus suture (after Phillips *et al.* 1976*).

the age of post-tectonic granites on the northwestern side of Iapetus. From data derived from magmatic rocks, it therefore seems that the ocean was quite narrow (less than 600 km) even during the Arenig, and that closure was a relatively slow process (0·65 cm a⁻¹).

The marked discrepancy of more than an order of magnitude between the values produced by the two independent studies serves as a reminder that even plausible assumptions must sometimes be in error.

8c Closure and the pattern of deformation

The Caledonides exhibit two types of structural zone: (a) the orthotectonic zone displaying intense folding and low- to high-grade metamorphism; and (b) the paratectonic zone displaying more simple folds and only low grades of metamorphism. Broadly, the early tectonic events − the Morarian(?), Celtic and Grampian − were orthotectonic, and the later events were paratectonic. If the deformation was indeed linked with the process of ocean closure, then two mechanisms of folding are implicated: (a) shearing caused by the underthrusting of the oceanic lithosphere beneath the leading margin of the continental lithosphere; and (b) deformation caused by the collision of the two continental plates. It is tempting to see the present-day exposures of the paratectonic Caledonides in the Southern Uplands, W. Ireland, Longford Down, the Lake District and Wales (Figs 8.2 and 8.3) as the relics of deformation caused by the collision, but this is almost certainly an oversimplification.

Consideration has already been given to the hypothesis of McKerrow *et al.* (1977*) that the Southern Uplands represent a deformed continental slope − trench sequence. If the analogy with modern trench sequences is correct, then the earliest phases of deformation, at least, were caused by underthrusting rather than by collision. Similar evidence has been accrued in the Lake District, on the southern side of the proposed zone of collision. Simpson (1967), noting the intensity of deformation in the Skiddaw Slates (Llanvirn), has argued that a strain sequence comprising three phases of deformation preceded the emplacement of the Borrowdale Volcanics in the Llandeilo. Dewey (1969b*) has pointed to the lower ductility of the Borrowdale Volcanics and declared Simpson's argument equivocal. But the major unconformity beneath the Caradocian Coniston Limestone is at least a clear demonstration of pre-Caradocian folding. The pattern of the strain sequence in the Lake District − steep cleavage (D1), wrench faulting, flat strain-slip cleavage (D2), steep strain-slip cleavage (D3) − is broadly comparable to that recorded in the Southern Uplands and Wales (Figs 8.2b and 8.3). Vertical extension (D1) was followed by axial extension (wrench faulting). This gave way to vertical shortening (D2), which was finally followed by axial shortening (D3).

In Wales early compression resulted in vertical extension represented by E−W (D1) and N−S (D2) structures. But the generation of these structures was partially synchronous. Two major domes, the Harlech and Berwyn, are prominent in N. Wales, but Dewey (1969b*) has considered these volcanotectonic centres rather than structures resulting from the interference of the two earliest phases of deformation (D1 and D2). R. M. Shackleton (in Dewey 1969b*) has attributed some of the buckling of Ordovician−Silurian sediments to the activity of steep faulting in the basement. During the phase of vertical shortening there was thrusting and the development of subhorizontal strain-slip cleavage (D3). Finally, axial shortening produced NW−SE strain-slip cleavage (D4) comparable to D3 in the Lake District and D5 (kink bands) in the Southern Uplands.

In summary, the paratectonic Caledonides generally exhibit four successive stages of deformation − vertical extension, axial extension, vertical

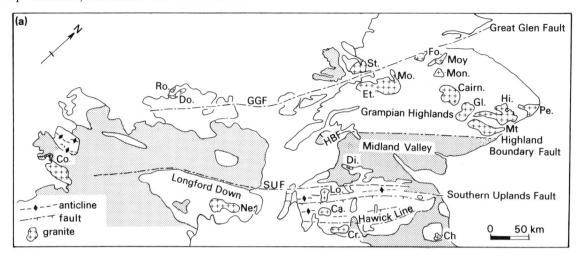

Figure 8.2a Caledonian paratectonic structures in the N. British Isles and the late Silurian−Devonian granites. Granites: Ca. = Cairnsmore of Fleet; Ch. = Cheviot; Co. = Connemara; Cr. = Criffel; Di. = Distinkhorn; Et. = Etive; Cairn. = Cairngorm; Do. = main Donegal; Fo. = Foyers; Gl. = Glen Cairn and Lochnager; Hi. = Hill of Fare; Lo. = Loch Doon; Mo. = Moor of Rannoch; Mon. = Monadliath; Moy = Moy; Mt = Mt Battock; Ne. = Newry; Pe. = Peterhead; Ro. = Rosses; St. = Strontian.

shortening and axial shortening − involving up to five (D1 to D5) phases of deformation. It seems probable that some of the deformation recorded in the remnants of the paratectonic Caledonides took place during the process of closure, but it is not yet clear exactly which phases of strain were caused by the collision of the continents.

8d Closure and the generation of strike-slip faults

Phillips *et al*. (1976*) have sought an explanation for the major sinistral strike-slip faults in Scotland. They have proposed that the NW-directed zone of subduction sloping beneath the Southern Uplands converged with an opposing SE-dipping zone. The latter is envisaged as a continental subduction zone, with crust-bearing Lewisian rocks, underthrusting the Moine and Dalradian pile along the Moine Thrust zone. Phillips *et al*. have constructed a vector diagram and demonstrated that it is improbable that the subducted slabs were directly opposed. They have reasoned that the convergence of the subducted slabs with their respective vectors was responsible for the initiation of the Great Glen Fault. A similar argument is employed to infer a dextral shear along the zone of the Iapetus Suture

(b)

BULK STRAIN	SOUTHERN UPLANDS Wigtownshire
AXIAL SHORTENING	granites
	D5 : N−S kink bands
VERTICAL SHORTENING	D4 : flat cleavage, late thrusts
AXIAL EXTENSION	D3 : sinistral activation of strike faults wrench faulting
VERTICAL EXTENSION	D2 : S-facing monoclines superimposed on D1 monoclines
	D1 : NW-facing monoclines; cleavage development and steepening of fold plunges; thrust faults

Figure 8.2b Paratectonic strain sequence in Wigtownshire, SW Southern Uplands (after Dewey 1969b*).

following the collision of the two leading continental margins (Fig. 8.1).

8e Late Caledonian magmatism

Throughout much of the long history of the Iapetus Ocean there was conspicuous igneous activity. Essentially, this activity took two forms. (1) There was emplacement of plutons and eruption of volcanics during two of the orogenic episodes: the late Precambrian Celtic Orogeny which affected the south,

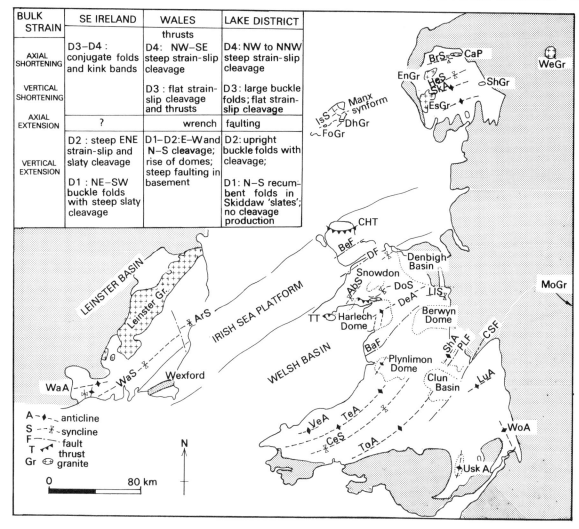

BULK STRAIN	SE IRELAND	WALES	LAKE DISTRICT
		thrusts	
AXIAL SHORTENING	D3–D4 : conjugate folds and kink bands	D4: NW–SE steep strain-slip cleavage	D4: NW to NNW steep strain-slip cleavage
VERTICAL SHORTENING		D3 : flat strain-slip cleavage and thrusts	D3: large buckle folds; flat strain-slip cleavage
AXIAL EXTENSION	?	wrench	faulting
VERTICAL EXTENSION	D2 : steep ENE strain-slip and slaty cleavage	D1–D2:E–W and N–S cleavage; rise of domes; steep faulting in basement	D2: upright buckle folds with cleavage;
	D1 : NE–SW buckle folds with steep slaty cleavage		D1: N–S recumbent folds in Skiddaw 'slates'; no cleavage production

Figure 8.3 Caledonian paratectonic structures and strain sequences in the S. British Isles, and the late Silurian–Devonian granites. AbS = Abersoch Syncline; ArS = Arklow Syncline; BaF = Bala Fault; BeF = Berw Fault; BrS = Braithwaite Syncline; CSF = Church Stretton Fault; CaT = Carmel Head Thrust; CeS = Central Wales Syncline; DeA = Derwen Anticline; DiF = Dinorwic Fault; DoS = Dolwyddelan Syncline; HeS = Helvellyn-Scafell Syncline; IsS = Isle of Man Syncline; LlS = Llangollen Syncline; LuA = Ludlow Anticline; PLF = Pontesford-Linley Fault; ShA = Shelve Anticline; SkA = Skiddaw Anticline; TT = Tremadoc Thrust; TeA = Teifi Anticline; ToA = Towey Anticline; UskA = Usk Anticline; VeA = Velindre Anticline; WaA (S) = Waterford Anticline (Syncline); WoA = Woolhope Anticline; CaP = Carrock Fell pluton; DhGr = Dhoon Gr; EnGr = Ennerdale Granophyre; EsGr = Eskdale Gr; FoGr = Foxdale Gr; MoGr = Mountsorrel Gr; ShGr = Shap Gr; WeGr = Weardale Gr (concealed). (Sources − Dewey 1969b*, Roberts 1977).

and the early Ordovician Grampian Orogeny which affected the north. These orthotectonic episodes were associated with regional metamorphism which locally attained high grades. (2) As subduction proceeded in Ordovician times, volcanoes flared up on the northern and southern continental margins. In the south, volcanic activity was greatest in the Lake District, Leinster Basin and Welsh Basin, areas subject to rapid subsidence. In the north, volcanism was rather less extensive, but eruptions occurred along a zone extending from W. Ireland northeastwards to the Southern Uplands. It is notable that emplacement of large scale plutons was not associated with this magmatism and accompanying paratectonic deformation.

However, in late Silurian to early Devonian times, after a short lull following the volcanism associated with subduction, igneous activity became important again. Granite plutons were widely emplaced, from N. Scotland, the Grampians and Southern Uplands in the north (the 'Néwer Granites'), to Ireland, the Lake District and Charnwood Forest (Mountsorrel) in the south (Figs 8.2a and 8.3). The 'Newer Granites' of Scotland are associated with a widely scattered suite of hybrid bosses, sills and dykes. Investigators of this appinite suite have discovered that, whatever the content of ferromagnesian minerals, the feldspars are always alkalic (sodium and potassium rich). Joplin (1959) has envisaged that hybridisation occurred in two phases. First the water and volatiles of a slowly rising granodioritic magma permeated and hybridised a mafic or ultramafic rock, producing a hornblende-rich suite. Later these first-stage hybrids were invaded by the granodioritic magma, resulting in the generation of tonalites and quartz monzonites. The calc-alkaline 'Newer Granites' commonly have a concentric arrangement of compositional zones. The outer dioritic or tonalitic margin, often displaying flow structures, passes inwards to a zone of granodiorites, which in turn passes inwards to an adamellitic core. Some compositional boundaries are gradational; others are sharp and exhibit chilled margins. The contacts with country rocks are either sharp and transgressive, or veined and sheeted. Sharp contacts are characteristic of the steep walls of stock-like granites such as the Criffell intrusion, or laccolithic forms like the Cairngorm Complex. The strike of the surrounding country rock is commonly deflected into conformity with the granite margin and indicates that emplacement was forceful. Some granites also show complex margins with granulites and schists riddled by granite, pegmatite and aplite sheets and veins. Contact metamorphism is common.

The arrival of the 'Newer Granites' preceded the passive emplacement of the 'last' granites in early Devonian times. Cylindrical blocks of country rock subsided along ring-shaped fractures and permitted granitic magmas to rise and fill in the space. This cauldron subsidence mechanism of emplacement operated at Glencoe and Ben Nevis in the Grampian Highlands. Lavas of basalt, andesite and rhyolite poured out over the terrain. Southwest of the Grampian Highlands there was further volcanism in the eastern part of the developing Midland Valley Graben. These rocks are now exposed in the Ochil, Sidlaw and Pentland Hills. In sharp contrast to the Ordovician volcanism of the south, tuffs were subordinate.

The late Caledonian magmatism was associated with mineralisation. Dagger (1977) has maintained that mineralisation at Coniston in the Lake District was controlled by the emplacement of granite. Early haematite mineralisation took place as the sedimentary pile suffered deformation and low-grade regional metamorphism. This was followed by a main phase of sulphide mineralisation. Iron pyrites, chalcopyrite, arsenopyrite and magnetite formed as granite was emplaced at depth. Finally there was carbonate mineralisation associated with uplift of the granite.

Clearly the upper Silurian to lower Devonian period was a time of great heat production. It is tempting to link this magmatism with crustal friction caused by continental collision. This is not, however, a fully satisfactory explanation, because the open style of the late Silurian folds does not appear to record the intense deformation that would be expected. It is also difficult to explain why the Caledonian plutons plugged the Grampian and Southern Uplands terrain but generally shunned the subsiding Midland Valley Graben. Thus the origin of this late Caledonian magmatism is far from being completely understood.

8f Summary

Although the various lines of evidence are consistent with the view that the hypothetical Iapetus Ocean gradually closed during the Lower Palaeozoic, culminating in continental collision in Wenlock to early Devonian times, there is much that is not understood. When did closure actually commence? What was the cause of the Celtic, Grampian and Lakelandian orogenic episodes, apparently occurring alternately on the northern and southern continental margins? What was the rate of the closure? (Estimates to date vary greatly and make gross

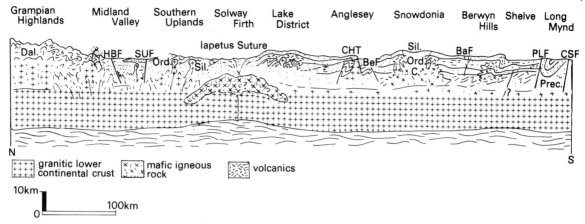

Figure 8.4 Hypothetical crustal section across the Caledonides in late Silurian times. BaF = Bala Fault; BeF = Berw Fault; CHT = Carmel Head Thrust; CSF = Church Stretton Fault; Dal. = Dalradian; HBF = Highland Boundary Fault; PLF = Pontesford-Linley Fault; SUF = Southern Uplands Fault; C. = Cambrian; Ord. = Ordovician; Prec. = Precambrian; Sil. = Silurian.

assumptions.) How much deformation was caused by the underthrusting associated with subduction, and what was the cause of the late Caledonian magmatism? To all these questions there are not yet satisfactory answers; but it is clear that, with the cessation of marine sedimentation and the subse-

quent deformation and magmatism, late Silurian to early Devonian times witnessed the end of an important chapter in the evolution of the British Isles. In essence, this was the growth of the Caledonides (Fig. 8.4).

Theme

After the Caledonian deformation in Britain a period of rapid uplift and erosion ensued in a semi-arid climate. In N. Britain the Old Red Sandstone (ORS) was deposited in internal molasse basins. In the Irish and Welsh Basins, and in Belgium and N. Germany, the ORS rivers meandered across subsiding coastal plains of alluviation and deposited some of their load on to branches of the complex shelf of the long-established Rheic Ocean and its marginal seas. A prominent deepening phase, accompanied by basic volcanism, occurred in the Devon and Rhenish areas at the beginning of the upper Devonian. This caused fragmentation and subsidence of middle Devonian carbonate platforms and might have been due to the development of small marginal ocean basins. After the Bretonic uplift during late Devonian times a major marine transgression, causing extensive carbonate deposition, flooded the margins of the ORS continent and gave rise to a number of interconnected gulfs during the early Carboniferous. The block—basin system that developed on the Carboniferous shelf might have been due to lower crustal hot creep directed towards the marginal basins to the south. During late lower Carboniferous times a major phase of uplift occurred in the Caledonian orogenic belt. This coincided with the Sudetic orogenic phase, caused by subduction of oceanic lithosphere, in the Hercynides. Major river systems spread into many basins, depositing the coarse Millstone Grit facies in N. England. By upper Carboniferous times vast deltaic plains, steaming under a humid tropical climate, covered much of N. Europe and eastern N. America. Thick peat deposits periodically accumulated in the delta backswamp. Final contractions of the Hercynian orogenic belt in late Carboniferous times gave rise to new upland areas, ushering in a new era of aridity.

Plate 3 shows a Coal Measures cycle exposed at Swillington Quarry, near Leeds, Yorkshire. The photograph shows a small area on the left of the large field sketch in Figure 11.20. The vertical tree trunk grew from the level of the top of the thin coal. The river or distributary channel deposits that surround most of the trunk were presumably quickly deposited around the (?dead) trunk which must have continued to have stood up proud from the channel bed. A prominent scour-like feature surrounds the trunk about half way up. (Photo by Andrew Scott)

9 The Old Red Sandstone: Caledonian molasse

9a The Old Red Sandstone continent

The landmass formed by the final welding together of a northern and a southern lithospheric plate in late Silurian to early Devonian times has been named the 'North Atlantic' or 'Old Red Sandstone' (ORS) continent (Fig. 9.1). The immediate post-orogenic history of the area during Devonian times may be pieced together using evidence from the distinctive ORS successions that crop out in many areas. The ORS rests in many northern areas of the British Isles with angular unconformity upon Lower Palaeozoic or Precambrian rocks. The great thickness and coarse grainsize of some ORS successions provides evidence for rapid differential vertical crustal movements within and adjacent to the Caledonian mountain chains, the analysis of palaeocurrent patterns and pebble and heavy mineral suites helping to deduce the location, magnitude and timing of uplift. The ORS is largely a continental succession of facies zoned by fish faunas and containing interesting soil horizons, analysis of which supports the palaeomagnetic evidence for tropical latitudes in Britain during Devonian times (Fig. 9.1).

9b Anglo-Welsh–Irish basins

The widespread outcrops of ORS in this area (Fig. 9.2) are increasingly deformed towards the south. Lower ORS deposits generally rest conformably

Figure 9.1 Approximate extent and palaeolatitudes determined by palaeomagnetism of the ORS continental landmass in upper Devonian times (partly after Woodrow *et. al.* 1973).

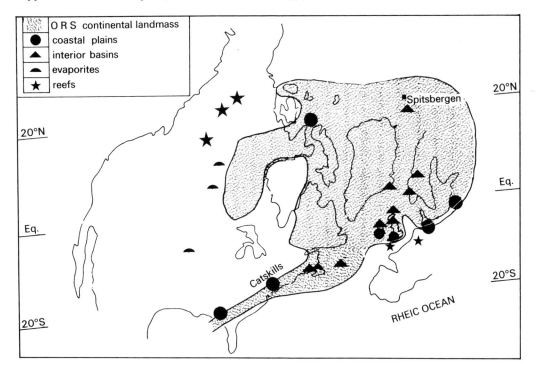

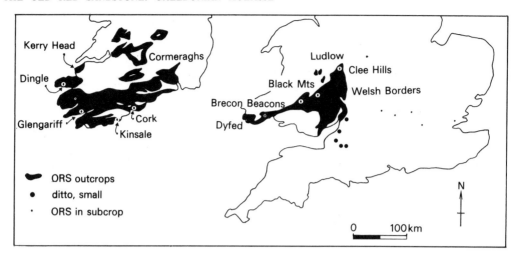

Figure 9.2 Main ORS outcrops in the Anglo-Welsh–Irish area.

IRELAND		WALES		ENGLAND	
DINGLE	SHERKIN	S DYFED	N BRECON	CLEE HILLS	
Glengarrif Harbour Gp 3000m	Old Head Fn (200m)	Skrinkle Sandstone (300m)	Grey Grits (60m)	Farlow Gp (0–150m)	upper ORS
	Toe Head Fn (350m)		Plateau Beds (45m)		
Kilmurry Fn (660m)		Ridgeway Conglom. (365m)			middle ORS
Inch Conglom. Fn (600m)	Castle Haven Fn (1000m)				
Ballymore Fn (1200m)		Cosherston Gp (3000m)	Brownstones (425m)	Woodbank Gp (235m)	lower ORS
Trabeg Conglom. Fn (265m)	Sherkin Fn (1200m)	Red Marl Gp (365m)	Senni Beds (300m)	Ditton Gp (450m)	
			Red Marl Gp (1220m)	Downton Gp(515m)	

Table 9.1 Main stratigraphic units defined in the ORS in Ireland, Wales and the Welsh Borders. Shading indicates absence of strata.

upon marine Silurian. However, in the Winsle Inlier, Dyfed, the contact between marine Silurian and the Red Marl Group of the ORS is an unconformity denoted not by obvious angular discordance, but by the progressive development of a distinctive discolouration in the marine Silurian rocks and an abrupt facies change. In most areas

middle ORS is absent, the upper ORS resting unconformably upon gently folded lower ORS (Table 9.1). There is a transition from terrestrial upper ORS to marine late Devonian or early Carboniferous rocks.

The Welsh Borderland outcrops around Ludlow (Fig. 9.2) have long been recognised as classic localities for the study of the lower ORS regressional episode. The basal Downton Group (Table 9.1) rests erosively, but conformably, upon underlying Silurian (Ludlovian) flagstones. The Ludlow Bone Bed at the base of the group (c. 0·15 m thick) is of variable lithology, ranging from well-sorted sandstones to mudstones with silty laminations (J. R. L. Allen 1974a*). The sandstones contain comminuted thelodont and ostracoderm fish debris, lingulid shell fragments and phosphatic pellets. Tops of sandy or silty units show both symmetrical and asymmetrical ripple forms. The Ludlow Bone Bed is considered analogous to present-day shelf lag concentrates produced by slow sedimentation and reworking. Subsequent shoreline regression led to the formation of well-sorted, flat to low-angle, laminated sandstones of suspected high-energy, coastal beach–barrier facies (Fig. 9.3).

Overlying formations of the Downton Group record further regression, with facies of shallow subtidal to intertidal aspects (J. R. L. Allen 1974a*). Lingulids frequently occur in life position in mudstones, and other vertical burrow systems are plentiful. These low-energy coastline facies are succeeded

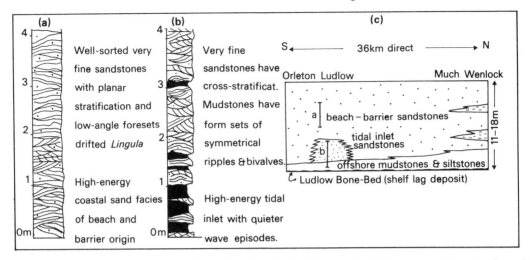

Figure 9.3 Graphic logs and restored stratigraphic section for the Downton Castle Formation (Downton Group) in the Ludlow area (after J. R. L. Allen 1974a*).

by a great thickness of repeated examples of fining-upwards cycles. The meandering river channels and floodbasins that produced the cycles were a persistent aspect of ORS geography on the broad coastal plains of alluviation.

Thus, in lower ORS times, the palaeogeographic picture in the Welsh Borderland is of a S-migrating strandline of beaches and barriers with sheltered lagoonal or tidal flat deposits, behind which were extensive alluvial plains with river channels and low floodplains. Evidence for shoreline regression at about this time is also seen in Dyfed and SW Ireland, including a thick sequence of conglomerates and mudstones of alluvial fan origin in the Ridgeway Conglomerate of Dyfed (Fig. 9.4). The thick succession of ORS deposits on the Dingle Peninsula, Ireland (Table 9.1), is particularly noteworthy because of evidence for periodic, very coarse-grained incursions attributed by Horne (1975) to braided stream and alluvial fan environments. Aeolian deposits have been recognised interbedded with fluvial facies in the Inch Conglomerate Group (Table 9.1).

A period of uplift and mild deformation ensued in middle ORS times with depositional episodes represented by the (supposed middle ORS) Ridgeway Conglomerate of Dyfed. In upper ORS times thick sequences of repeated fining-upwards cycles were again deposited. There is an upward change to marine late Devonian to early Carboniferous

Figure 9.4 Graphic logs through: (a) part of the lower ORS Red Marl Group at Freshwater West, Dyfed to show repeated alluvial fining-upwards cycles (after J. R. L. Allen 1963); and (b) the middle ORS Ridgeway Conglomerate at Skrinkle Haven, Dyfed to show alluvial fan facies (after B. P. J. Williams 1971).

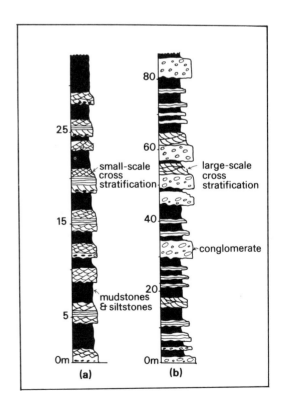

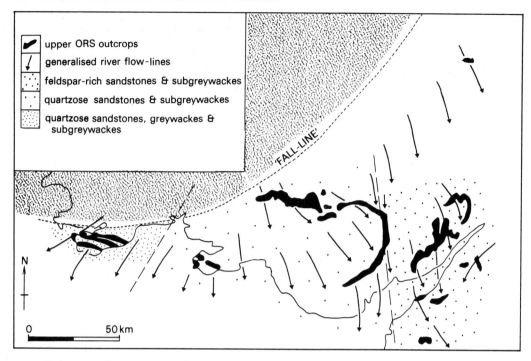

Figure 9.5 Outcrops, palaeocurrents and sandstone types of the upper ORS of S. Wales (after J. R. L. Allen 1965*).

deposits in S. Ireland and Dyfed. This is well shown in the coastal sections southwest of Cork, Ireland, where fining-upwards cycles pass up into grey units with mud drapes or flasers in sandy ripple troughs, lenticular bedding, bipolar small-scale cross-stratification (herringbone structure) and frequent wave-formed ripples. The topmost Devonian marginal deposits of the Skrinkle Sandstone in Dyfed are succeeded by shallow-water bioturbated mudstones and marine limestones, the latter becoming dominant upwards.

In the Welsh Borderland, J. R. L. Allen (1974a*) has found convincing palaeocurrent evidence for SE-directed river flow throughout much of ORS times. Heavy minerals and exotic rock fragments of high-grade metamorphic types arrived in floods early on in lower ORS times, but these decreased in abundance upwards until in the upper ORS, igneous and sedimentary rock fragments and igneous heavy minerals dominate. The only suitably garnetiferous metamorphic sourcelands to the northwest, up the presumed ORS palaeoslope, are the Moinian–Dalradian metamorphic rocks of the Caledonides in NW Ireland and the Scottish Highlands. Evidently the length of the ORS drainage

basin at this time was in excess of 400 km. Severance of the 'metamorphic connection' as ORS times proceeded is attributed to late orogenic uplift in the paratectonic regions of the orogenic belt. The igneous and sedimentary hinterlands of later ORS deposition lay in the central to N. Welsh (Fig. 9.5) region, where a great variety of greywacke- and volcanic-rich strata may be matched closely in the ORS pebble assemblages.

As well as this northern sourceland during ORS times, there is increasing evidence for a southern 'Bristol Channel' landmass. Thus B. P. J. Williams (in Owen *et al.* 1971) has found evidence of N-directed currents and N-decreasing clast size and angularity in the Ridgeway Conglomerate (middle ORS) of Dyfed, attributed to deposition in N-prograding alluvial-fan systems. Clasts are of sandstones with ?Cambro-Ordovician fossils, abundant and often large angular fragments of vein quartz and, towards the top, much phyllite. J. R. L. Allen (1975) has noted exotic clasts of poorly rounded acid–intermediate lavas and tuffs, various lithic sandstones and pink quartzites in the lower ORS Llanishen Conglomerate around Cardiff. He has argued that the assemblage is of southern provenance, with the volcanic clasts indicating 'a modest westerly extension of early Silurian volcanics from Tortworth and the Mendip Hills'. This may imply that the ORS facies in the dominantly marine Devonian of Devon (Ch. 10) was deposited in a separate depositional basin from that of the main Anglo-Welsh alluvial plains to the north.

9c The Midland Valley of Scotland, and Ireland

ORS sedimentary and volcanic rocks crop out extensively in the Midland Valley Graben and in the Scottish Borders area to the south. Sediments of the same age also occur in the extension of the rift in Ireland (Fig. 9.6). Middle ORS rocks are absent from the area; the upper ORS, as determined by fish faunas, usually rests upon the lower ORS with angular unconformity (Table 9.2). Midland Scotland, along with much of the Anglo-Welsh area, was deformed and faulted during middle ORS times.

Lower ORS sequences are usually thick and associated with abundant calc-alkaline andesitic, rhyolitic and basaltic volcanics (Fig. 9.7). The thickest and most extensive area of present outcrop

BORDERS	MIDLAND VALLEY	CAITHNESS	SHETLAND	
upper O R S (250m)	upper O R S (200–900m)	Dunnet Fn (600m)		upper O R S
		John O'Groats Sandstone Gp (630m)	Melby Fn (760m)	middle O R S
		Caithness Flagstone Gp (3850m)	Walls Fn (9000+ m)	
Cheviot Lava Gp	Strathmore Gp (2000m) Garvock Gp (1525m) Arbuthnott Gp (2100m) Crawton Gp (670m) Dunnotar Gp (1660m) Stonehaven Gp (1550m)	Sarclet Gp (430m)	Sandness Fn (1350–3000+ m)	lower O R S

Table 9.2 Main stratigraphic units defined in the ORS in Scotland. Shading indicates absence of strata.

Figure 9.6 Distribution of ORS sedimentary and volcanic rocks and early Devonian plutons ('Newer Granites') in Midland Scotland, S. Scotland and Ulster.

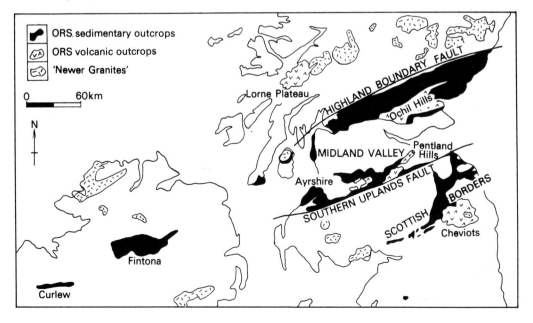

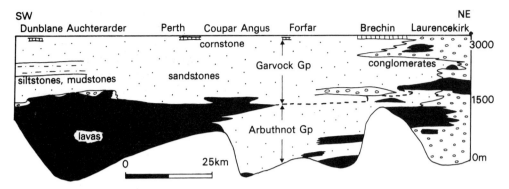

Figure 9.7 Stratigraphic section for the upper part of the lower ORS along the N. Midland Valley. Note: the lateral variations in lithology are due chiefly to the proximity of the Laurencekirk area to the syndepositional Highland Boundary Fault (abundant conglomerates) and to the proximity of the Dunblane area to a major andesitic lava depocentre (after Armstrong & Paterson 1970).

occurs in the Strathmore area along the northern fringe of the Midland Valley, bounded by the major Highland Boundary Fault to the north (Fig. 9.7). Thicknesses gradually decrease from Stonehaven southwestwards to Arran and hence to Cushendall, Ulster. Successions comprise thick members of conglomerates, pebbly sandstones, sandstones and siltstones. Trends towards decreasing grainsize outwards from the Highland Boundary Fault often exist. A. C. Wilson (1971) has recognised facies of sheetflood, braided stream and ephemeral-stream floodplain types, and postulated the former presence of interfingering alluvial-fan systems draining from source areas in the present-day Highland Borders region north of the Midland Valley (Fig. 9.9). Over the whole area the commonest clasts in conglomerates are of volcanic (andesite, basalt), metaquartzite and quartz types, with garnets prominent among heavy mineral assemblages which may also be rich in the less stable silicates such as hornblende and pyroxene. In the northeastern outcrops, where the succession is thickest, clasts include a greater proportion of high-grade metamorphic material such as sillimanite and staurolite gneisses, particularly towards the top of the succession. Rapid vertical variations in lithology may be correlated with events of alluvial fan aggradation and dissection, possibly due to synchronous movements along the Highland Boundary Fault. Lateral variations along strike (Fig. 9.7) reflect changing hinterland palaeogeology and the location of major depocentres of volcanic lavas. A second area of subsidence, sedimentation and volcanism occurred along the southern margin of the Midland Valley Graben in the area of the present-day Pentland Hills. Although little is known about the sedimentary succession, the detritus is thought to have been derived from the Southern Uplands Massif to the south.

Pebble types, heavy minerals, facies changes and palaeocurrent measurements all indicate that the Midland Valley was a true sedimentary basin surrounded by extensive high hinterlands undergoing rapid weathering and erosion to the north and south. Hinterland relief and uplift rates were probably highest adjacent to northeastern areas, where the greatest sediment thicknesses are now seen and where present day outcrops of the highest-grade Dalradian metamorphic rocks occur. This area of maximum uplift was probably separated from the subsiding graben by an active fault line scarp, now seen as the Highland Boundary and Ochil Faults. To the southwest, in Kintyre, Arran and Fintona, there is no firm correlative of the Highland Boundary Fault, and the basin margin must have been a broader zone of contemporaneous downwarp or a hingeline. North of these areas remnants of lower ORS are seen preserved in volcanic vents and under the Lorne Plateau Lavas.

The onset of sedimentation in the Midland Valley Graben was coincident with the late Silurian to early Devonian deformation of the Southern Uplands to the south and with the intrusion of the very extensive 'Newer Granite' swarm of calc-alkaline granites and granodiorites into both the Highlands belt and the Southern Uplands (Fig.

9.6). Rapid uplift ensued around the Midland Valley.

It is worth stressing that the andesitic lava cover in the Highland Borders area was probably far more extensive than that seen today (e.g. the Lorne Plateau Lavas) because of the very great abundance of lava pebbles seen in outcrops of the lowest ORS. Being the high level equivalents of the 'Newer Granite' plutons, their position became unstable due to great post-extrusive uplift of the majority of the Highland Borders area. In other areas, such as the Cheviots to the south, where uplift was presumably much less, great thicknesses of andesitic lavas are still present above and around the Cheviots granitic pluton.

Upper ORS successions are generally both thinner and finer-grained than those of the lower ORS. It is possible that some of the upper ORS may be early Carboniferous in age, since plant spores indicative of the Devonian/Carboniferous boundary do not exist. The middle Devonian deformation threw the lower ORS into a broad series of open folds which were much denuded prior to upper ORS deposition. Data derived from the upper ORS indicate the continued existence of the Midland Valley as a depositional alluvial basin, with fluvial palaeocurrents indicating drainage out from the Southern Uplands and Highlands land areas (Fig. 9.9). However, several source areas also existed within the basin.

In the Firth of Clyde area, Bluck (1967) has recognised an association of conglomerates, sandstones and siltstones indicative of alluvial fan (debris flow and stream flow deposits), braided stream and alluvial floodplain origin. Stratigraphic sections (Fig. 9.8) show an upward reduction in grainsize, together with a change in depositional environment from fan mudflow to fan stream to braided stream to floodplain. Such features suggest the diminishing effects of the source areas through time at any one locality, probably due to the erosion and planation of newly emergent, fault-bounded uplands.

In the northern outcrop strip of upper ORS around Stirling, repeated fining-upwards cycles

Figure 9.8 Graphic logs to show: (a) coarse-grained alluvial fan facies from the upper ORS of the Clyde area (after Bluck 1967); (b) fining-upwards cycles with prominent cornstone horizons from the upper ORS Cornstone Formation of the Stirling area (after Read & Johnson 1967); and (c) interior-basin flood facies from the upper ORS of the Scottish Borders (after Leeder 1973).

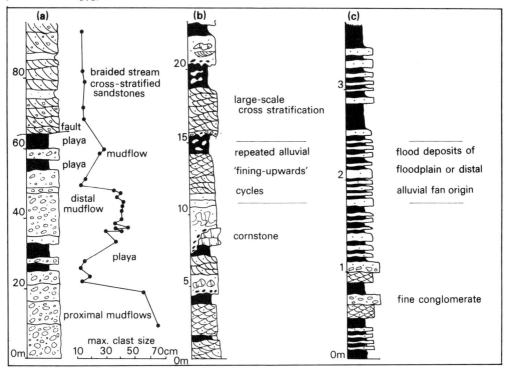

occur (Fig. 9.8), the coarse members of arenitic to subarkosic sandstones being predominant. Together with palaeocurrent evidence, these indicate deposition in meandering stream channels draining from the Highland Borders zone to the north. A formation rich in cornstones of soil origin underlies the first undoubted lower Carboniferous fluvio-lacustrine cementstone facies. In the Lomond area, Chisholm and Dean (1974) have recently interpreted well-sorted bioturbated sandstones with bimodal current directions as being tidal in origin, thus implying a temporary marine transgression eastwards into the Midland Valley.

In the southern outcrop of upper ORS in Ayrshire, Burgess (1961*) has recognised a development of thick cornstones (Ch. 9e) within sandstone-dominated fining-upwards cycles. The sandstone members show current directions from the Southern Uplands area towards the south and west.

An interesting discovery of ORS of unknown age occurs as the reservoir in the Buchan field in block 21/1 off Aberdeen (Fig. 9.10). About 500 m of thin fining-upwards cycles make up the succession, each cycle being about 1 m thick and containing conglomeratic cornstones as clasts immediately above scoured surfaces. The succession bears some similarity to waning-flow sheetflood sequences and may record persistent location in a distal alluvial-fan environment. Whether the succession is Midland Valley or Orcadian Basin in context is unknown, but the sourcelands were presumably of acid or intermediate igneous composition since zircon and hornblende are common heavy minerals, and garnet is comparatively rare.

The upper ORS of the Scottish Borders region is a wholly fluviatile succession. The basal unconformity with deformed Silurian flysch, lower ORS being absent, shows evidence of relief (more than 150 m) before burial by a sequence of pebbly sandstones, sandstones and siltstones of braided and meandering channel origin (Leeder 1973). Palaeocurrents indicate stream flow from the northeast and southwest (Fig. 9.9) into the Jedburgh area, where fine-grained siltstones and thin sandstones are interpreted as basin-centre floodplain facies (Fig. 9.8). It is debatable whether the basin was entirely of internal drainage type or whether some of the streams, together with those of the Midland Valley, drained into the North Sea Basin (Fig. 9.9). Some support for the latter view is given by the recent discovery of marine middle Devonian reef limestones in the Auk oilfield, indicating a connecting seaway from the Hercynian ocean into the ORS continent. Such a seaway might have existed in upper ORS times and been the route of the brief marine incursion into the upper ORS of the Midland Valley noted above. At any rate, deposition in the Scottish Borders Basin was terminated by the development of a regional cornstone horizon in the topmost upper ORS (Leeder 1976a), prior to the eruption of a sequence of Carboniferous alkali basalts, the Birrenswark–Kelso Lavas.

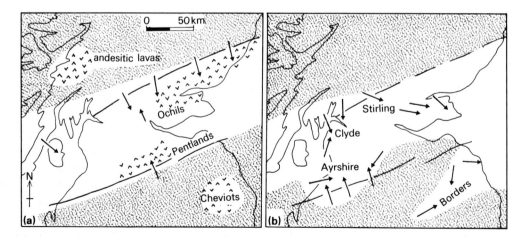

Figure 9.9   Generalised palaeocurrent directions for (a) the lower ORS and (b) the upper ORS of the Midland Valley and S. Scotland.

9d The Orcadian Basin of NE Scotland

The extensive outcrops of ORS in NE Scotland, the Orkneys and the Shetlands (Fig. 9.10) are mainly of middle Devonian age (Table 9.2). Over 5 km of strata are present in Caithness, with possibly up to 10 km in Shetland. The basal parts of the succession are usually conglomeratic and rest unconformably upon a metamorphic Moinian and intrusive granitic basement. The major part of the middle ORS in Caithness and the Orkneys consists of the Caithness Flagstone Group (over 3·8 km thick) and its correlatives, which contain an abundant fauna of fishes at several horizons. The famous Rhynie Chert occurs

Figure 9.10 Outcrops of the ORS and palaeocurrents in the Orcadian Basin and Shetlands (after Donovan et al. 1976) and restoration of strike-slip fault displacements in the Shetlands (after Mykura & Phemister 1976). Note: the Buchan field contains ORS which may be either Midland Valley or Orcadian in context.

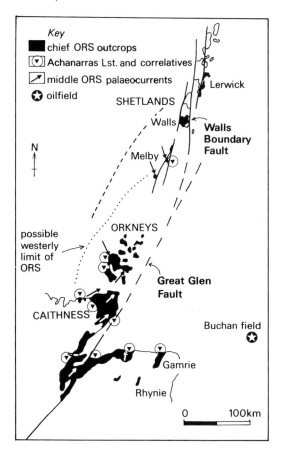

in a basement group facies in an inlier in Aberdeenshire, and consists of silicified peats with primitive vascular land plants, fungi and the earliest known insect. The upper ORS rests unconformably upon the main middle ORS succession.

At many localities there is clear evidence of considerable relief beneath the basal unconformity, with carbonate-cemented breccias and arkosic sandstone tonguing outwards from buried landscape hills and knolls into surrounding fine-grained flagstone lithologies. The unconformable surface may be coated by successive generations of algal stromatolite growth forms, with calcareous tufa penetrating into joints in the basement. The breccias show imbrication and thinning trends away from the basement hills and contain locally derived clasts. Some marginal carbonates, up to 4 m thick, contain non-stromatolitic laminations and abundant spar-filled fenestrae or birds-eye structures attributed to shrinkage during desiccation.

The main Caithness Flagstone Group and its correlatives, the Stromness—Rousay Flagstones in the Orkneys, consist of a large number of distinctive sedimentary cycles. Laminites or varved sediments make up much of the sequence (Rayner 1963*, Donovan 1975*, Donovan et al. 1974). Some show 0·1−1 mm laminae of alternating micritic calcite or dolomite and clastic siltstones, sometimes with phosphates. Where carbonate laminae are dominant, the clastic laminae are reduced to 0·1 mm organic streaks, as in the Achanarras Limestone where fish faunas are abundant. Non-carbonate laminites show alternations of siltstone and mudstone or of coarse and fine siltstones. Subaqueous and subaerial shrinkage-crack casts are abundantly present on the bases of many coarse laminae, and the tops of thicker laminae show preserved symmetrical wave-formed ripples. Thicker, sharp-based and sometimes erosive sandstones overlie mudstone units. These contain current-produced sedimentary structures. Overlying and underlying mudstone units may show subaerial desiccation cracks. Fish remains are generally scarce in all units apart from the carbonate-rich laminites. Upward-coarsening cycles involving the above lithologies may often be correlated for many kilometres along strike, where exposures permit. There is a general upward trend from the various 'flagstone' lithologies to a coarser cyclical arrangement of alluvial fining-upwards cycles in the Caithness area (Fig. 9.12). Palaeocurrents indicate generally western to southern derivation in the Orcadian Basin (Fig. 9.10).

The above sedimentary features may be interpreted as due to the interplay between a large nonmarine lake and streams draining into the lake margins. The carbonate–clastic laminites are 'nonglacial varves', probably resulting from a process of annual lacustrine rhythm involving (a) the growth of warm and dry season algal blooms causing increased pH and carbonate precipitation, alternating with (b) the wet season introduction of clastic silt and mud from streams.

Many modern lakes show a marked stratification involving an upper warm oxygenated water mass (epilimnion) and a lower cool dark anaerobic water mass (hypolimnion). The bounding thermocline acts as both an ecological and a chemical barrier (Fig. 9.11). Such stratification has been suggested for the Orcadian lakes (Donovan 1975*), because fossil fish within laminites are preserved undisturbed and the laminites show no signs of bioturbation or wind-induced sedimentary structures. However, these are more in evidence as the former lake margin is approached where, presumably, the stratified waters were mixed due to increased turbulence. The dolomitised and silicified marginal carbonate-rocks probably formed in these more permanently warm waters away from areas of clastic input.

The coarser clastic laminites were probably deposited in increasingly closer proximity to areas of river or stream influx. Periodic minor density currents resulted as river water debouched into the lake, causing the sharp-based sandstone beds with current-produced sedimentary structures. Minor, probably seasonal, fluctuations in water level resulted in extensive subaerial desiccation of the lake margin flats. The fluvial fining-upwards cycles that increasingly dominate the succession in Caithness indicate a major fluvio-lacustrine regression with time (Fig. 9.12).

The relative roles of long-term climatic fluctuations and tectonism in producing lake margin advance and retreat are difficult to assess. The basic pattern of cyclic sedimentation within the middle

Figure 9.11 The inferred relationships of lithologies at a lake margin site abutting on to a palaeohill in the Caithness area: local scree and alluvial clastics deposited during a period of low lake level; lacustrine facies deposited after a quick transgressional episode (after Donovan 1975*).

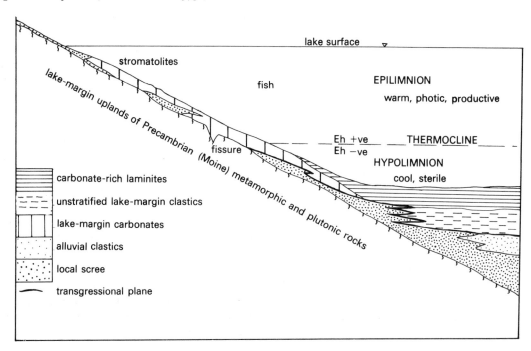

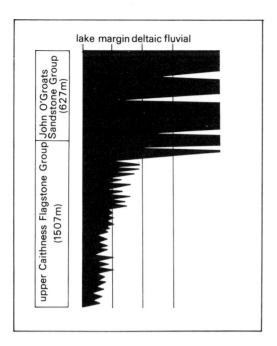

lake margin deltaic fluvial

John O'Groats Sandstone Group (627m)

upper Caithness Flagstone Group (1507m)

Figure 9.12 The upward change to dominantly fluvial conditions in the Caithness area during late middle ORS times (after Foster 1972).

ORS of the Orcadian areas probably resulted from the effects of lake level rise and fall combined with the consequent rapid retreat and slow advance of lake-margin delta systems.

The Orcadian lake basin was probably intermontane (cf. modern Alpine lakes) and separate from the marine Devonian sea that transgressed northwards into the present-day North Sea area at this time. Basinal subsidence was rapid, since over 5 km of sediment accumulated in only 10 million years.

In the Shetlands, three distinct basins of ORS sedimentation appear to have been juxtaposed by dextral wrench faulting (Mykura & Phemister, 1976). A restoration of these basins in relation to the main Orcadian Basin is shown in Figure 9.10. The Melby Formation, bounded by the Melby Fault, shows similarity with the Orkneys and Caithness outcrops in that fish-bearing cycles occur with a fauna similar to that of the Sandwick and Achanarras horizons. This may imply continuity with the main intermontane Orcadian lake system. The Walls Basin, bounded to the east by the Walls Boundary Fault, a presumed offshoot of the Great Glen Fault system, contains 10 km or more of ORS sediments and volcanics. The basal Sandness For-

mation (over 3 km), dated lower to middle ORS age by contained plants, represents a coarse-grained, mainly fluvial infill of a pre-Devonian landsurface with considerable topography. The formation increases in thickness southwestwards, which was also the direction of stream flow. The overlying Walls Formation contains sandstone—shale rhythms, the latter with calcareous mudstones and limestones, 0·75—20 m thick. Fish are recorded from several horizons, and the whole formation is regarded as of deep lacustrine origin since evidence of emergence is not seen. In the eastern Shetlands the Nesting Fault bounds the most northern ORS basin where present outcrops only show coarse basin—margin alluvial facies thought to be derived from a mountainous hinterland to the west. These Shetland basins show many of the depositional characters expected in an active wrench-fault terrain. There was also abundant contemporaneous volcanism in the area, the Clousta Volcanics in the Walls Basin comprising a suite of calc-alkaline andesitic and basaltic lavas, ignimbrites and abundant tuffs and agglomerates.

9e Old Red Sandstone soils and climate

The account of ORS deposition given above frequently mentions the occurrence of carbonate-rich horizons known as cornstones which occur within fluvial siltstone and sandstone horizons. In the field these carbonate units range from decimetres up to 10 m thick, although the majority are 1—2 m thick. The carbonate occurs as scattered nodules, coatings or filaments in red siltstone facies of floodbasin origin or in sandstone facies of river channel origin. Horizons of coalesced nodules, sometimes arranged in vertical pipes, define more massive carbonate beds, and these include delicately laminated horizons (Fig. 9.13). The carbonate beds show evidence of penecontemporaneous brecciation fabrics, cavity fills and distinctive pisolitic concretions. Rarely a catenary-shaped fold profile is observed within massive or nodular horizons. Several thicker carbonate units show a well-marked vertical profile, passing from small scattered nodules at the base, through increasingly closer packed and larger nodules, up to a massive coalesced nodule horizon with a capping of laminated carbonate at the top.

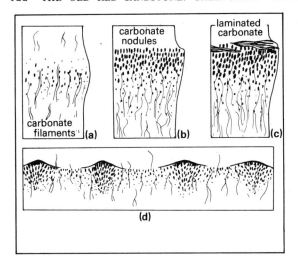

Figure 9.13 The development of (a) youthful to (c) mature carbonate palaeosols (cornstones). (d) The development of expansion folds as seen in certain cornstones of the Welsh ORS (all after J. R. L. Allen 1974a*).

In addition to these undoubted *in situ* concretionary-type carbonates, carbonate clasts occur intermixed with other intraformational and exotic clasts in the basal parts of the coarse members of fining-upwards cycles. In thin section the carbonate shows a variety of fabrics that indicate *in situ* growth within a clastic host. Carbonate replaces clastic grains, often producing a 'floating-grain' fabric of embayed quartz grains set in a micritic calcite or dolomite groundmass.

All the features noted above show great similarity to textures and morphologies observed in Tertiary and Holocene calcareous soil horizons known variously as caliche (USA), calcrete (Africa), nari (Israel) or kunkar (India and Pakistan). These form by carbonate precipitation and replacement above the water table in the C soil horizon, usually on static depositional or erosional surfaces, the carbonate usually being derived from surface levels by percolation. The pedogenic carbonates thicken with time, so that thick profiles with a massive, brecciated or laminated upper zone give radiocarbon evidence for formation over periods in excess of 10 000 years.

Modern calcretes are restricted to semi-arid climatic zones with mean annual temperatures in the range of 16−20°C and a seasonally distributed, but not excessively peaked, annual rainfall of 100−500 mm. Such conditions occur today in latitudes of 35° or less. Such low latitudes are, in fact, indicated by palaeomagnetic data (Fig. 9.1) for ORS times.

The occurrence of thick calcretes must also indicate periodically lowered alluviation rates over parts of the ORS alluvial plains, since active alluviation by river flooding would prohibit the preservation of stable surfaces for the lengthy time required for mature calcrete formation. Stabilisation is best explained by periodic river-entrenchment phases, causing low terrace areas to become isolated from flooding processes. The causes of entrenchment may be tectonism, climatic variations or eustatic sea-level changes. In the Scottish Borders the regionally extensive calcrete horizons in the topmost upper ORS (Ch. 9c) have been attributed to the combined effects of crystal upwarp and lava eruption, causing incision and diversion of the drainage system (Leeder 1976a). The widespread and characteristic red colouration of much of the ORS, in particular the siltstone horizons, may be attributed to the release of ferric iron minerals by the oxidative diagenesis of unstable ferromagnesian minerals in the vadose zone.

9f   Old Red Sandstone faunas and boundary problems

Most of the ORS is of Devonian age, but there are several difficulties regarding correlation both within the ORS and between the ORS and laterally equivalent marine successions in SW England and Europe.

The fish, which provide the chief means of dating and correlation within the ORS, include the jawless ostracoderms (pteraspids, cephalaspids), the acanthodians, antiarchs and dipnoans. The last, the lung-fish, are thought to be responsible for the suspected oestivation (hibernation) burrows found in some areas in fluvial sediments. In alluvial coastal plain and interior basin environments, the fish are usually preserved in channel facies as the disarticulated remains of a large number of individuals of the same species or of mixed species, strongly water sorted. Rarer faunas are known with well-preserved skeletons in suspected floodbasin facies. An example from near Mitcheldean, in the Welsh Borderland, contains a number of disarticulated, but not substantially broken, skeletons. These may represent a shoal of pteraspids that died during a flood in a channel and whose neutrally buoyant carcasses were transferred into the floodbasin and rapidly buried. In the lacustrine Orcadian Basin fish are usually to be found in well-defined bands within carbonate-rich laminites (e.g. the Achanarras Limestone). These well-preserved skeletons probably result from their sinking after death into the anoxic hypolimnion of the stratified lake. Here the carcass would have rotted slowly with minimum disarticulation and little subsequent disturbance by burrowing carrion faunas.

The base of the Devonian in the Welsh Borderland has been traditionally placed at the horizon of the Ludlow Bone Bed, which is higher locally than the last Ludlovian graptolite zone of *Monograptus leintwardinensis*. However, in central Europe the top of the Silurian is drawn at the base of the still higher graptolite zone of *Monograptus uniformis*. Evidently, the massive ORS regression due to uplift and denudation of the Caledonian orogenic belt began at some time during the late Ludlovian as defined in Europe. The British Isles is therefore not a suitable locality for defining the Silurian/Devonian boundary stratotype. In the absence of cosmopolitan faunas such as goniatites, conodonts and graptolites, it is apparent that the base of the Devonian in the British Isles must lie *within* the ORS. Much of the previously defined Downton Group is now included in the Silurian.

Similar problems arise at the top of the ORS, where it is known that fluvial red-bed facies of the ORS type continued to develop locally into early Carboniferous times, as in the Midland Valley of Scotland and the Northumberland Basin.

9g The Old Red Sandstone as Caledonian molasse

The discussion above has shown that the ORS is largely a continental assemblage of facies derived from uplifted areas of the Caledonian orogenic belt. It will now be useful to consider briefly the development of similar sedimentary facies in more recent orogenic belts.

In the Alpine–Pyrenean orogens, deep sedimentary basins flank the deformed orogenic sequences. They contain very thick successions of continental and shallow marine facies and represent the terminal sedimentary products of orogenesis. Palaeocurrents and clast types indicate a rather local provenance from adjacent mountain areas. Facies of this type have come to be known as 'molasse'. In the Alpine area, molasse sedimentation closely followed the main lower Oligocene deformation when rapidly subsiding flanking basins ('foredeeps') developed soon after orogenic uplift. In the northern molasse basin, between the Helvetic nappes and the folded Jura, nappe gravity gliding was still operative during molasse sedimentation. As the nappe thrust front overrode the molasse basin, it gave rise to the sub-Alpine molasse. It is estimated that about 80 per cent of the material eroded from the Alps since the lower Oligocene has been trapped in the northern, Po and Hungarian Plain molasse basins. A maximum of 6 km of molasse is recorded, but thicknesses are usually in the range 1–3 km. These thicknesses equal or exceed the total pre-orogenic fill to the Alpine 'geosynclines'.

Molasse basins are widespread within and marginal to many other recent orogenic belts. The Siwalik Basin in the Himalayas is of molasse type and contains several kilometres of Tertiary molasse. Also, 10 km or more of molasse are reported from certain intra-Cordilleran molasse basins of late-Tertiary age in the Chilean–Bolivian Andes, where accompanying calc-alkaline volcanics are frequent.

The main conclusions gained from studying the sequence of late orogenic events in orogenic belts are:

(a) Orogenic uplift, due possibly to stress relaxation, closely follows the termination of large scale deformation and regional metamorphism due to continent–continent collision or lithosphere subduction.
(b) Massive injection of magma (often calc-alkaline) as plutons and volcanics may closely preceed, coincide or closely postdate this postorogenic uplift.
(c) Rapid subsidence of areas marginal to or within the rising orogen gives rise to drainage from the orogen into the basins. This is the molasse phase of sedimentation.

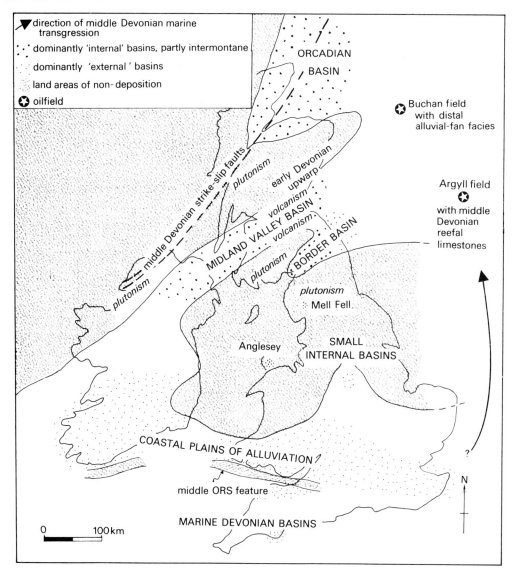

Figure 9.14 Major palaeogeographic features during deposition of the ORS in the British Isles.

(d) Molasse basins may be *internal* or *external* in type. External basins may pass out into shelf or oceanic areas, causing the molasse of one orogenic phase to be the lateral equivalent of the flysch of a future phase (e.g. the Ganges molasse transported offshore on to the Bengal fan in the Indian Ocean; the Po Basin molasse transported into the Adriatic Sea).

It seems that the ORS exhibits all the major sedimentary features of molasse facies. Basins such as the Midland Valley and Orcadian emerge as

internal molasse troughs showing the greatest subsidence and sedimentation rates (Fig. 9.14). The Anglo-Welsh−Irish basins emerge as *external* molasse basins, with their rivers transferring detritus into the northern fringes of the supposed Hercynian continental margin (Fig. 9.14).

The onset of molasse sedimentation in Devonian times was closely coincident with the intrusion of the 'Newer Granite' swarm of plutons. In the Midland Valley to Highland Borders area much andesitic volcanism occurred. Maximum sediment accumulation, and therefore probably maximum vertical crustal movements, occurred in the lower ORS in most areas. Middle ORS times were marked by mild intramolasse phase deformation and strike-slip faulting (Great Glen and Shetland Faults), coincident with the Acadian orogenic phase in Maritime Canada and the Appalachians.

Armed with this interpretation of the ORS as a molasse deposit, it is now possible to reflect upon the wider implications of other molasse basins within the ORS continent (Fig. 9.1). The Norwegian, Greenland and Spitsbergen Basins all contain great thicknesses of ORS facies and are of internal type. The external coastal plains of alluviation are thought to have extended eastwards from the Anglo-Welsh–Irish basins across S. England into Belgium, Germany, Poland and the Baltic. The upper Devonian Catskill facies of NE USA emerge as the mirror image of the Ango-Welsh–Irish external alluvial plains. These fluvial facies (c. 2·5 km thick) fine and intertongue westwards into shallow marine facies, and they show palaeocurrent and clast evidence for derivation from the newly uplifted (middle Devonian) Appalachian extension of the Caledonides.

9h   Summary

The ORS is predominantly a clastic red-bed sequence of continental facies deposited as a post-orogenic molasse after uplift in the Caledonian mountain chains. Zonation of the ORS is by fish faunas. Severe chronostratigraphic problems arise in accurately placing both the Silurian/Devonian and Devonian/Carboniferous boundaries within such a non-marine sequence. Depositional environments that may be recognised include meandering and braided alluvial plains, alluvial fans and freshwater lakes. Clast types, heavy minerals and palaeocurrent data indicate stream drainage into broad alluvial coastal plains in S. Ireland, S. Wales and S. England and into interior drainage basins in the Scottish Borders, Midland Valley and Orcadian Basin and Shetland areas. Generally, coarsest detritus and inferred greatest subsidence occurred in the interior drainage basins. Abundant soil carbonates of calcrete type within the ORS indicate semi-arid climates and periodic lowered deposition rates. A low latitude position for the ORS continent is proven by palaeomagnetic data.

10 Devonian Southwest England

10a Introduction

Only very recently has the beginning of a structural and stratigraphic synthesis emerged for SW England. A virtually complete succession of marine Devonian is present, with intercalations of Old Red Sandstone facies. Carboniferous sequences, referred to as 'Culm facies' in the older literature, lie in a complex synclinorial structure in central Devon and N. Cornwall (Fig. 10.1, Ch. 11b). The great Permo-Carboniferous Cornubian granite batholith, with its associated mineralisation and dyke swarms, is intruded into both Carboniferous and Devonian sediments.

Major problems of stratigraphic definition occur in the area because of rapid facies changes (Fig. 10.2), upon which is superimposed a variably complex history of structural deformation, often polyphase, with some major local inversions. The most important zone fossils in marine facies are goniatites and conodonts. House (1963*) has reorganised the full range of Devonian goniatite zones (Table 10.1), based upon the less deformed German successions. Goniatites occur in bands mainly in 'basinal' quiet-water marine mudstone sequences and in condensed pelagic limestones. Conodonts are widespread in most limestone successions, and coral zonation is possible in the Devonian reefs of S. Devon. However, much further zonal study is needed before good correlation between all successions in the area can be established.

Figure 10.1   Generalised geology of SW England with subdivision of the area discussed in the text (partly after Dearman 1971).

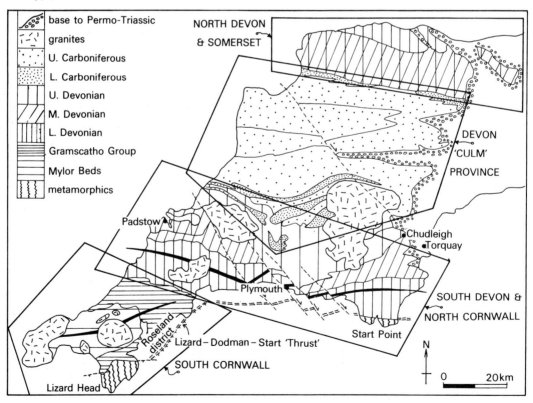

Table 10.1 Devonian stages and ammonoid zones (after House 1975a*).

	Stages	Stuffen	Ammonoid Zones
UPPER	Fammenian	Wocklumeria	Prionoceras sp. Cymaclymenia euryomphola Wocklumeria sphaeroides Kalloclymenia subarmata
		Clymenia	Gonioclymenia speciosa Gonioclymenia hoevelensis
		Platyclymenia	Platyclymenia annulata Prolobites delphinus Pseudoclymenia sandbergeri
		Cheiloceras	Sporadoceras pompeckji Cheiloceras currispina
	Frasnian	Manticoceras	Manticoceras sp. Crickites holzapfeli Manticoceras cordatum Pharciceras lunulicosta
MIDDLE	Givetian	Maenioceras	Maenioceras terebratum Maenioceras molarium Cabrieroceras crisoiforme
	Eifelian	Anarcestes	Pinacites jugleri Anarcestes lateseptatus
LOWER	Emsian Siegenian	Mimosphinctes	Sellanarcestes wenkenbachi Mimagoniatites zorgesis Anetoceras hunsrueckianum
	Gedinnian	Graptolite Zones	Monograptus hercynicus Monograptus uniformis

Figure 10.2 Schematic section to show Devonian successions and facies in SW England (after House 1975a*).

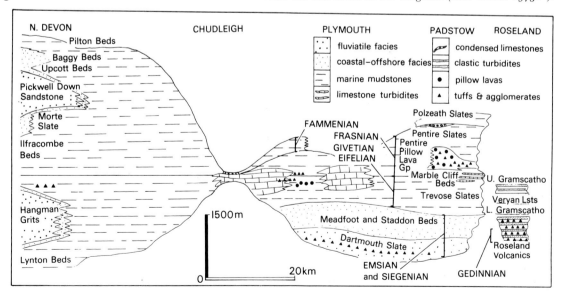

10b N. Devon and Somerset Devonian

Up to 5 km of Devonian may be present in N. Devon and Somerset (Figs 10.2 and 10.3) but, despite magnificent coastal exposures, thicknesses are difficult to measure because of structural repetition, particularly in more incompetent units like the Lynton Beds, Morte Slates and Pilton Beds. It is also difficult to zone the sequence according to the Devonian marine stages since goniatites are absent.

The Lynton Beds, possibly of Emsian age, may be of the order of 450 m thick. The main lithologies are cleaved mudstones with siltstone laminae and thin fossiliferous sandstones (Fig. 10.3). The coarser laminae may show isolated current ripples, some with symmetrical wave-rounded crests. Streaky siltstone laminae are more common, with abundant bioturbation mostly attributable to *Chondrites*. Faunas are rich in individuals but of low diversity, and occur as current-sorted death assemblages rich in spiriferid brachiopods and bivalves such as *Modiomorpha*, *Pterinea*, *Limoptera* and *Actinodesma* (Goldring *et al.* 1967). Crinoid debris is also common.

The above features suggest deposition in a muddy marine shelf environment dominated by long periods of quiescence and mud influx, with the development of extensive burrowing infaunas of soft-bodied types. Periodic introduction of sands and silts may have been due to storm processes, perhaps superimposed on normally weak tidal currents. Wave-induced modification of ripples and dunes also occurred, but in general the environment was below the effective wave base.

The Hangman Grits (possibly 1 km thick) are an

Figure 10.3 (a) Generalised Devonian succession in N. Devon, with sketches to show representative shallow marine facies of: (b) and (c) the Lynton Beds, (d) the Ilfracombe Beds, (e) the Baggy Beds and (f) the Pilton Beds.

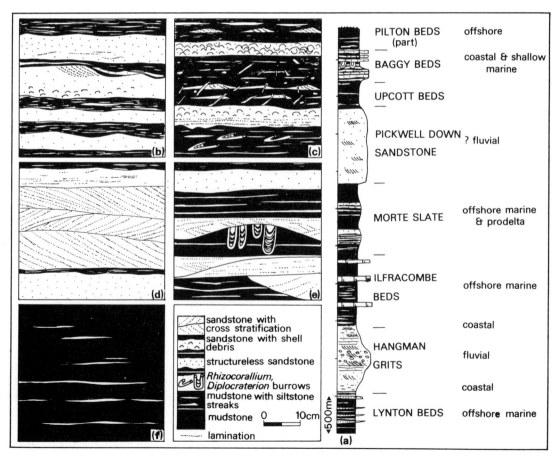

incursion of mainly continental facies into the area. Tunbridge (1976) has recently studied the whole group; his findings, supplemented by some of his unpublished information, are as follows. The basal Hollowbrook Formation passes up conformably from the Lynton Beds; it comprises plane-bedded sandstones with erosive longitudinal runnels and associated bioturbated siltstones and ripple-laminated sandstones with mudstone flasers. Sub-aerial desiccation shrinkage cracks are recorded. Deposition appears to have been in sublittoral to intertidal coastal environments with wave action producing the plane-bedded sandstones. The overlying Trentishoe Formation consists predominantly of unfossiliferous planar-laminated sandstones with scoured basal surfaces, signs of channelling and intraformational conglomerates. The Rawns Beds follow abruptly, with a notable increase in grainsize. Abundant exotic pebbles occur, many of which are tuffaceous. Fining-upwards units with large-scale cross-stratification seem to be of braided-stream channel origin, with certain persistent units of conglomerates possibly marking former sheetflood episodes. The exotic clasts cannot be matched with any known outcrops in Wales, but their coarseness and angularity indicate a relatively close, northern source. This may be the Bristol Channel landmass alluded to in Chapters 9 and 11 (Fig. 10.8). The topmost Sherry Combe and Little Hangman Formations show more marine influences, with flaser bedding and *in situ* beds of the bivalve *Myalina*. These eventually pass into the basal bioturbated sediments of the shallow-marine Ilfracombe Beds.

The Ilfracombe Beds have been dated as Givetian to Frasnian age by coral faunas. They are dominantly mudstone with sandstone becoming common towards the base and top (Fig. 10.3). Four thick coralliferous limestones are present, with numerous thinner examples. Overall, the clastic members are very similar to the clastic facies of the Lynton Beds. The thin limestones are sharp-based bioclastic types with abundant crinoid debris and bivalve and brachiopod fragments. Thicker examples — such as the Roadwater Limestone of the Brendon Hills and its supposed correlative, the Jenny Start Limestone of Ilfracombe — contain abundant *in situ* tabulate and rugose coral heads and stromatoporoids, forming small patch reefs. Ooliths are also recorded from the Jenny Start Limestone.

These features suggest that, after marine transgressions over the Hangman Grits coastal plain, the Ilfracombe Beds were deposited in a dominantly muddy offshore shelf environment. Webby (1966*) has described the facies as shallow prodelta and delta platform deposits, but it seems that tidal and storm processes were more effective than direct fluvial input of sediment. The thicker coralliferous limestones indicate periodic clear shallow conditions over large areas. Webby has noted lateral changes in faunas from lower energy assemblages of solitary rugose corals in the west to higher-energy nearshore assemblages of tabulate corals and stromatoporoids in the east.

The Morte Slates (Fammenian age, *c*. 400 m thick) overlie the upper Ilfracombe Beds conformably. The group is composed largely of cleaved mudstone in N. Devon but coarsens somewhat eastwards, so that in Somerset siltstones and sandstones are more common. Major coarsening-upwards sequences (50 m thick) may be seen at Woolacombe, where pure mudstones pass upwards to mudstone with thin siltstone streaks and lenses, with abundant bioturbation. Shelly faunas are extremely rare. Likely environments of deposition include prodelta to delta slope, according to Webby (1966*).

Very little information is available concerning the nature of the overlying Pickwell Down Beds, save that they comprise alternating purple cross-stratified sandstones and mudstones of presumed fluvial origin.

These grade upwards into the fine-grained Upcott Beds (*c*. 250 m thick) which, according to Goldring (1971), comprise olive to greenish-grey cleaved mudstones and graded siltstones with occasional 10–20 m thick cross-laminated sandstones. Pockets of comminuted shells occur locally, but none of the fragments have proven identifiable. Goldring has noted that the Upcott Beds may be backswamp (but no rootlets or coals) or shallow lacustrine facies.

The basal sandstone member of the overlying Baggy Beds (Fammenian age, 440 m thick) rests sharply upon the muddy Upcott Beds. Goldring (1971*) has recognised nine separate facies in the magnificent coastal sections around Baggy Point, with the bulk of the sandstones occurring in the lower parts of the succession. Penecontemporaneous erosion due to wave action seems to have been the dominant control upon these non-cyclical marine facies. The commonest facies are mudstones, siltstones and very fine sandstones with the U-shaped dwelling burrows of *Diplocraterion yoyo* (Fig. 10.3). These U-shaped traces migrated upwards and downwards in response to local erosional and depositional episodes. Other sandstone facies are thought to represent a spectrum of sublittoral and nearshore environments. Lagoonal or restricted bay origins are postulated for bioturbated graded siltstones and mudstones with patches of *Lingula*, but cross-stratified and parallel-laminated channellised sandstones with plant fragments may be of delta distributory origins. In conclusion, Goldring has suggested that the Baggy Beds were deposited in a spectrum of shallow marine environments that were subjected to a moderate tidal range and to frequent episodes of intense wave action, and situated close to a delta or coastal plain. The overlying Pilton Beds contain the Devonian/Carboniferous boundary, as revealed by trilobite faunas. This is dominantly a mudstone succession but contains occasional horizons of thin sharp-based fine sandstones with internal parallel laminations or low angle scours (Fig. 10.3). The contact with the topmost Baggy Beds is quite sudden and follows after a prominent slumped horizon. The Pilton Beds indicate a dominantly muddy offshore shelf environment with rare sand introduction due probably to storm processes.

To summarise: the deposits in N. Devon record a series of regressions and transgressions (Figs 10.2 and 10.3). Southward advance of river and delta systems from the ORS continent (Hangman Grits, Pickwell Down Beds) was followed by marine transgressions depositing predominantly muddy shelf and nearshore sandy facies. During middle Devonian times, coral limestone facies extended into the area, probably from the south where, as will be shown, a major carbonate build-up occurred.

10c S. Devon and N. Cornwall Devonian

The earliest-known deposits in this area are the Dartmouth Slates (Fig. 10.2). These comprise several hundred metres of sandstones, siltstones and mudstones exposed in the core of the Dartmouth Anticline in S. Devon. Dineley (1966) has recognised fluvial fining-upwards cycles of ORS aspects from the succession, with non-marine faunas of ostracoderm and placoderm fish fragments and plant remains, which indicate a lower Devonian (Siegenian age) for the group. Tuffs are present in parts of the succession. Although palaeocurrent data are not available, it seems reasonable to suppose that the Dartmouth Slates represent the furthest southward advance of the coastal plains of alluviation of the ORS continent. Continental equivalents of the Dartmouth Slates are not seen in N. Devon, but they may lie below the Lynton Beds discussed in Chapter 10b (Figs 10.2 and 10.4).

The Meadfoot and Staddon Beds sharply overlie the Dartmouth Slates, their lithology and marine faunas indicating that they are transgressive shallow-marine clastic sequences. Around Plymouth Sound, Harwood (1976) has recognised facies comprising (a) highly fossiliferous calcareous sandstone, mudstone and thin limestones, (b) black mudstone with rare calcareous horizons, and (c) sparsely fossiliferous sandstone—mudstone alternations. The fossiliferous sandstones contain large thamnoporoids, bryozoans, crinoids, bivalves and brachiopods.

The overlying Jenny Cliff Shales in S. Devon comprise fossiliferous mudstones and spilitic lavas, often associated with thin limestone beds. The end of mudstone deposition in the S. Devon area heralded a major change in deposition between here and the N. Cornwall area. Subsequently, in S. Devon extensive carbonate build-ups dominate the middle Devonian scene, whereas in Cornwall very thick, dominantly mudstone facies with periodically extensive volcanicity occur and continue up into the upper Devonian.

The Chudleigh—Torquay—Plymouth limestones are the only exposed British examples of the Devonian reefs so splendidly developed in Belgium and, more spectacularly, in the subsurface of the Canadian plains, where they form important oil reservoirs. Dineley (1961) and Scrutton (1975, 1977*) have noted that in the Givetian of the Torquay area a stromatoporoid bank developed. The massive to laminar stromatoporoid growths are up to 2–3 m wide and 1·5 m high. Scrutton (1977*) has postulated that these were linked to form a reefal barrier, restricting circulation to the north and giving rise to

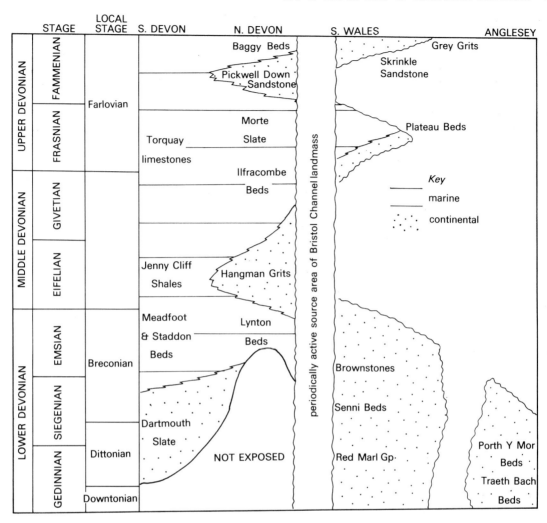

Figure 10.4 Interrelationships between the non-marine facies of ORS aspect in Devon and those of the ORS in S. Wales (after J. R. L. Allen 1974b).

a sheltered backreef environment supporting the solitary coral *Amphipora* in the Newton Abbot area. The stromatoporoid bank was succeeded in Torquay by a mixed coral fauna in bioclastic limestones during the late Givetian, which Scrutton (1977*) has attributed to bank drowning due to transgression. Deeper-water goniatite-bearing mudstones of Frasnian age succeed the limestones in the Torquay area. To the north, carbonate deposition persisted till late Frasnian times, culminating in a phase of *Frechastrea−Phillipsastrea* bioherm growth.

Braithwaite (1967) has declined to accept that the Torquay limestones are reefal, and feels that at no time did the corals or stromatoporoids form a growth lattice having a topographic expression comparable to modern reefs. He has noted that, although patches of *in situ* bioherm occur, the dominant limestone-building process was due to variable current winnowing around these bioherms causing the production of spreads of skeletal-lime sands and muds. It should be noted here that work in progress on the structure of the Torquay area by M. Coward and K. McClay indicates that much of the Torquay limestones are allochthonous thrust sheets, derived from the south. This indicates the need for some radical rethinking concerning the middle Devonian palaeogeography of S. Devon.

The subsequent Devonian history of S. Devon is

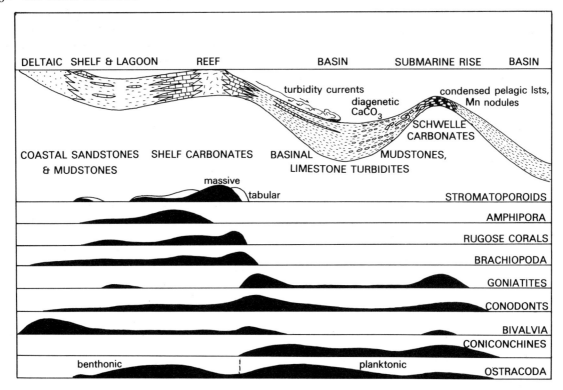

Figure 10.5 Hypothetical reconstruction to illustrate facies and faunal variations in the Devonian of SW England (after House 1975b*).

of great interest. In the Chudleigh area Scrutton (1969) has shown that the coral faunas in the early Frasnian limestones indicate gradual deepening. The overlying non-reefal Mount Pleasant Series is less than 60 m thick and contains a full Fammenian sequence of ammonite and conodont zones. It is thus a truly condensed deposit (House 1963*), since in N. Devon the Fammenian may exceed 1·5 km in thickness. Such condensed carbonate–mudstone deposits are thought to have been deposited over relatively deep-water submarine rises as pelagic sediment. These rises, widely present in the German Devonian, are termed 'schwellen' (Fig. 10.5). Tucker and van Straaten (1970) have shown that two facies occur over the massive reefal limestones at Chudleigh. The lower Dunscombe Goniatite Bed comprises 1·7 m of greyish-pink micritic limestones, with many pressure solution planes and very thin irregular mudstone laminae. A rich and varied, dominantly pelagic fauna occurs with goniatites, trilobites and bivalves, the bed occupying a whole conodont zone. The Kilnwood Beds comprise 9 m of

black and dark grey micaceous mudstones, with siliceous horizons and thin limestones containing mostly disarticulated small bivalves and crinoid ossicles brought in by bottom currents. The facies indicate that deposition occurred on a localised deeper part of the upstanding *schwelle*. The remaining Mount Pleasant Series consist of nodular limestones and mudstones, the nodules being considered as being of early diagenetic origin.

In the Torquay area, early Frasnian 'reefal' limestones are overlain by several hundred metres of dominantly mudstone facies containing goniatite bands, abundant ostracods and volcanic horizons. Again by analogy with interpretations placed upon German successions, the environment of deposition is considered to have been a deep water basin or 'becken' (Fig. 10.5). Some good evidence that shallower *schwellen* were nearby is seen in the Saltern Cove Beds, where palaeontological and sedimentary evidence have suggested to van Straaten and Tucker (1972) that the unit is a sedimentary slump deposit or 'mélange', since reworked Frasnian and

Fammenian clasts are randomly mixed within Fammenian mudstones. Slope failure on an adjacent, but unexposed, *schwellen* flank is indicated.

The N. Cornwall middle and upper Devonian successions around Padstow (Fig. 10.2) contrast greatly with the S. Devon successions discussed above. More than 1600 m of beds have been noted by Gauss and House (1972) in the Trevone area, these being dominated by becken-type mudstones with pelagic faunas of goniatites in bands, current-aligned styliolinids, ostracods and posidoniid bivalves. Graded tuffs and agglomerates abound in the Long Carrow Cove Beds, which are the lateral equivalents of the Pentire Pillow Lava Group (450 m). The Gravel Caverns Conglomerate (90 m) in the Pentire area contains prominent debris-flow conglomerates. Considerable significance may be attached to the Marble Cliff Beds (70 m) of Trevone, which lie on the inverted limb of a large recumbent fold. They comprise sharp-based beds of crinoidal limestone (*c.* 0·6 m thick), alternating with clastic mudstones. They show grading and small sets of large-scale cross-stratification. Typical flutes are rare, but broad grooves and channels are present on bed bases. Tucker (1969) has interpreted these beds as limestone turbidites derived from an adjacent schwelle populated in part by crinoid thickets.

Volcanic rocks are abundant, particularly in the middle to upper Devonian and Carboniferous of the area (Fig. 10.6). Spilitic submarine pillow-lavas, tuffs and agglomerate are very common, as at Pentire and in S. Devon (Gauss & House 1972, Middleton 1960). Potassium-rich trachytes are common in the lower Carboniferous of the area, and high level sills, sheets, dykes and minor diapirs of generally basic composition occur widely. Floyd (1972*) has noted that many of the spilitic lavas show evidence of autometasomatism. His chemical analyses reveal that the basalts are sodium-rich, with major and minor element characteristics of alkali basalts (Fig. 10.6), not of modern oceanic tholeiites. They might have been derived by a small degree of partial melting from the deep upper mantle, but they do not support claims that they were formed as oceanic tholeiites related to any mid-European or Hercynian ocean (Ch. 12).

Figure 10.6 (a) Outcrops of basic igneous greenstones in SW England. (b) Total alkalies versus silica plot to show the dominantly alkaline nature of the greenstones (after Floyd 1972*).

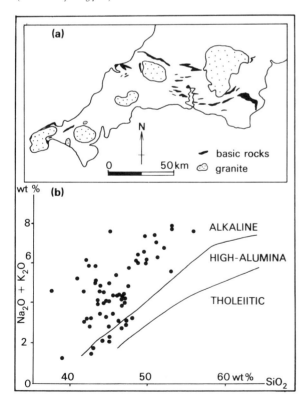

10d S. Cornwall Devonian

The age, stratigraphy and interpretation of rocks within this area (Fig. 10.1) have long been controversial. North of the Lizard−Dodman−Start 'Thrust', now not regarded as a single structure (Sadler 1974*), three main outcrop areas are loosely defined: that of the Mylor Beds, that of the Gramscatho Beds, and the belt of supposed thrust breccia shown on IGS Sheet 353 (Mevagissey).

The Mylor Beds remain a puzzle, comprising a completely unfossiliferous, cleaved, dominantly mudstone lithology. They are generally thought to underlie the Gramscatho Beds. The Gramscatho Beds outcrop over a broad area, but only in the Roseland area (Figs 10.2 and 10.7) has a detailed stratigraphic and chronological scheme been established for them and for the underlying Roseland Volcanics and Veryan Beds. Sadler (1973*), by

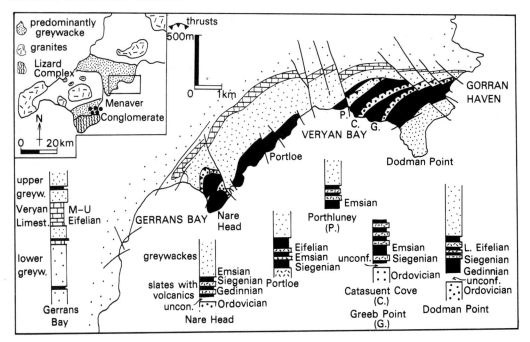

Figure 10.7 Devonian outcrops, structure and stratigraphy of the Roseland area of S. Cornwall. Note especially the thrust sheets containing Ordovician quartzites of Armorican type (after Sadler 1973*).

studying both structural and faunal aspects, has established the successions in Figure 10.7. The ages of the units are the result of conodont zonation of included limestones in the successions. The basal Roseland Volcanics (*c.* 400 m thick) — comprising spilitic lavas, lava breccias, agglomerates, conglomerates and black mudstones — are of Gedinnian to Emsian age. They unconformably overlie Ordovician (Llandeilan) quartzites, dated by trilobites, which are repeated by low-lying thrust sheets in Gerrans and Veryan Bays. The conglomerates contain both intraformational pebbles and exotic metamorphic and coarse crinoidal-limestone clasts. The overlying Gramscatho Beds are divided into a lower (*c.* 600 m thick) and an upper (*c.* 350 m thick) group, separated by the Veryan Limestones. In Gerrans Bay they comprise dominantly mudstones but with frequent, thin, sharp-based, graded and highly immature greywackes of coarse sand to silt grade. At Gunwalloe Fishing Cove the greywackes range up to 1·5 m thick and show turbiditic internal structures. The Veryan Limestones (*c.* 370 m thick) comprise clastic mudstones with numerous thin sharp-based recrystallised limestones. Black cleaved mudstones with thin beds and laminae of radiolarian chert occur at the base of the sequence. Conodont faunas from the limestones suggest a middle Devonian age.

The Lizard−Dodman−Start line was originally proposed by Sedgwick as a single 'mineral axis' bringing up pre-Devonian rocks in S. Devon and Cornwall. Later authors have considered the breccias bounding the northern margin of the Lizard to be tectonic, regarding them as occurring in a zone of intense deformation (Meneage Crush Zone) produced by northward thrusting of the Lizard basement, with the northern Lizard boundary marking the trace of a major thrust plane. Similar crush breccias have been recognised in the Roseland area; but, as mentioned above, Sadler (1973*) has established that a normal, though much deformed, sedimentary succession is present here, so that this part of the crush zone concept is clearly discredited. The thrusts seen in Roseland, which bring up the Ordovician quartzites of Armorican affinities and their Devonian cover, do not seem to be major thrust planes. They do not bring up metamorphic basement of Lizard type, and Sadler (1973*) has interpreted them as décollement structures off a basement high perhaps only 2−3 km to the south.

At Start there is no evidence of any significant thrust, the northern limit of the Start Schists being a steep N-dipping fault. This 'basement' is faulted against a region of upright to steeply inclined, polyphase folds in the Meadfoot Beds.

J. L. M. Lambert (1965) has found no evidence to support the idea that the Lizard 'thrust breccias' are actually fault breccias. He has noted that, between Nare Head and Porthallow on the eastern side of the Lizard, four mappable divisions may be recognised within the 'crush breccia': (a) siltstone breccia, (b) 'slate with phacoids of greywacke', (c) conglomeratic pebbly slate, and (d) conglomerate. Fragments of siltstone within the breccia were present before cleavage development; some of the graded greywacke phacoids are inverted although others are the right way up, and the conglomerates have boulders and matrix differing in composition. Lambert has concluded that, although the succession is deformed tectonically, the lack of stratification in parts may be due to a previous phase of subaqueous slumping. Sadler (1974*) has noted that the slate-with-greywackes unit may be equivalent to the Gramscatho greywackes of the Roseland area and that locally occurring volcanics may be equivalent to those in Roseland. The coarse Menaver Conglomerate (see below) that outcrops north of the 'crush zone' are definitely older than the main Gramscatho outcrops to the north, and Sadler has pointed out their striking resemblance to the conglomerates in the Roseland Volcanics. Most significantly, Sadler has found one locality where the Menaver Conglomerate rests directly upon Lizard basement, although as well as being a normal unconformable contact the field evidence does not preclude a flat-lying fault contact.

The Menaver Conglomerate (Fig. 10.7) is a strikingly polymictic breccio-conglomerate with clasts locally up to 1 m in diameter. The clasts occur in a dominantly coarse sandy matrix, although mud-supported clasts occur at Flushing. Thin, coarse and rarely graded sandstone interbeds may occur. There are records of a wide range of metamorphic (phyllite, schist, mylonite), igneous (granophyre, dolerite, spilite) and sedimentary (sandstones, chert) clasts, which are mostly angular. The unit thins rapidly from Nare (65 m) towards Flushing (6 m). The whole deposit indicates very minimal transport abrasion and sorting, with high transport power. If the deposit is marine, which is deduced only from its presumed stratigraphic position under the Gramscatho turbidites, then it may indicate high hinterland gradients, short and fast stream transport and a rapid process of debris flow to emplace the deposit into a deep marine environment.

Thus it is concluded that an important sedimentary basin existed in S. Cornwall during lower and middle Devonian times. This basin was infilled by sediments that are quite distinct from those seen in S. Devon. Palaeocurrent data are not available, but the evidence of the Menaver and Roseland exotic clasts indicates derivation from the south, with evidence for a Precambrian to Lower Palaeozoic hinterland in the English Channel—Brittany area. We must conclude with Sadler (1974*) that the northern limit of the Gramscatho outcrop represents a major tectonic *and* palaeogeographic discontinuity, juxtaposing as it does the above succession with the totally different S. Devon sequence.

10e   The European context

Devonian SW England is usually included in the Rheno-Hercynian zone of the Hercynian mobile belt as defined in the Ardennes and Rhenish Schiefergebirge (Fig. 12.4). In these latter areas there is a sequence of Devonian shallow-marine clastic deposits derived from the north, overlain by an assemblage of late middle to upper Devonian becken and schwellen facies similar to those seen in the S. Devon area. Pelagic limestones in the German outcrops occur above volcanic swells, reefs and upfaulted basement ridges. Basic volcanism in Europe reached its peak around the middle/upper Devonian boundary, again as seen in S. Devon. Clearly, a major extensional phase with subsidence and volcanism was widely established in the Rheno-Hercynian zone during late middle Devonian times, lasting until the lower Carboniferous.

It should not be thought that the Rheno-Hercynian zone was a continuous arc-like trough, since individual sedimentary basins may die out laterally to the west and east. They have intimate connections with adjacent crystalline continental crust uplifts, with evidence that the loci of greatest subsidence in individual basins shifted progressively northwards during early Devonian to late

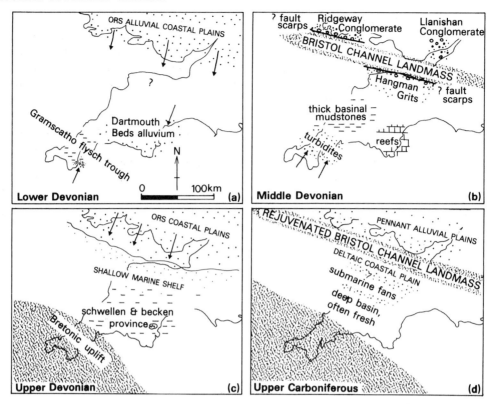

Figure 10.8 Summary maps to illustrate the upper Palaeozoic palaeogeographic evolution of SW England. Further discussion in Chapter 12.

Carboniferous times. The central German block (Mittel Deutsche Schwelle) was a most persistent feature of Upper Palaeozoic palaeogeography in central Europe and lay north of the Saxo-Thuringian zone, where the various sedimentary basins and blocks were more-or-less stationary with time. Sediment thicknesses in the basins (c. 700 m for Devonian) were much less than those seen in the Rheno-Hercynian zone before the onset of the early Carboniferous flysch phase.

Partly continental successions in S. Belgium and N. France show northward thinning on to the London–Brabant Massif, with northward overlap due to transgressional phases. Prominent Frasnian reefs occur in the Couvin area of the Dinant Basin. In Britanny, as in Cornwall, Silurian strata are unknown; the early Devonian Schistes et Quartzites, a flysch-like deposit thought to be derived from a hinterland in the English Channel or Western Approaches, is the earliest depositional event (Renouf 1974). The Roseland Basin of S. Cornwall

may thus have existed north of this hinterland. Subsequent marine transgression deposited shallow marine sands as far north as Finistère and the Cotentin. In middle and upper Devonian times throughout Armorica there was general regression, due possibly to upwarp associated with the Bretonic 'deformational' phase (Ch. 12). There is, in fact, little direct evidence of deformation at this time in Brittany, although late Devonian K−Ar dates in S. Cornwall have been widely accepted as indicating a Bretonic deformational phase (Ch. 12).

It may be concluded that the SW England area shares many features in common with Belgian and German Devonian successions. The palaeogeographic evolution of SW England is summarised in Figure 10.8.

10f Summary

SW England lay on the southern margin of the Old

Red Sandstone continent in Devonian times and formed the most westward extension of the Rheno-Hercynian zone of N. Europe. In N. Devon, successions indicate S-flowing rivers that deposited coastal plain facies of Old Red Sandstone type. These alternated with nearshore to offshore shelf facies deposited after northward transgressional episodes. Similar events are recorded in the Belgian and N. German areas. In S. Devon, early Devonian successions also contain shelf and coastal plain facies, but middle to upper Devonian successions become increasingly distinct, with the development of reefal carbonate platforms and then extensive deep-water basins containing locally upstanding ridges or schwellen with condensed pelagic deposits. Again, many similarities with N. European successions exist, especially since the phase of subsidence was accompanied by extensive, mildly alkaline, basaltic volcanism in many areas. The S. Cornwall area contains quite distinct basin-infill sequences of pelagic mudstones, clastic turbidites and slump breccias with evidence of a southern provenance. It is doubtful whether the Lizard—Dodman—Start line, which presently forms the southern boundary to the area, is a laterally continuous thrust as has previously been thought.

11 Carboniferous environments and basin evolution

11a Introduction and correlation

During late Devonian to early Carboniferous times the southern margin of the Old Red Sandstone continent was transgressed by a major marine invasion, which subsequently spread gradually northwards. On the resulting Carboniferous shelf, from Northumberland to the Rhine and to Kansas, the distribution of positive blocks and rapidly subsiding basins is critical for our understanding of facies distributions, source areas and thickness trends during much of Carboniferous times.

Carboniferous faunas and floras have been investigated for almost 100 years with a view to establishing workable zonal schemes. Table 11.1 shows a current compendium of the various proposed schemes. Major problems arise in the Dinantian (lower Carboniferous), where facies influences, provincialism and homeomorphy among the shelly macrofaunas (mainly brachiopods and corals) have created many problems. In recent years these problems have faded somewhat with the increasing use of microfossils (conodonts, foraminiferans, spores) for zoning purposes. Zonally useful species of these forms may be much more widespread geographically and more abundant than macrofaunas. Spores are particularly valuable zonal indices in non-marine facies, although problems remain in the correlation of spore zones with conodont and foraminiferal zones established in marine facies.

The most recent and novel correlation aid in Dinantian sequences has been provided by Ramsbottom (1973*), who has recognised six major cycles of sedimentation, recently given stage names (George *et al.* 1976*). These major cycles, thought to have been controlled by eustatic sea-level changes, are defined by transgressive basal members and regressive top members (Table 11.2). Such major cycles are much easier to define at basin margins than in basin centres, since changes in sea level in the latter have a much reduced influence upon facies

types. Whether these major cycles were caused by the waxing and waning of ice sheets, or the changing volumes of the oceans due to sea-floor-spreading effects, is at present an open question.

A particularly good zonal scheme exists for Namurian sequences, based upon the marine nektonic goniatites, as first devised by Bisat. The goniatite stages so defined may be recognised over the whole northern hemisphere. About sixty diagnostic goniatite bands are recognisable in the British Namurian. Adopting a figure of *c.* 12 Ma for the duration of the Namurian, Ramsbottom (1969) has calculated that on average a new goniatite fauna arose every 200 000 years. Such a high quality of time correlation between the thin bands is comparable with that achieved in the Jurassic where ammonities are used. Ramsbottom (1977) has also recognised a number of major eustatic cycles in the British Namurian.

The traditional threefold classification of the British Coal Measures into lower, middle and upper divisions has been standardised by placing the boundaries at major marine incursions, where goniatites occur. Further subdivision is based upon non-marine bivalves. A complementary zonation based upon miospores has recently been developed, seven assemblage zones being recognised in the British Westphalian.

11b Culm Province Carboniferous

In SW England the passage between the Devonian and Carboniferous is always conformable and transitional. Within the Pilton Beds of N. Devon (Ch. 10) there is no facies break; the dominantly offshore shelf mudstone facies with thin sandstone and limestone beds is uninterrupted. Relatively deepwater mudstone facies predominated during earliest Carboniferous times, continuing the Devonian becken facies described in Chapter 10c. In addition,

AGE (Ma)	SERIES	Stages	Major Cycles	Coral–Brachiopod	Goniatite	Non-Marine Bivalve
280	STEPHANIAN					
						prolifera
290	WESTPHALIAN	D				tenuis
		C				phillipsi
		B			A	similis–pulchra
		A			G2	modiolaris
						communis
315	NAMURIAN	Yeadonian	11		G1	
		Marsdenian	9, 10		R2	
		Kinderscoutian	6, 7, 8		R1	
		Sabdenian	4, 5		H	Spores
		Arnsbergian	2, 3		E2	
325		Pendelian	1		E1	
	DINANTIAN				P2	
		Brigantian	6	D2	P1	VF, NC
		Asbian	5	D1	B	TC, NM
		Holkerian	4	S2		
		Arundian	3	C2, S1	Pe	PU
		Chadian	2	C1, C2, S1		
360		Courceyan	1	K, Z	Ga	VI, CM

Table 11.1 Carboniferous stages, major cycles and main zonal subdivisions.

radiolarian cherts were also deposited during Dinantian times. These may consist of radiolarian tests set in a structureless cryptocrystalline matrix, or they may show primary depositional structures with clastic carbonate silica grains. Prentice has regarded these as 'deep water' cherts, analogous to modern abyssal radiolarian oozes. However, it is difficult to define the actual depth; deposition may have occurred in a mud-free 'basin' just a few hundred metres deep. That the cherts are deep water and not lagoonal is indicated by the fact that they are overlain by turbiditic sandstones of the Crackington Formation. In the structurally complex Boscastle and Okehampton areas, tuffs, agglomerates and lavas of the Tintagel Volcanic

Formation and its correlatives occur within the Dinantian sequence, which is about 200 m thick. Elsewhere thicknesses are very difficult to determine, but overall the Dinantian of the Culm province is very much thinned compared to the carbonate deposits of the Bristol area only some 100 km to the northeast.

11c Dinantian of the Southwestern Province

Dinantian shelf sequences in the Bristol–Mendips area and S. Wales appear to have been deposited on the southern flanks of the low-lying emergent hinterlands of a block area known as St George's Land

Environment	Facies	Faunas and Floras
Supratidal	karstic surfaces sabkha evaporites calichification some dolomitisation	
Intertidal	early cementation some micrites	stromatolites
Agitated shallow subtidal	oölites algal clasts	oncolites *Linoprotonia*
Quiet shallow subtidal	micrites and pelletal micrites	stromatolites, athyrids, *Modiolus*, rhynchonellids
Shallow offshore	pale, thick-bedded, bioclastic limestones	costate productids, spiriferids, colonial rugose corals
Deep offshore	dark, thin-bedded, bioclastic limestones	spinose productids, solitary rugose corals, phillipsoid trilobites
Deep muddy basins	calcareous mudstone and mudstone	smooth spiriferoids, nuculoid bivalves, phillipsoid trilobites, goniatites, pectinoid bivalves, radiolarians, pelagic ostracods

Table 11.2 Depth-related lithological features and faunas definable in the Dinantian. Not exhaustive (modified after Ramsbottom 1973*).

(Fig. 11.1). This block probably continued across the Irish Sea to join up with the Leinster block in Eire. It is not known whether the Bristol Channel landmass, present during Devonian and late Carboniferous times (Chs 9 and 11), was present during the Dinantian.

The major early Dinantian marine transgression over the low-lying alluvial plains of the Old Red Sandstone continent is recorded in the lower Limestone Shale Group of the Bristol–S. Wales area and in the basal Kinsale Group of Eire. In the Bristol area, fining-upwards alluvial cycles are succeeded by thin limestones with crinoidal and bryozoan debris and clastic siltstones. Limestones with stromatolites, phosphate nodules, vermetid gastropods and haematised shell debris follow. Some limestones show sharp erosive bases and channel forms and contain flaser bedding indicative of tidal- or storm-influenced coastal deposition.

In SW Eire the transgression is entirely within a clastic sequence, with reddened fluvial fining-upwards cycles passing upwards into rapid alternations of sandstone and mudstone containing bidirectional cross-laminations and mud flasers indicative of tidal action (Kuijpers 1971, J. R. Graham 1975). Mudstones with goniatites comprise the remainder of the Dinantian here (Fig. 11.1) and indicate the persistent presence of a sheltered offshore basinal environment. Further north, at Hook Head, upper Old Red Sandstone fluvial facies are again succeeded by tidally deposited facies, with the eventual formation of extremely fossiliferous muddy limestones of low-energy shelf origin.

Subsequent Dinantian sedimentation in most areas was dominated by carbonate deposition, with major cycle boundaries being well developed in the Bristol and Mendips sections (Fig. 11.2). Non-sequences developed at the top of some of the regressive facies. A period of extended non-deposition is indicated at the top of the Black Rock Limestone of the Bristol area, where there is extensive associated dolomitisation. Similarly to the north of Cardiff, extensive dolomitisation is present beneath the disconformity associated with the overstep of the Caswell Bay Oölite over the underlying Tears Point

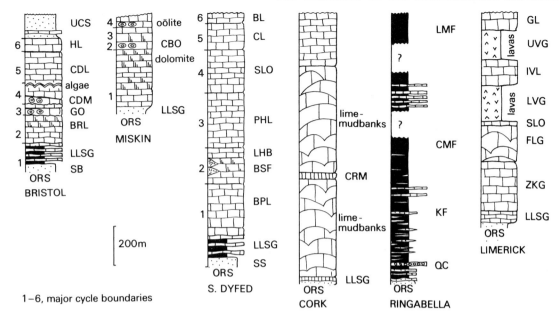

1–6, major cycle boundaries

200m

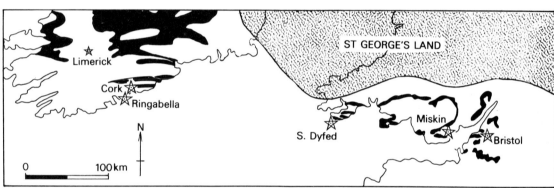

Figure 11.1 Dinantian outcrops (shaded black), representative sections (mudstones in black) and generalised palaeogeographic features of the Southwestern Province (sections and cycle boundaries modified after George *et al.* 1976* and references cited therein).

Bristol section: SB = Shirehampton Beds; LLSG = lower Limestone Shale Group; BRL = Black Rock Limestone; GO = Gully Oölite; CDM = Clifton Down Mudstone; CDL = Clifton Down Limestone; HL = Hotwells Limestone; UCS = upper Cromhall Sandstone.

Miskin section: LLSG = lower Limestone Shale Group; CBO = Caswell Bay Oölite.

S. Dyfed section: SS = Skrinkle Sandstone; LLSG = lower Limestone Shale Group; BPL = Blucks Pool Limestone; BSF = Berry Slade Formation; LHB = Linney Head Beds; PHL = Pen-y-Holt Limestone; SL = Stackpole Limestone; CL = Crickmail Limestone; BL = Bullslaughter Limestone.

Cork section: LLSG = lower Limestone Shale Group; CRM = Cork Red Marble.

Ringabella section: QC = Quartz Conglomerate; KF = Kinsale Formation; CMF = Courtmacsherry Formation; LMF = Lispatrick Mudstone Formation.

Limerick section: LLSG = lower Limestone Shale Group; ZKG = *Zaphrentis konincki* Group; FLG = *Fenestrellina* Limestone Group; SLG = *Seminula* Oölite Group; LVG = lower Volcanic Group; IVL = Inter-Volcanic Limestone; UVG = upper Volcanic Group; GL = *Gigantella* Limestone.

ORS = Old Red Sandstone.

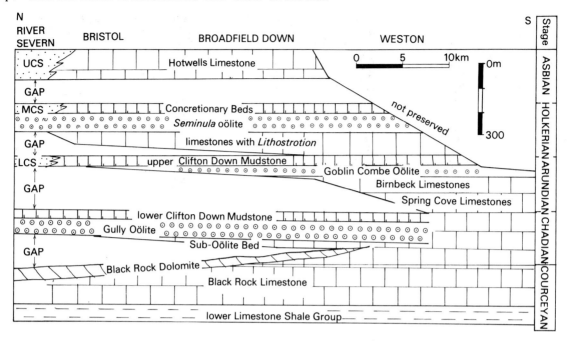

Figure 11.2 Cross-section to illustrate Dinantian thickness and facies changes due to onlap on to the Bristol Shelf fringing the southern margin of St George's Land. Note the increasing hiatuses that develop northwards at cycle boundaries (after Ramsbottom 1977).
LCS, MCS, UCS = Lower, Middle and Upper Cromhall Sandstone respectively.

Limestone. The former contains much evidence for nearshore and tidal flat deposition, with cross-stratified oölitic limestone, pelletal calcite mudstone and stromatolites with calcite pseudomorphs after gypsum. The dolomitisation seems to be mainly confined to micritic limestones. Bhatt (1976) has postulated that dolomitisation was due to the reflux of dense brines on the floor of a lagoon bounded on its seaward side by a reef barrier, as seen today in Bonaire, East Indies, or the Coorong, S. Australia. A disadvantage of the hypothesis is that there is no actual evidence for the reefal barrier in present outcrops.

In Dyfed, northward thinning and associated off-shore to nearshore facies changes occur as the presumed shoreline of St George's Land is approached. The thinning is partly a product of multiple over-step, due to the periodic northern rise of St George's Land relative to the more uniformly subsiding southern area.

In most areas, late Dinantian sequences are dominated by offshore marine limestones. A significant exception is the clastic sequences laid down in the Bristol and Forest of Dean areas (e.g. Cromhall and Drybrook Sandstones). Those in the Forest of Dean are clearly of fluvio-deltaic origin and have a presumed northern provenance from the southern flanks of St George's Land. The Bristol examples are thinner and show evidence of marine reworking under wave-influenced conditions.

11d Dinantian of the Central Province

Generally, basin deposits in this area consist of thick deposits of limestone and mudstone, with major cycles 1–6 all developed. Surrounding blocks may show an incomplete number of cycles (Fig. 11.3).

A major feature of interest in the area are the so-called 'reef knolls', which outcrop widely. Similar reef knolls occur in the Dinantian of Belgium and in the Mississippian of the USA.

Pioneer work by Tiddeman on the Clitheroe reef knolls established that they were primary depositional structures, although some later authors have doubted whether the knoll forms were original, preferring to regard them as tectonic modifications of original lime sheets. In view of the variation in

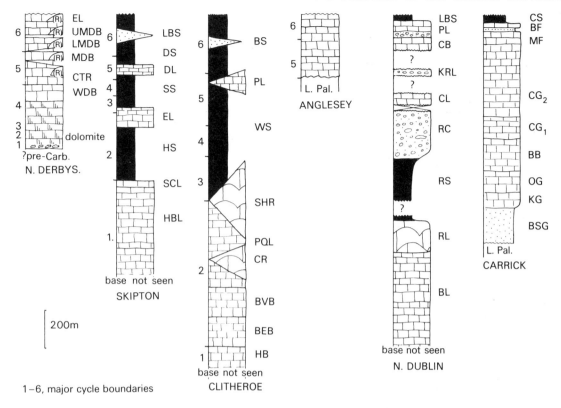

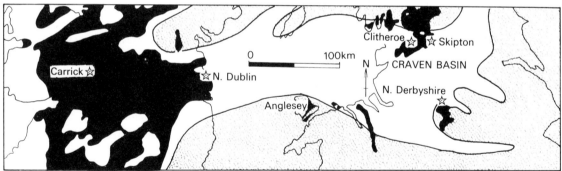

Figure 11.3 Dinantian outcrops (shaded black), representative sections (mudstones in black) and generalised palaeogeographic features of the Central Province (sections and cycle boundaries modified after George *et al.* 1976* and references cited therein).

N. Derbyshire section: WDB = Woo Dale Beds; CTR = Chee Tor Rock; MDB = Millers Dale Beds; LMDB, UMDB = lower and upper Monsal Dale Beds respectively; EL = Eyam Limestone.

Skipton section: HBL = Haw Bank Limestone; SCL = Skipton Castle Limestone; HS = Halton Shales; EL = Embsay Limestone; SS = Skibeden Shales; DL = Draughton Limestone; DS = Draughton Shales; LBS = Lower Bolland Shales.

Clitheroe section: HB = Horrocksford Beds; BEB = Bankfield East Beds; BVB = Bold Venture Beds; CR = Coplow Reef; PQL = Peach Quarry Limestone; SHR = Salthill Reef; WS = Worston Shales; PL = Pendleside Limestone; BS = Bolland Shales.

N. Dublin section: BL = Bedded Limestone; RL = Reef Limestone; RS = Rush Slates; RC = Rush Conglomerate; CL = Carlyan Limestone; KRL = Kate Rocks Limestone; CB = *Cyaxonia* Beds; PL = *Posidonomya* Limestones; LBS = Loughshinny Black Shale.

Carrick section: BSG = Boyle Sandstone Group; KG = Kilbryan Group; OG = Oakport Group; BB = Ballymore Beds; CG₁ = Croghan Group; CG₂ = Cavetown Group; MF = Meenymore Formation; BF = Bellavally Formation; CS = Carraun Shale.

L. Pal. = Lower Palaeozoic; pre-Carb. = pre-Carboniferous.

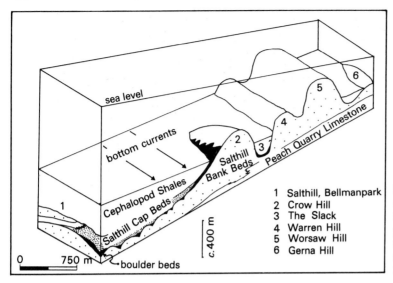

1 Salthill, Bellmanpark
2 Crow Hill
3 The Slack
4 Warren Hill
5 Worsaw Hill
6 Gerna Hill

Figure 11.4 Presumed topography of the knoll lime-mudbanks in the Clitheroe area of Lancashire after a phase of intra-Dinantian uplift, and erosion indicated by the presence of boulder beds (after Miller & Grayson 1972*).

morphology and size of the knolls, Lees (1964*) has introduced the term 'lime-mudbanks' as an alternative.

Within the knoll lime-mudbanks of the Clitheroe Limestone Complex, Miller and Grayson (1972*) have recognised four main sedimentary facies:

(a) The low-energy bank facies, with depositional dips, which comprises white to grey, poorly bedded, micritic limestones containing stromatactis cavities and occasional bryozoans, brachiopods and crinoids.

(b) The bank margin facies, with depositional dips, which comprises well-bedded limestones that are very rich in crinoidal debris. A higher energy of deposition than for the bank facies is postulated, but the currents were still of low velocity, since on death crinoids often toppled over in the current direction with only minor disarticulation.

(c) The interbank facies which comprises thinly bedded and lenticular argillaceous limestones alternating with thin calcareous mudstones. Faunas include zaphrentid corals, crinoids and bryozoans.

(d) The cover mudstone facies which comprises crinoid-rich mudstones interbedded with the thinner crinoidal limestones.

The occurrence of fissures in the lime-mudbanks, infilled with later debris and thick bank-derived boulder beds, is taken as evidence for a phase of subaerial erosion that locally accentuated the primary moundlike nature of the mudbanks (Fig. 11.4). Problems still remain concerning the origin of the

mudbanks, due to the rarity of framebuilders. Local production of lime-mud might have occurred with extensive crinoid thickets around the bank edges acting as sediment baffles, a role probably shared by the bryozoan fronds that are so common in many Waulsortian mudbanks. Stabilisation of mounds must have required either rapid and early carbonate lithification and/or the presence of gelatinous algal mats as sediment binders. The latter are present in quantity as stromatolites only in the Castleton, Derbyshire, apron reefs. A relatively early diagenetic age for cementation is indicated by blocks of mudbank facies in associated talus breccias and limestone turbidites. This is well illustrated by the Draughton and Pendleside Limestones of the Craven Basin (Fig. 11.3), which are thought to have been derived from the Cracoe reef knolls to the north.

Lime-mudbanks reached their maximum thickness and extent in the Irish Carboniferous of the Limerick−Cork area, where they coalesced to form a sheet up to 1500 m thick (Fig. 11.5). Scattered single mudbanks of knoll type occurred widely in the Irish Midland Plain area, where they are closely associated with lead−zinc mineralisation, and the Dublin Basin (Fig. 11.5). In the English part of the Central Province, many of the knoll lime-mudbanks occur along the southern margin of the Askrigg block along the Craven Fault system. The Derbyshire reefs similarly fringe the carbonate shelf area from gulfs and basins to the north and west (Fig.

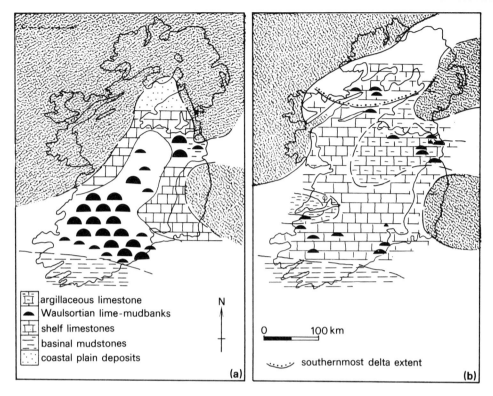

Figure 11.5   Distribution of (a) early and (b) late Dinantian facies in Ireland (after MacDermot & Sevastopulo 1973*).

Figure 11.6   Location of blocks and basins in central England as determined by outcrop, borehole and gravity surveys (partly after Falcon & Kent 1960 and Kent 1967).

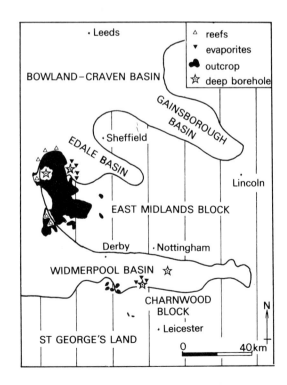

11.6). The older Clitheroe mudbanks occur within the main basin area, where they may have formed upon more positive axes of uplift coincident with what were to be later structural anticlinal axes.

Figure 11.6 shows a number of narrow basinal areas separated by stable blocks around the southeastern and eastern margins of the main central English basin. These have been delimited largely by geophysical work during the search for oil in the Midlands area. Numerous boreholes have helped to delimit block/basin boundaries and, in the basins, have proven facies that show evidence for hypersaline conditions. The Widmerpool gulf is most noteworthy, since the Hathern no. 1 borehole at the gulf margin has revealed regressive sabkha evaporites of late major cycle 1 (Courceyan), with nodular 'chicken-mesh' anhydrite. A previously unsuspected southern extension of the Edale gulf has recently been proven by the Eyam borehole, which passed through nearly 2000 m of Dinantian rocks, mainly limestones. The late major cycle 1 deposits here also contain nodular anhydrites and dolomite of sabkha aspect. On the Derbyshire block to the west, many different types of bioclastic limestones are recorded, with several associated horizons of alkali basalt, lavas and tuffs. Of especial interest is Walkden's (1974) discovery of karstic horizons, providing good evidence for minor cyclicity on this late Dinantian shelf.

The northern margin of St George's Land seems to have stood up as a low-relief land area during much of Dinantian times, until it was transgressed across in major cycle 5 (Asbian) times, as evidenced by marine limestone successions in N. Wales, the northern Welsh Borderland and Leicestershire. Following the transgression, a phase of quite rapid subsidence in N. Wales allowed over 1000 m of mostly shallow offshore bioclastic limestones to accumulate in places. Here again, though, minor cyclicity was apparent, with formation of karstic surfaces and spectacular infilled potholes (as at Lligwy Bay, Anglesey). Much thinner sequences are present further east in the Midlands, but nowhere is there evidence for significant clastic input into the main basin from St. George's Land.

11e Dinantian of the Northern Province

This area (Fig. 11.7) has come to be regarded as the classic area in Britain for the study of blocks and basins on the Carboniferous shelf. In early Dinantian times, block areas were land and basins took the form of geomorphic gulfs with quite distinctive sedimentary facies and faunal assemblages. Let us begin by examining gulf evolution in the area.

The Stainmore gulf, a westward extension of the E. Yorkshire Basin, was bounded by the Alston and Askrigg blocks to the north and south respectively. The form of the Stainmore Basin itself was first delimited by gravity surveys. Surface outcrops are restricted to the head of the supposed gulf at Ravenstonedale, where a complete Dinantian succession is exposed. The basal Pinskey Gill Beds and their correlatives seem to represent locally derived fluvio-marine deposits. They rest unconformably upon the Lower Palaeozoic. The overlying Stone Gill and Coldbeck Beds contain micritic limestones (often dolomitised), calc-arenitic limestones, algal stromatolites and calcareous mudstones. Although corals are present at certain horizons, the fauna is usually very restricted. Transgression at the top of these major cycle 1 (Courceyan) deposits is indicated by the Scandal Beck Limestone, which retains its bioclastic character with open marine faunas over a wide area. Subsequent transgressions during major cycles 3 (Arundian) and 4 (Holkerian) times gradually changed the gulf-like form of the basin into a non-uniformly subsiding open marine area.

The Northumberland—Solway gulf (Fig. 11.8) was bounded by the Cumbrian—Alston and Southern Uplands blocks. Along the northern basin margin is seen an upward passage from red-beds of fluvial origin, presumed to be upper Old Red Sandstone in age, via the alkali basalt Birrenswark—Kelso Lavas (5—200 m), to the Lower Border Group deposits of major cycle 1 (Courceyan) times. Rivers draining off the Southern Uplands land area deposited the Whita and Annan Sandstones on the narrow coastal plains of the gulf sea (Leeder 1974a*). On the north Solway coast at Rerrick, thick coarse immature breccio-conglomerates and sandstones of alluvial fan aspect make up the entire Dinantian succession (Deegan 1973). They contain local detritus derived from the rapid weathering and erosion of the Criffell pluton (dated at 390 Ma) and give good evidence for penecontemporaneous movements along the N. Solway Fault. In Berwickshire, coastal-plain fluvio-lacustrine facies of the Cementstone type (see below) conformably succeed fluvial red-beds of Old Red Sandstone aspect bearing evidence of S-flowing rivers.

Periodic advances of a delta system flowing southwestwards into the Northumberland gulf are recorded in the Border Group. During periods of delta absence a wide variety of limestones with stromatolite assemblages were deposited. Evidence for hypersalinity in high intertidal environments, such as calcite pseudomorphs after gypsum, abounds in the Scottish outcrops close to the former northern gulf margins. Normal marine faunas are rare in Lower Border Group limestones — especially corals, foraminifera and conodonts — which seems

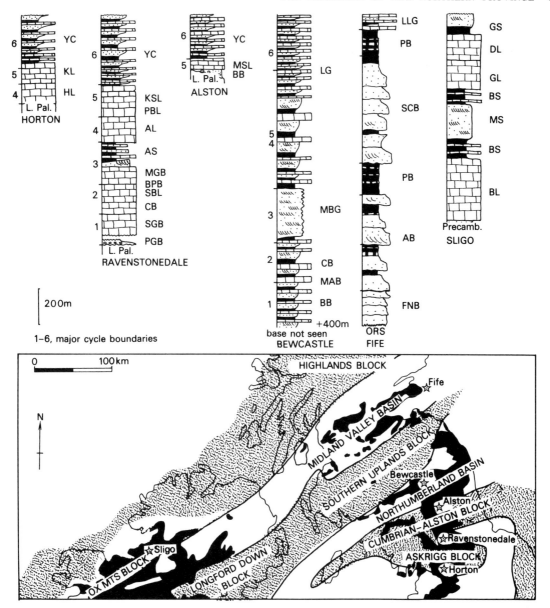

Figure 11.7 Dinantian outcrops (shaded black), representative sections (mudstones in black) and generalised palaeogeographic features of the Northern province (sections and cycle boundaries modified after George *et al.* 1976* and references cited therein).

Horton section: HL = Horton Limestone; KL = Kingsdale Limestone; YC = Yoredale cycles.

Ravenstonedale section: PGB = Pinsky Gill Beds; SGB = Stone Gill Beds; CB = Coldbeck Beds; SBL = Scandal Beck Limestone; BPB = Brownber Pebble Beds; MGB = *Michelinia grandis* Beds; AS = Ashfell Sandstone; AL = Ashfell Limestone; PBL = Potts Beck Limestone; KSL = Knipe Scar Limestone; YC = Yoredale cycles.

Alston section: BB = Basement Beds; MSL = Melmerby Scar Limestone; YC = Yoredale cycles.

Bewcastle section: BB = Bewcastle Beds; MAB = Main Algal Beds; CB = Cambeck Beds; MBG = Middle Border Group; LG = Liddesdale Group.

Fife section: FNB = Fife Ness Beds; AB = Anstruther Beds; PB = Pittenweem Beds; SCB = Sandy Craig Beds; PB = Pathhead Beds.

Sligo section: BL = Ballyshannon Limestone; BS = Bundoran Shale; MS = Mullaghmore Sandstone; BS = Benbulben Shale; GL = Glencar Limestone; DL = Dartny Limestone; GS = Glenade Sandstone.

Precamb. = Precambrian; L. Pal. = lower Palaeozoic; ORS = Old Red Sandstone.

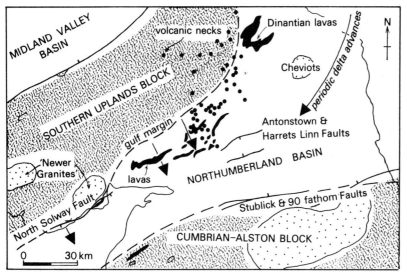

Figure 11.8 Approximate limits of the Dinantian Northumberland gulf and the distribution of volcanics on the northern gulf margin (after Leeder 1974b).

to indicate the development of a special faunal province, due perhaps either to geographic isolation or, more likely, to slight hypersalinity.

Middle Border Group rocks include the Fell Sandstone, now regarded as the sheet sand deposit of a succession of major braided river courses (Hodgson 1978). Upper Border Group successions include the Scremerston Coal Group, which marks a major fluvio-deltaic regressional episode in the gulf and, in Berwickshire, contains the oldest workable coals in the European Carboniferous. The transgressive Dun Limestone which overlies the Scremerston Coal Group marks a marine transgression that also spread over the Alston block to the south, depositing the Melmerby Scar Limestone. From this time onwards the gulf-like character of the Northumberland Basin was no more. Major cycle boundaries in the Northumberland Basin are very difficult to uphold, since minor cyclicity due to delta advances and retreats is predominant.

The Southern Uplands block seems to have effectively separated the Midland Valley Basin from the Northumberland Basin during much of Dinantian times. There is a conformable upward change from fluviatile red-beds of upper Old Red Sandstone type to the early Dinantian Cementstone Group in the western outcrops of the Midland Valley. In the Stirling area, fluvial red-beds with cornstones reappear above a development of the Cementstone Group, illustrating that the local Devonian/Carboniferous boundary merely marks a temporary

facies change. The Cementstone Group comprises alternating calcareous mudstones, siltstones, sandstones and dolomitic limestones. Some of the latter are nodular and clearly diagenetic, but others are well bedded with internal laminations and must be primary (Belt *et al.* 1967*). Pseudomorphs after halite, desiccation polygons, gypsum lenses and a rare restricted fauna of ostracods, bivalves and gastropods all point to a periodically hypersaline, probably lacustrine, environment of deposition for the dolomites of the group. Interbedded erosive-based sandstones with cosets of large-scale cross-stratification seem to be of fluvial channel origin. Very thick, dominantly carbonate−mudstone successions, as at Ballagan Glen, seem to record long-standing lacustrine areas far from the coastal plain rivers.

Large scale eruption of basaltic lavas occurred during the early Dinantian of the Midland Valley (Fig. 11.10), causing rapid lateral variations in sediment thickness around the area of lava spreads (Fig. 11.9). After the eruption of the Clyde Plateau Lavas, for example, a barrier was established between a western basin and an eastern basin in the Lothians, where organic-rich oil-shale facies were deposited (Fig. 11.9). Eastwards to Fife it appears that Cementstone facies die out, for the early Dinantian Calciferous Sandstone measures comprise at least 1500 m of fluvio-deltaic facies deposited by S- and SW-flowing distributaries. The overlying lower Limestone Group everywhere shows evidence for

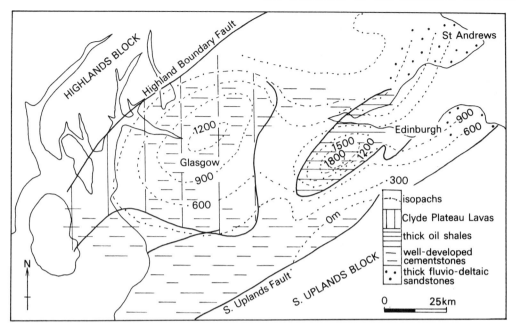

Figure 11.9 Major features of Dinantian palaeogeography in the Midland Valley (after George 1958*).

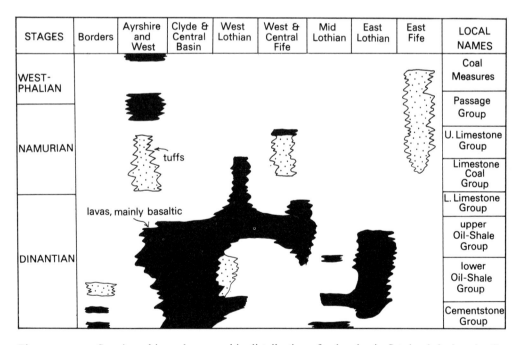

Figure 11.10 Stratigraphic and geographic distribution of volcanics in Scotland during the Carboniferous (after Francis 1967).

Yoredale-type cyclicity (Ch. 11f), with very rapid lateral and vertical changes of facies and thickness due to non-deposition over the site of maximum thickness of the Clyde Plateau Lavas in Ayrshire. The basal Hurlet transgression seems to have established uniform shallow marine conditions over much of the Midland Valley area.

The sedimentary basins and gulfs discussed above were separated by block areas that were initially land. In general the relief and area of these blocks probably increased northwards, for clastic deposits of local derivation increase in frequency and thickness northwards. The granite-based Alston and Askrigg blocks (Bott 1967) were reasonably mature surfaces coated by fluvial pebbles when a marine transgression deposited the Great Scar and Melmerby Scar Limestones upon them in major cycle 5 (Asbian) times. These limestones are uniformly open marine bioclastic calcarenites, with occasional crinoid banks. Schwarzacher (1958) has shown that within the Great Scar Limestone of the Askrigg block eight prominent bedding planes, usually marked by mudstone in underground exposures, indicate minor cyclicity due probably to periodic emersion, since potholed surfaces of presumed karstic origin frequently underlie the mudstones. Succeeding Dinantian beds comprise a number of Yoredale cycles (Ch. 11f) that continue up into the Namurian over the whole area. The much-thinned Dinantian successions on the Southern Uplands block at Sanquhar (10 m) and Thornhill (15–45 m) are both of late Dinantian age, and suggest that at this time transgression had covered at least part of the block.

In Ireland, the southwestward extensions of the Southern Uplands and Highlands blocks are both present, as the Longford Down and Donegal–Connaught blocks. However, these exerted much less influence upon sedimentation than their British equivalents. A marine transgression in late major cycle 1 (Courceyan) times deposited offshore bioclastic limestones on the southwestern fringe of the Longford Down block, while fluvial coastal-plain deposits, often coarse, were laid down to the west and north. The only example of Cementstone facies in Ireland occurs at Cultra on the southern shore of Belfast Lough, where 180 m of Old Red Sandstone facies are overlain by 300 m of Cementstone facies of middle Courceyan age. In major cycle 2 (Chadian)

times, northward transgression deposited offshore limestones over much of Sligo, Roscommon, S. Mayo, Galway and N. Clare (MacDermot & Sevastopulo 1973*). In later Dinantian times, fluvio-deltaic sandstones of N–NW derivation were deposited (Fig. 11.5). In Clare, however, great thicknesses of carbonates accumulated in shallow shelf seas, for in the Burren more than 600 m of bioclastic limestones are known. In Leitrim and Cavan, extensive deposits of intertidal and supratidal facies with stromatolites and sabkha-type evaporites alternate with fluvial sandstone facies.

11f Yoredale cyclothems in N. Britain

Yoredale cyclothems are distinctive rhythms of alternating clastic and carbonate members (5–80 m per rhythm) seen in the late Dinantian and early Namurian of N. Britain. Similar cycles also occur in the Dinantian Border Group of the Northumberland Basin. R. C. Moore and later workers have recognised essentially similar cycles in the Mississippian of the eastern USA.

In N. Britain, the clastic members of cycles are of presumed deltaic origin, with a variety of subenvironments recorded. Above the limestone members (Fig. 11.11), coarsening-upwards usually signifies delta progradation. Further upward coarsening into a sandstone member may be accompanied by signs of wave reworking, indicating spit formation marginal to the distributary mouth-bar front. Coals (usually thin) or rootlet horizons with thin sandstones and mudstones of ondelta origin complete the cycles. Other sandstone members are sharp and erosive based and may occupy channel forms or show lateral-accretion cross-stratification indicative of point-bar deposition. These members may cut down to almost any level within the cycle. Good evidence for periodic delta abandonment may be seen in the occurrence of calcareous well-sorted sandstones containing marine faunas and low-angle beach-type cross-stratification. Such sandstones are analogous to those in the Mississippi delta that form due to lateral switching of an active delta lobe to another location (e.g. the Chandeleur Isles). Traced on to the former active delta, these sandstones pass laterally into coals, recording active peat accumulation over the now defunct alluvial plain (Elliot

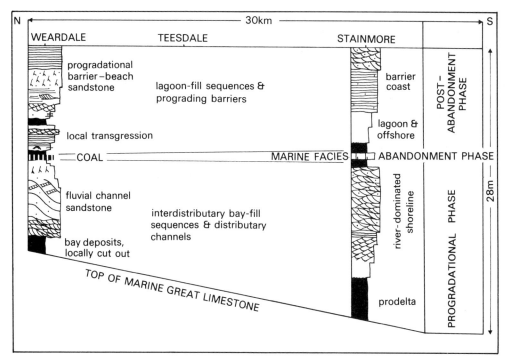

Figure 11.11 Lateral and vertical facies changes in the clastic member of the Great Limestone cyclothem from the N. Pennines (after Elliott 1975).

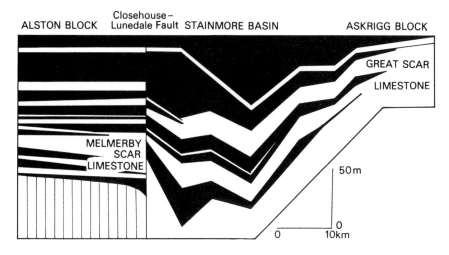

Figure 11.12 Facies and thickness changes as Yoredale cycles are traced from the Askrigg and Alston blocks to the Stainmore Basin (after Burgess & Mitchell 1976).
Clastic members shaded; vertical lines indicate lower Palaeozoic basement.

1974*). It is thus apparent that a variety of different clastic members may evolve according to the precise evolution of a delta lobe at any one particular site.

Carbonate members overlie the deltaic clastics abruptly or with rapid transition. The limestones are usually of grain-supported bioclastic facies with comminuted brachiopod and crinoid debris. Faunas include coral heads, productid, chonetid and rhynchonellid brachiopods, crinoids, foraminifera and conodonts. A winnowed, clear, shallow-water shelf environment with normal marine salinity is indicated.

Most Yoredale cycles are remarkably persistent when mapped out over the area, although thickness increases and limestone 'splitting' (Fig. 11.12) are prominent when cycles are traced from blocks to basins. The origin of the cyclicity is the source of continuing debate. Tectonic hypotheses advocate periodic hinterland upwarp causing rivers to advance basinwards until an equilibrium is reached, with the subsequent marine transgression causing carbonate deposition over the abandoned delta pile. Eustatic hypotheses advocate delta drowning and carbonate deposition during sea level rises, with delta advance following sea level falls. Purely sedimentary hypotheses advocate periodic delta-lobe abandonments or switches (as seen in the pre-Recent Mississippi delta), so that carbonate sedimentation resumes over large areas. It is quite possible that each of these mechanisms may influ-ence Yoredale cyclicity in particular areas, yet there is at present no detailed study based upon firm evidence that may lead us to understand the process fully.

11g Dinantian environments beyond Britain

Outcrops and boreholes in Belgium, Holland and Germany north of the Ruhr reveal the existence of a broad carbonate platform on the southern and eastern fringes of the Brabant Massif, which we can now identify as the most eastward extension of St George's Land (Fig. 11.13). Belgian Dinantian carbonate environments were established following marine transgression over the sandy shelf and littoral deposits of the upper Devonian. Numerous regressional–transgressional phases are recognised, some of which are correlated with the major cycle boundaries of the British Dinantian (Conil, in Bless et al. 1976). A major phase of lime-mudbank growth began in upper Tournaisian times, giving rise to the Waulsortian reefs now seen in outcrop. It is evident that both the Irish and Belgian mudbank provinces were initiated at roughly the same time.

In the Hercynian 'geosyncline' east of the Rhine, Dinantian sequences record the existence of a deeper water basin undergoing slow pelagic mud and chert deposition, with periodic influxes of limestone turbidites from the 'Brabant' carbonate plat-

Figure 11.13 Generalised Dinantian palaeogeography of NW Europe and N. America. 1 = Rhenish 'Geosyncline'; 2 = Mittel Deutsche Schwelle; 3 = Cantabria; 4 = Maritime Graben Province; 5 = Appalachian Basin; 6 = Eastern Interior Basin; 7 = Michigan Basin.

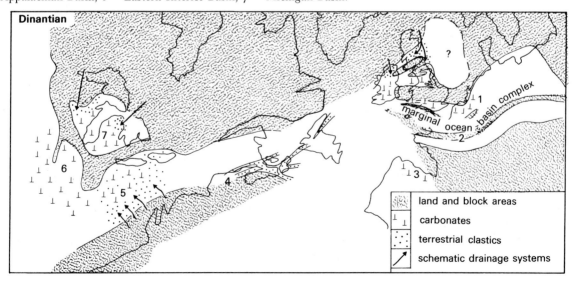

form and from drowned Devonian reef pinnacles within the deep water basin (Meischner 1971). The above discussion has already shown how similar the Devonian Rhenish 'geosyncline' was to parts of the Devonian Culm basin of SW England, and it is evident that these similarities continued into Dinantian times. As Dinantian times proceeded in Germany there were increasing influxes of clastic turbidites from a rising basement high to the south, which were eventually to fill in the whole basin during Namurian and Westphalian times. Again obvious parallels exist with the infill of the Culm basin (Ch. 11j) by turbidites fed from major river channels issuing from the rising Bristol Channel landmass.

In Brittany, coarse clastic facies of fluvial molasse aspect dominated early Dinantian successions. There is considerable evidence that late Devonian (Bretonic) uplift was appreciable, causing a great reduction in the areas undergoing marine sedimentation. A phase of shallow-marine carbonate deposition followed in later Dinantian times, particularly in the Laval Basin (Renouf 1974).

In N. America the Dinantian is roughly equivalent to the Mississippian. In the Maritime Provinces of Canada (Belt 1968) there are well-defined narrow basinal areas containing thick Mississippian sequences that are separated from block areas by a complex system of faults, which have been interpreted as dominantly dextral wrench faults. The narrow sedimentary basins show many of the features attributable to basins formed as pull-apart structures between wrench faults. Thick conglomerates and sandstones of alluvial fan and fluvial aspects interdigitate basinwards with fine-grained mudstones and limestones identical to the Scottish Midland Valley Cementstone facies (Belt et al. 1967*). Marine transgression during Viséan times is recorded in the Windsor Group, containing a suite of carbonate facies, stromatolites and evaporites of shallow-marine, intertidal and sabkha aspects.

In the eastern USA (see Briggs 1974*) there existed a number of large sedimentary basins mostly initiated in earlier Palaeozoic times, such as the Michigan, Appalachian and Eastern Interior Basins. These were separated by positive 'arches' or blocks, which acted as sediment source areas (e.g. the persistent Piedmont area) or as more stable platforms with reduced thicknesses of sediment.

Over 600 m of Mississippian sediments occur in the Michigan and Eastern Interior Basins and more than 1400 m in the Appalachian Basin. For much of Mississippian times, shallow water limestones were the most widespread facies in the eastern USA following the early Mississippian transgression. Persistent influxes of fluvial sediment are seen on the eastern margin of the Appalachian Basin, with rivers draining off the Piedmont block, and on the northern and northeastern flanks of the Eastern Interior Basin, with rivers draining off the Wisconsin−Kunkakee 'arches'. Late Mississippian carbonate−clastic cycles of Yoredale aspect are well developed in the Chester Series of the Eastern Interior Basin, with the periodic delta influxes coming from hinterland sources areas north of the Great Lakes area. A major period of regression occurred over much of the eastern USA at the end of Dinantian times, the stratigraphic implications of which will be discussed in Chapter 11i. Ramsbottom (1973*) has briefly noted, in fact, that the major cycles recognisable in the European Dinantian may also be present in the N. American successions.

11h Namurian (Millstone Grit) rivers and deltas

Sedimentation during Namurian times in the S. Pennines area gave rise to a maximum of over 2000 m of clastic deposits (Fig. 11.14), which are by no means entirely coarse sandstones. The pioneer work of Gilligan (1920), who recognised abundant fresh microcline feldspars and metamorphic heavy minerals within the coarse clastic horizons, led to a hypothesis for the origin of the Millstone Grit as a major river deposit whose hinterlands included extensive metamorphic terrain in the Scottish−Scandinavian Caledonides. Much sedimentological work in recent years has confirmed and amplified Gilligan's conclusions and led to the recognition of a number of environments of deposition. The most significant discovery (J. R. L. Allen 1960; R. G. Walker 1966; Collinson 1969*, 1972) is that turbidity currents issuing from submarine fan channels were responsible for a significant portion of the succession in Derbyshire and Yorkshire (Figs 11.15 and 11.16). Laterally persistent sandstones, such as the Kinderscout Grit, are deduced to be

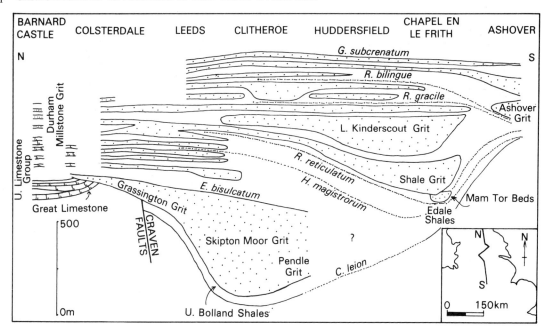

Figure 11.14 Section to show thickness trends and stratigraphy of the Namurian in the Central Pennines Basin (after Ramsbottom 1966). Inset shows line of section.
U. = upper; L. = lower.

truly fluviatile deposits. The sheet-like nature of such members, together with an obvious lack of meandering stream facies, indicates that wide braided channels were dominant. Their frequent floodplain wanderings caused the sheet-like form of the sandstones now seen in outcrop. Some idea of the magnitude of the river channels may be gained from McCabe's (1977) interpretation of certain 35 m sets of cross-stratification as being due to channel bar migration, thus implying at least 35 m water depth during bankfull stages.

Correlation within Namurian Millstone Grit facies is based entirely upon the goniatites that occur in thin marine mudstone bands. These rapidly evolving forms have enabled correlations to be made from Yorkshire to Russia in single marine bands. They thus imply a eustatic origin for the marine transgressions. Figure 11.14 shows that in N. England the location of the greatest thickness of sediments moved southwards during Namurian times. Exceptions to the usual northern or northeastern derivation occur in the Chatsworth and Ashover Grits in Derbyshire, which seem by petrographic and palaeocurrent evidence to have been

derived from the south, presumably from a freshly rejuvenated St George's Land.

North of the Craven Fault system, on the Askrigg and Alston blocks, the early Namurian is still represented by a Yoredale-type facies association (Upper Limestone Group; Fig. 11.14). However, the thin marine limestone members gradually die out upwards, and a thinned Millstone Grit facies develops in most areas.

Very coarse Millstone Grit lithologies are absent from the Midland Valley of Scotland. Namurian deposits of the Limestone Coal Group (60–550 m) are mainly of deltaic clastic facies, with numerous thick workable coals and marine mudstone bands. Volcanism is restricted to the Bo'ness district of W. Lothian and to parts of Fife and Ayrshire. The overlying mid-Namurian Upper Limestone Group is a reversion back to more typical Yoredale cyclothems. As deposition of the late Namurian Passage Group commenced, the sedimentation pattern was changed abruptly by a sudden influx of coarse-grained clastic sediments of fluvial origin, due perhaps to tectonic upwarp at the basin margins. Peat swamps rarely developed in the group,

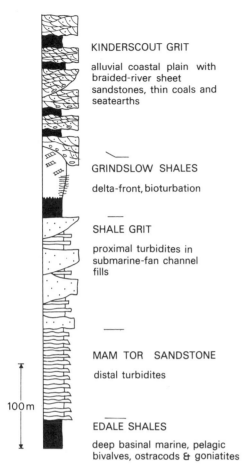

KINDERSCOUT GRIT

alluvial coastal plain with braided-river sheet sandstones, thin coals and seatearths

GRINDSLOW SHALES

delta-front, bioturbation

SHALE GRIT

proximal turbidites in submarine-fan channel fills

MAM TOR SANDSTONE

distal turbidites

100 m

EDALE SHALES

deep basinal marine, pelagic bivalves, ostracods & goniatites

Figure 11.15 Summary log through a Namurian basin infill succession in the Central Pennines Basin (after Selley 1970 and data from references cited therein).

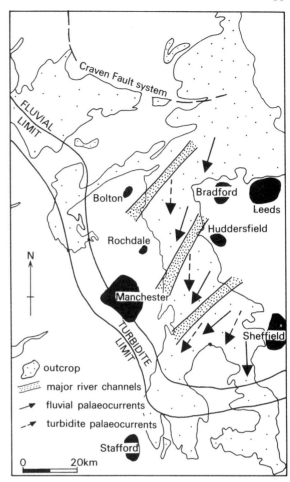

Figure 11.16 Schematic palaeogeography for Namurian R2 times in N. England (from an unpublished compilation of J. Collinson).

presumably due to low floodplain water tables or to excessive sedimentation by river flooding. Periodic and rapid marine incursions are indicated by the numerous marine mudstone bands. In Ayrshire, basaltic lavas make up most of the Passage Group and contain bauxitic horizons, which are also present in equivalent sequences in Arran and in the Sanquhar−Thornhill Outliers where the succession is much thinner (15−25 m).

In S. Wales there is growing evidence for a late Dinantian regressional phase with carbonate sediments being deposited in successively smaller areas. Associated uplift of the southern flanks of St George's Land and the intra-Carboniferous Usk axis caused the early Namurian basin to be a considerably shrunken remnant of the Dinantian basin.

Northern outcrops around the S. Wales coalfield reveal evidence of a sub-Namurian unconformity of variable magnitude. Fluvial input into the S. Wales Basin in early and middle Namurian (E and R zones) times was via S-flowing delta distributary systems in northern areas and N-flowing rivers in the south (Table 11.1; Kelling 1974). This pattern persisted into late Namurian (C2) times and hence into the Westphalian (Ch. 11j).

In Ireland, outcrops of the Namurian are restricted, the chief being in Clare to the north and south of the Shannon Estuary, where Rider (1969) has established a pattern of basin infilling similar to that noted above for the S. Pennines Basin. The sequence contains a lower, bedded, turbiditic sandstone and euxinic black-shale group; a middle, very slumped, dominantly fine-grained transition group; and an upper cyclothemic group. Each group is *c.* 600 m thick, and the whole sequence represents a complete basin full succession. The upper cyclothemic group (cycles 100−130 m), superbly exposed along cliff sections, contains many facies of deltaic origin that have indicated to Rider that the deltas were of birdsfoot type, like the modern Mississippi delta. A spectacular development of sedimentary growth faults in one cyclothem finds parallels in those of the Mississippi and Niger deltas, where the rapid deposition of delta front sands over prodelta muds and silts causes failure along arcuate rotational slip planes. Also present are spectacular sheet slumps due to the movement of material above sedimentary faults. These have preserved sand volcanoes on their upper surfaces.

11i Namurian environments beyond Britain

It has been shown above that Namurian strata in the British Isles record the advent of large new fluvial drainage systems, probably initiated by major uplift of hinterland and basin margin areas in late Viséan to early Namurian times. Similar stories emerge from other European and N. American areas (Fig. 11.17).

In Brittany (Renouf 1974), fresh influxes of fluvial clastics, the Schistes de Laval, poured into the Laval Basin, possibly following fresh uplift of the Mancellian batholith area. In the Ardennes−Rhenish area there was a great northward spread of immature clastic detritus from a southern rising hinterland, the Mittel Deutsche Schwelle, carrying on the trends noted above for the Dinantian. The earliest greywackes are highly feldspathic and still turbiditic, but they grade upwards into Westphalian alluvial and coastal deposits of molasse aspect (Meischner 1971).

In Maritime Canada, Cansoan stage rocks are equivalent to the Namurian of Europe and continue the rift−horst story of Dinantian times. Perhaps the most spectacular scenario in the USA is seen at the end of Chesterian times, with the occurrence of a major regressive phase that has perhaps been best documented in the Illinois section of the Eastern Interior Basin. Following deposition of the Yoredale-like Chesterian Group in this area, a major regression enabled a linear system of SW-flowing streams to entrench themselves into the Chesterian Group (Bristol & Howard 1974). The valleys so produced are up to 32 km wide and 150 m deep. Cross-sectional forms range from narrow and V-shaped to wide and flat bottomed, depending on the nature of the rock types involved. The large

Figure 11.17 Generalised Namurian palaeogeography of NW Europe and N. America. For key and main locations see Figure 11.11.

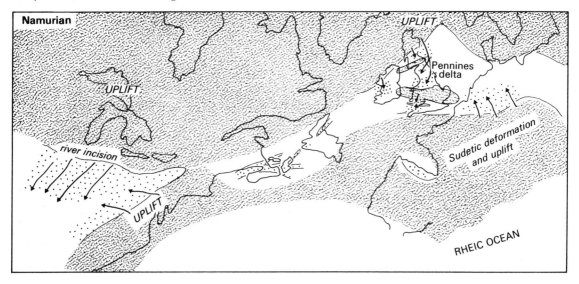

valleys are 160–240 km long, spread 32–60 km apart. Spectacular slump blocks which have slid from valley sides along rotational faults are observed. These palaeovalleys were subsequently infilled with fluvial clastic facies of early Pennsylvanian age. Similar, although less spectacular, evidence for pre-Pennsylvanian erosion is seen in the northern part of the Appalachian Basin.

We may conclude from this brief survey that, over extensive areas of the N. American–European 'shelf' north of the supposed Hercynian continental margin (see Ch. 12d), a major phase of sedimentary regression, probably due to vertical crustal movements causing rejuvenation of upland areas, took place during late Dinantian to early Namurian times.

11j Westphalian (Coal Measures) rivers, lakes and peat swamps

The Westphalian is the major coal-bearing stage in Britain and over much of NW Europe. The major British coalfields and their concealed extensions are shown in Figure 11.18, together with the sub-Permian outcrops of the Coal Measures as revealed by North Sea hydrocarbon exploration. The base of the Coal Measures, which in most areas is conformable upon the Namurian, is taken at the *Gastrioceras subcrenatum* marine band (Fig. 11.19). The upper limit is more problematical, since there is a major, but diachronous, upwards facies change towards 'barren' fluviatile red-beds in many coalfields. Some idea of the continuing crustal subsidence on the Carboniferous shelf is given by the fact that 1060 m of Coal Measures are preserved in the Midland Valley of Scotland, 3050 m in the Lancashire–N. Staffordshire area, 2440 m in S. Wales and 760 m in Kent.

Sections through Coal Measures rocks reveal a variable cyclicity of lithologies. Marine bands are usually fine-grained, dark and carbon-rich mudstones. Some nineteen widespread examples are known, the most complete series being found in the

Figure 11.18 Distribution of coalfields in NW Europe, together with the area of the southern North Sea underlain by Coal Measures, as proven by hydrocarbon exploration (after Calver 1969* and Eames 1975).

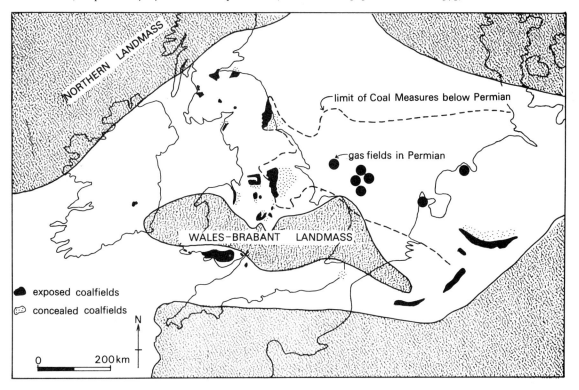

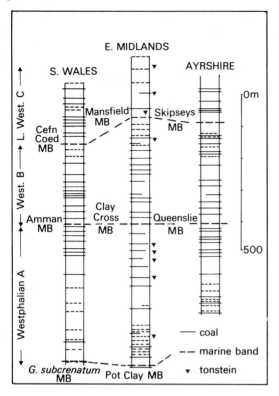

Figure 11.19 Stratigraphic columns for three British coalfields (after Calver 1969*).
G. = *Gastrioceras*

central parts of basins. Many marine bands may be correlated from S. Wales to Ayrshire and southeastwards into Europe; they must therefore represent marine transgression due to eustatic mechanisms. Faunal stages may be recognised within marine bands, corresponding to the advance,

climax and retreat of the marine incursion (Calver 1969).

Other clastic facies may be broadly divided into coarsening-upwards and fining-upwards units, as in the Yoredale cycles (Ch. 11f). The former usually begin with black carbon-rich mudstones containing siderite concretions and non-marine bivalves. The upward coarsening takes place over an interval of 2–30 m. The sequence may be succeeded by a rootlet horizon (seatearth) or coal, or may be cut out at any level by a sharp- and erosive-based, fining-upwards unit. The coarsening-upwards units may be interpreted as due to the gradual infilling of an ondelta lake body by a nearby river or delta distributary channel or by minor crevasse-splay from the partly breached bank of a major channel. A more varied marine fauna in the basal mudstone member possibly suggests that an interdistributary or marine prodelta water body was being infilled. The sharp-based fining-upwards sandstone units (1–20 m thick) sometimes occupy channel forms or define ribbon-shaped outcrops. They usually show cross-stratification, with certain examples showing lateral accretion sets indicative of deposition in meandering stream channels (Fig. 11.20; Plate 3). Such facies are of fluvial or delta distributary-channel origin; they are the 'wash-outs' of the older coalfield literature.

Structureless siltstones with abundant rootlets and occasional *in situ* tree-trunk bases are termed 'seatearths'. They frequently, but not always, underlie coal seams. In a deltaic environment a large number of sedimentary facies may be colonised by plants, including point-bar tops, levees,

Figure 11.20 Lateral variations in Coal Measures facies at Swillington Quarry, near Leeds.

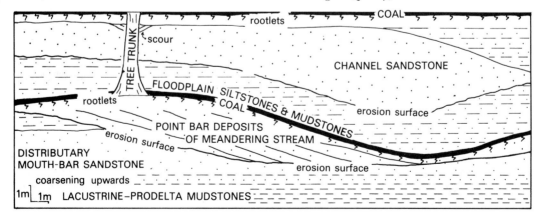

abandoned channel courses, lake margins and low-lying swamplands. Coals, of course, are the compacted and devolatilised remnants of swamp peat. It is important to realise that up to 80 per cent reduction in thickness due to compaction may be involved in the formation of a bituminous coal from peat. All peats seem to require water-saturated and low Eh (reducing) conditions in order to allow for plant preservation and peat development, as opposed to oxidation and bacterial decay.

Tonsteins are mudstones that are low in illite and high in kaolinite. They usually occur as light-coloured, poorly laminated layers less than 10 cm thick. Their microscopic and chemical characteristics indicate volcanic ash origins, and, indeed, many examples may be correlated over large areas.

Regarding the fossil floras found in the Coal Measures, Scott (1977), in his study of Westphalian B successions in the Yorkshire coalfield, has found that, although no one species is restricted to a particular sedimentary facies, each facies can be characterised by variations in plant cover, state of preservation, species association, diversity and variability. For example, prodelta lake clays have low plant cover and diversity with an assemblage dominated by *Lepidodendron* leafy shoots. Distributary mouth-bar coarsening-upwards sequences contain comminuted plant debris with rare *Calamites*, sometimes *in situ*. Swamp coals contain abundant *Lycopod* spores, with *Lepidodendron* and *Sigillaria* macrofloral remains. Floodplain and crevasse-splay deposits, which may give rise to 'roof shale floras' with a very variable and often high degree of plant cover, are dominated by *Neuropteris*, often as fairly large fronds. Scott has concluded that the degree of transport of plant remains after death is clearly the dominant control upon the final make-up of a given fossil floral assemblage.

Turning now to the vexed and much discussed problem of cyclicity in Coal Measures sequences, we may first of all deduce from the above discussion that random or stochastic processes must be a dominant control upon the kinds of sedimentary processes that lead to the formation of a cycle. River or distributary channel positions were the dominant controls upon subenvironments of deposition, these being controlled by random factors. Once established in a given area, however, the advance of a delta into a lake body would give rise to a predict-able vertical sequence of facies. The depth to which the channels cut would then control the completeness of a cycle (Fig. 11.20). Attempts to define an 'ideal' cycle thus seem to be pointless. In fact, Duff and Walton (1962), defining cycles as intervals between successive rootlet horizons in their statistical analysis of 1200 cycles, have shown that there are 320 different kinds of cycles. All that we may legitimately do in such cases is to define modal (most frequently occurring) cycles. The concept of rigorously identifying facies types as an aid to cycle analysis is a relatively recent trend, and it should have a beneficial effect upon the problem of cyclicity when individual cycles can be correctly interpreted in terms of depositional environments.

Generalised sequences of British Westphalian rocks are shown in Figure 11.19. The features common to all regions are: (a) a lower group of marine bands with thin coal seams and sparse non-marine bivalve horizons, although the *Gastrioceras subcrenatum* band is not recognised in Scotland; (b) a rarity of marine bands in top Westphalian A and lower Westphalian B times, with the development of the thickest and most numerous, economically valuable coals; and (c) an upper series of marine bands and thin coals below the Top Marine Band that marks the final marine incursions during Westphalian C times. Above the Top Marine Band, fluviatile red-beds indicative of well-drained and oxygenated floodplains appear in Scotland and along the northern margin of St George's Land. Some of the reddening in Scotland and in other areas of N. Britain is attributable to secondary oxidation from below the sub-Permian unconformity. The volcanic episodes in Midland Scotland during Coal Measures times should not be forgotten. Westphalian tuffs occur off the Fife coast in the Firth of Forth. The lavas that occur in Ayrshire and in the Sanquhar−Thornhill Basins are now thought to be Permian in age. More than sixty necks, probably feeders to these volcanics, occur, around the Mauchline Basin.

South of St George's Land, coal-forming environments continued after Top Marine Band times, although thick fluvial sandstone members of the Pennant Group dominate in S. Wales. In fact, the S. Wales coalfield basin is quite distinctive in many ways since, as in Namurian times, rivers drained into the area from both south and north. N-flowing

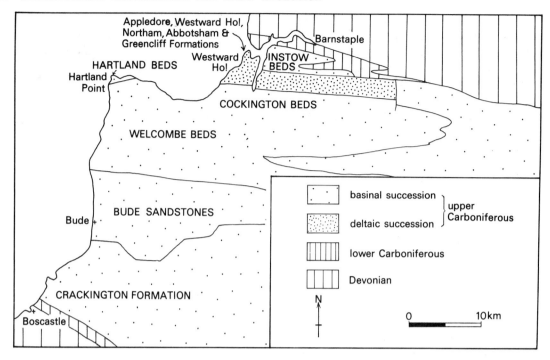

Figure 11.21 Distribution of basinal turbidite and deltaic coastline facies in the Culm synclinorium of SW England (after Burne & Moore 1971).

Table 11.3 Major stratigraphic units defined in the Namurian and Westphalian of the Culm province in Devon (after Burne & Moore 1971). L. = lower, M. = middle, U. = upper. Shading indicates absence of strata.

SERIES	GONIATITE ZONES	BASINAL SUCCESSION					DELTAIC SUCCESSION		NON-MARINE BIVALVE ZONES
		WEST COAST	BARNSTAPLE–BIDEFORD REGION						
			Westacott Cliff	Cockington Beds	Tawstock–Fremington	Instow Beds	Westward Ho!	Appledore Formation	
WESTPHALIAN	G₂	Bude Sandstones (700m)					Greencliff Formation (50m)		*Lenisulcata comm.*
		Hartland Beds (140m)					Abbotsham Formation (360m)		
							Northam Formation (430m)		
		Welcombe Beds (640m)					Westward Ho! Formation (500m)		
			Cockington Grits (100m)		upper (120m)		U. Sandstone Member (c.500m)		
		Black Mudstone Beds (100m)	Mudstones (30m)		middle (80m)		M. Mudstone Member (c.35m)		
	G₁		Cockington Flags (100m)		lower (150m)		L. Sandstone Member (c.200m)		
NAMURIAN	R	Crackington Formation (c.300m)			Limekiln Beds (?)				

SERIES — WEST COAST; *GONIATITE ZONES*; *NON-MARINE BIVALVE ZONES*

rivers dominated sedimentation by late Westphalian times, with much of S. Wales being
covered by alluvial plains. Kelling (1974) has found
that the Rhondda Beds sandstones contain a variety
of rock fragments, including spilitic and phyllitic
clasts of obvious southern provenance. In fact, these
late Carboniferous facies may be considered as
N-directed molasse, flowing in from a rising southern
landmass, the Bristol Channel portion of which was
active during Old Red Sandstone times (Ch. 9b).

A large area of upper Carboniferous (Fig. 11.21)
rocks, previously called the middle and upper
'Culm measures', outcrops in the Culm basin in SW
England (Fig. 10.1). Recent detailed mapping and
correlation by Burne and Moore (1971), using rare
goniatite horizons and non-marine bivalves, have
established the existence of two major contrasting
successions (Fig. 11.21, Table 11.3).

In the Westward Ho! area over 2 km of clastic
mudstones, siltstones and sandstones occur. In
classic detailed analysis of the Northam and Abbot-
sham Formations, de Raaf et al. (1965) distinguished six distinct sedimentary cycles; each cycle
passes gradually upwards from black mudstones
into coarser beds, the tops of which are sharply
separated from the black mudstones of the succeeding cycle. Turbiditic sandstones occur in the silty
mudstones above the black mudstones. The topmost sandstone members of two cycles comprise
erosive-based fining-upwards cycles which are succeeded by finer sediments with non-marine
bivalves, but coals are notably absent. As shown in
Figure 11.22, these major cycles are interpreted as
regressive sequences due to delta advance into moderately deep basins, the delta front area being
fringed by low gradient cones receiving occasional
turbidity currents due to high river discharge or to
delta front slumping. Palaeocurrents indicate a
northern derivation, most likely from the Bristol
Channel landmass that was also shedding
N-flowing rivers into the S. Wales area at the same
time. The onset of active fluvial clastic input into the
area, following the quiescent basinal early Dinantian sequences discussed above, suggests major
Namurian uplift in the Bristol Channel hinterlands.

The second major type of depositional framework
is seen mainly in the upper Carboniferous outcrops
south of Bideford, Devon (Fig. 11.21), named by
Burne and Moore (1971) as basinal successions.

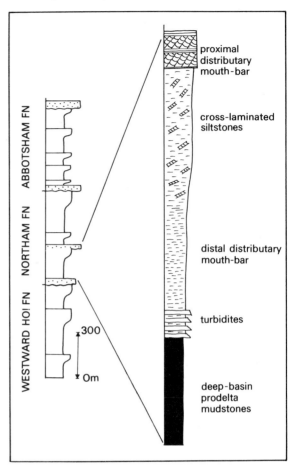

Figure 11.22 Coarsening-upwards cycles of deltaic
origin from the Culm synclinorium (after Elliot 1976, de
Raaf et al. 1965 and R. G. Walker 1970).

Here, alternations of mudstone and turbiditic sandstones are characteristic, in a sequence more than
2 km thick. Succeeding the Dinantian cherts and
mudstones is the Crackington Formation, a
Namurian sequence of turbiditic sandstones and
mudstones showing palaeocurrent evidence for
transport towards the west and southwest. The
overlying Black Mudstone, Welcombe and Hartland
Beds (total c. 900 m thick) again comprise alternations of graded sole-marked turbiditic sandstones
with black mudstones and other mudstones yielding
goniatites. Palaeocurrents indicate a dominantly
eastward transport direction. The overlying Bude
Sandstones comprise a variety of both proximal and
distal turbidites, together with interbedded mudstones and siltstones. There is palaeocurrent

evidence for southward transport in the turbiditic sandstones and E-travelling bottom currents in the interbedded non-turbiditic siltstones. Burne (1969) has proposed a model based upon the former presence of large submarine channels in a fan complex to explain the various facies present in the group. It is notable that marine faunas are present only at the very top of the Bude Sandstones, and that the presence of the freshwater *Planolites* and limulid (king-crab) tracks indicates a truly fresh- or brackish water environment for much of the time.

Faunal evidence (Table 11.3) indicates that the deltaic and basinal successions discussed above are, in part, time equivalent. Much of the basinal detritus was probably introduced into the Culm basin by turbidity currents issuing from submarine fan channels that cut the northern delta slope areas.

11k   Coal Measures environments beyond Britain

In Europe (Fig. 11.23) there were two distinct types of Coal Measures basins. Evidence from the Polish, Russian, German, Dutch, Belgian and N. French coalfields indicates that river and delta systems were fed from hinterlands to the south. As previously shown, northward clastic input, often turbidi-tic, from a rising Hercynian orogenic belt was a continuing feature of both Dinantian and Namurian times. The Coal Measures in these areas record the final pulsatory infilling of these so-called Hercynian foredeep areas to sea level, the term 'pre-orogenic molasse' being applied to such facies. The Namurian–Westphalian deposits of Devon, S. Wales and Kent were also situated near the northern boundary of the rising Hercynides, the succession in the Culm basin of Devon being the nearest British analogue to the successions seen in the Rhenish basins of Carboniferous age in N. Germany. In many of the above coalfields marine bands may be traced over vast areas, indicating that periodic marine transgressions over the low-lying Coal Measures alluvial plains were almost certainly of eustatic origin.

In contrast to the above basins, there also existed depositional basins within the uplifted Hercynides that contain no evidence for marine incursions. Examples include the Saar and Walbrzych coalfields (Fig. 11.23) and the coarse clastic deposits seen preserved in synclines within the basement massifs of the Alps (e.g. Salvan Syncline). These clearly represent intermontane basins and are analogous to the internal molasse basins discussed above with reference to the Caledonide orogenic belt in Old Red Sandstone times. Up to 5 km of this

Figure 11.23   Generalised Westphalian palaeogeography of NW Europe and N. America. For key and main locations see Figure 11.13.

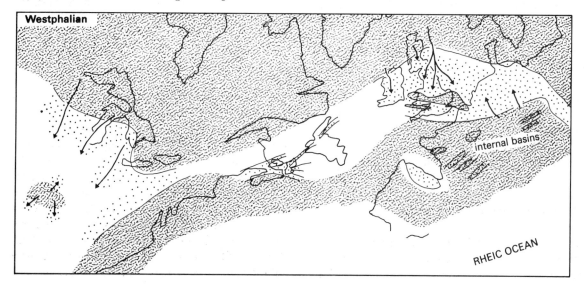

molasse sediment accumulated in the Saar Basin, with locally developed coals up to 10 m thick.

In Maritime Canada, the graben system noted above for Dinantian times was present through the early upper Carboniferous. A period of deformation occurred during Westphalian B (Cumberlandian) times. This was followed by the influx of very coarse fluvial facies over basin and block areas alike during Westphalian C to Permian times. By the end of the Carboniferous, the early Carboniferous graben system had lost much of its identity.

In the eastern USA, early Pennsylvanian deposits include the fluvio-deltaic and beach−barrier deposits of southern and northern clastic systems feeding a marine area centred on northern W. Virginia, E. Ohio and Pennsylvania (Fig. 11.23). The northern fluvial system was dominant in the Michigan Basin, draining a Great Lakes hinterland. Pennsylvanian deposits in the Eastern Interior Basin comprise more than 800 m of fluvio-marine cycles, fifty-one separate delta advances being recorded in all. In the Appalachian Basin, the southern fluvial systems advanced northwards in middle to upper Pennsylvanian times to cover W. Virginia with ondelta and alluvial facies in which coals were formed. Separate delta lobes advanced into Ohio at the same time. Up to 1800 m of Pennsylvanian rocks were deposited in Kentucky, although thicknesses usually range from 200 to 700 m in the Appalachian coalfield areas. The Pennsylvanian is clearly thickest and coarsest grained along the eastern side of the Appalachian Basin, indicating a major eastern hinterland in addition to the northern Great Lakes hinterland mentioned above.

This brief survey illustrates the tremendous persistence of coaliferous facies in upper Carboniferous times. Regional upwarp during early Namurian times, with the establishment of major fluvial drainage systems in northern and southern hinterlands, continuing subsidence and a hot humid climate (cf. Dinantian times), must all have contributed towards the preservation of such sequences.

11l The Carboniferous climate

During Old Red Sandstone times, a hot semi-arid climate prevailed over much of the N. Atlantic area, and the latitude of the British area was approximately equatorial (Ch. 9a). The presence of cornstones in early Dinantian fluvial red-beds in Scotland (Ch. 11e), together with the palaeomagnetic data of Turner and Tarling (1975), clearly indicates that these climatic and latitudinal conditions persisted into the Dinantian. The distinctive sabkha-type anhydrites noted previously from several Dinantian sequences support this conclusion and further, by analogy with modern sabkhas, indicate mean annual temperatures in excess of 30°C. Sabkha-type anhydrite has recently been discovered in the Belgian Dinantian also.

It has long been surmised that the increasing prevalence of coal-bearing deltaic facies in later Carboniferous times indicates an increasingly humid tropical climate over much of NW Europe. Peat swamps require a more-or-less continuously water-saturated environment in which to form; otherwise oxidation and bacterial destruction of organic material will predominate. Support for increasingly humid tropical climates comes from the occurrence of truly bauxitic soils developed within Namurian lava sequences in Ayrshire and Sanquhar, Scotland. The upper Carboniferous humid tropical climate is traditionally thought to have ended in late Westphalian to Stephanian times, when the evidence of red-beds and oxidised coal seams indicates the onset of the lengthy Permo-Triassic arid period in NW Europe (Ch. 13).

11m Subsidence, volcanism and mineralisation on the Carboniferous shelf

Basin and block areas persisted throughout Carboniferous times from their apparent initiation in late Devonian to early Carboniferous times, although their effects upon subsidence rates decreased somewhat with time in some areas. Dinantian block/basin boundaries may be recognised by distinctive facies changes and thickness trends. Many examples of syndepositional faults and hingelines are known at the margins of basins, the great majority being normal tensional features with NE−SW Caledonide trends. It is thus clear that a regional NW−SE-directed tensional-stress regime was operative over the Carboniferous shelf. Some modifications to this basic theme of crustal subsidence under a tensional stress regime must have occurred in Namurian times, since the spread of detritus from the Bristol Channel landmass, St George's Land and the Scotto-Scandinavian landmass indicates uplift and rejuvenation of these hinterland areas.

Alkali basalts, and rarer peralkaline lavas and intrusives, were particularly important in the Midland Valley of Scotland and in the Northumberland Basin. In both areas there is an upward change towards dominantly pyroclastic activity. Recently, Macdonald *et al.* (1977) have proposed that the Carboniferous—Permian alkaline magmatism was related to two thermal cycles, each of which produced a similar trend of magma compositions, towards increasing silica undersaturation, with time. One cycle was early Dinantian to early Namurian in age, the other middle Namurian to early Permian. The cycles are tentatively thought to represent mantle partial melting at progressively greater depths with time. Such a model, however, still leaves unexplained the important late Carboniferous quartz dolerite suite (Ch. 12b). As noted above (Ch. 11e), the Midland Valley volcanics had a profound effect upon sedimentary environments and palaeogeography. In the Northumberland Basin the volcanics are concentrated along the northern Scottish margin and seem to have been initially extruded just prior to basin initiation (Leeder 1974b). Elsewhere, volcanism was generally of less importance, but notable occurrences are seen in the Isle of Man, Derbyshire and around Limerick, Ireland.

The numerous base metal deposits, believed to be of syngenetic or early diagenetic origin, that have recently been discovered in the Irish Waulsortian mudbank province (Ch. 11d, Fig. 12.2) have led that country to the top of the European league of producers. The close link between mineralisation and lithology in many deposits has led to the theory of mineralisation based upon ascending hot brines carrying the metals to receptive early-diagenetic hosts within or adjacent to the mudbank complexes. Russell (1972) has suggested that since some of the major deposits lie along N—S lines, which he has termed 'geofractures', the mineralisation may have resulted from brine circulation over hotspots located where NE—SW-trending faults intersect the postulated geofractures. This is still, however, a controversial hypothesis. The origin of basement subsidence and volcanism on the Carboniferous shelf will be discussed in Chapter 12.

11n Summary

A major transgression during late Devonian to early Carboniferous times flooded the southern margin of the Old Red Sandstone continent. Block and basin areas were quickly established, with many basins taking the form of geomorphic gulfs during the early Dinantian. Carbonate sediments dominated Dinantian deposition in most areas, with major, eustatically controlled cyclicity responsible for regional facies changes. Extensive lime-mudbank accretion occurred in many basinal areas, with a major phase of associated 'stratiform' lead—zinc mineralisation in Ireland. Northern areas show evidence of repeated introductions of fluvio-deltaic clastic detritus and abundant alkali-basaltic volcanism. An extensive hinterland-uplift phase in late Dinantian times, recognisable from Silesia to Illinois, was followed by the copious introduction of clastic detritus, seen in the English Pennines as the Namurian Millstone Grit. By upper Carboniferous times an enormous, low-lying fluvio-deltaic plain lay over much of NW Europe, causing many coal-bearing cycles to form and establishing the raw materials needed for the future industrial prosperity of the area.

12 The Hercynides: a problematic orogenic belt

12a Introduction

The last three chapters have traced the evolution of the British Isles and adjacent areas in the 100 Ma or so subsequent to the major final deformation of the Caledonides. The task of this chapter is to examine the nature of the Hercynian Orogeny, first in the British Isles and then in Europe and N. America. We will then be in a better position to examine critically some of the numerous plate-tectonic schemes that have been proposed in recent years to account for the Hercynides.

12b Structure and plutonism in SW England

The basic structural pattern of SW England is a complex synclinorium pierced along a NE−SW axis by a major granite batholith. Much structural analysis in recent years has enabled a reasonably coherent regional deformation pattern to emerge. Sanderson and Dearman (1973*) have defined a series of twelve structural zones between N. Devon and S. Cornwall (Fig. 12.1). This complex zonation is due both to the original variation of primary structures and to later refolding events. The structural zones may be divided into two major groups. Northern zones 1−6 in the Culm trench show a divergent-facing trend centred on the upward-facing upright folds of zone 2. Southern zones 7−12 contain primary folds facing the north or northwest. These two major zones are separated by a zone of facing confrontation. The southern folds that face north have a slaty cleavage that dies out northwards. The beds north of Polzeath were not affected by this early phase of deformation. These unfolded beds were later deformed into S-facing recumbent folds, the local F_1 phase, with a well-developed axial planar slaty cleavage. This later deformation refolded the earlier N-facing recumbent folds, where

the northern slaty cleavage is developed as a crenulation cleavage.

Stratigraphic evidence can only suggest times of folding of later than upper Devonian and Westphalian for the early and late fold phases respectively. K−Ar dating of micas in the slates (Dodson & Rex 1971*) has given valuable evidence for minimum ages of mica recrystallisation during cleavage developments (Fig. 12.1). South of the line of facing confrontation, away from areas reheated by granite intrusion, ages are mostly in the range 340−320 Ma. North of the line these are significantly different, being in the range 300−280 Ma. Thus there is evidence for major deformation phases at the end of Dinantian times (Sudetic phase) and during late Carboniferous times (Asturic phase). In addition, in S. Cornwall there are K−Ar ages of 365−345 Ma, indicating a late Devonian (Bretonic) event in the area. These figures indicate that the uplift of areas associated with deformation occurred at progressively later ages towards the north in SW England.

The status and significance of the Lizard Complex are of major significance, since every plate-tectonic scheme has proposed the complex as a thrust ophiolite mass. The complex comprises the metamorphosed muds, sands, tuffs and basalts of the Old Lizard Head Series and Landewednack Hornblende Schists. The Man-of-War Gneiss appears to have intruded the Old Lizard Head Series in a migmatitic fashion. The largely serpentinised peridotite mass has mainly faulted boundaries and was further intruded by a large mass of gabbro and gabbro dykes. Basalt dykes swarms cut both the gabbro and serpentinite, and younger acid injection gneiss and non-foliated microgranites occur in the serpentinite, marking the final significant events in the complex. A number of K−Ar dates from the various members of the complex cluster in the range 371−350 Ma ($n = 6$), the three oldest dates being 391, 442 and 492 Ma. It is reasonable to

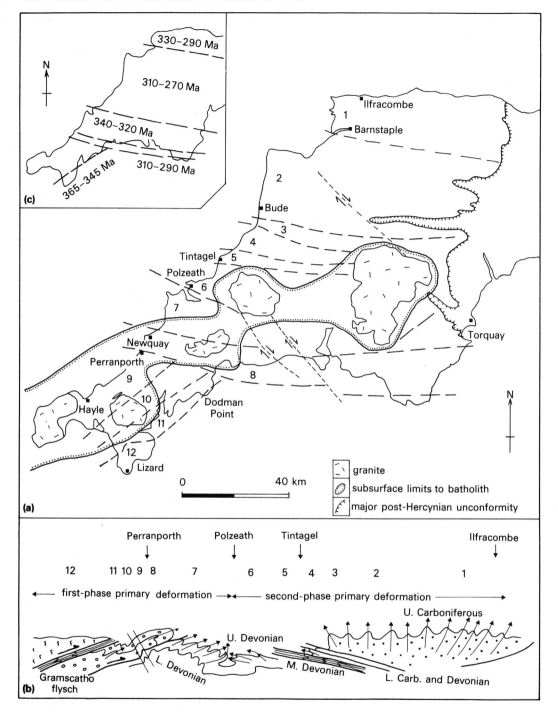

Figure 12.1 (a) Structural zones defined in the Hercynian fold belt of SW England. (b) Structural sketch section. (c) The significant groupings of K−Ar ages within the area (after Sanderson & Dearman 1973* and Dodson & Rex 1971*). L. = lower, M. = middle, U. = upper.

assume that the youngest dates represent only the last igneous, metamorphic or structural events to affect the Lizard. Miller and Green (1961) have noted that the oldest ages may be due to the sampled hornblende schist and hornblende granulite being partly insulated from late thermal effects, and that they may reflect partial argon loss from Precambrian or Cambrian minerals. It thus seems possible that part of the Lizard is a portion of late Precambrian basement. Indeed, the Landewednack Hornblende Schists have been compared to the gneisses and hornblende schists of the Sark (Channel Isles) Cadomian. Tectonic emplacement of the Lizard Complex during late Devonian times need not have involved much northward travel, since large areas of strongly magnetic basement lie at shallow depths in the English Channel off SW England.

Recent geochemical results (Floyd *et al.* 1976) tend to rule out a simple comagmatic oceanic crustal origin for the various members of the Lizard Complex. Three distinct geochemical trends exist for (a) the Landewednack Hornblende Schists, (b) the gabbro and associated dykes, and (c) the pillow lavas and other lavas. All the basaltic units are tholeiitic or subalkaline, the hornblende schists being the only group to show affinities with ocean floor basalts. As noted above, it is precisely these rocks that give the 'old' ages. Clearly, the Lizard shows a complex, and perhaps lengthy, pre-Hercynian history.

Undoubtedly the major economic feature in the area is the great Cornubian granite batholith, cupolas of which now rise above the main mass of subsurface granite (revealed by gravity surveys) to outcrop as separate plutons. The granites, actually potassium-rich adamellites, mostly yield K–Ar ages of 280–270 Ma, i.e. of clearly Permian age. The batholith is a multiple intrusion, with sharp outward contacts against hornblende-hornfels aureole rocks. Seismic refraction data indicate that the granitic material persists to a depth of 12 km, where it grades into lower crustal material of presumed intermediate composition. Since the granite cuts the primary folds and also some of the late phase folds, it is evident that primary deformation was complete before emplacement. Beds showing an intense F3 deformation lie close to and above the batholith and have a ubiquitous flat-lying crenulation cleavage, generated in response to the

flattening deformation due to the rise of the batholith.

Mineral deposits associated with the batholith include huge china-clay deposits and the tin–copper–tungsten veins that follow fractures within the granites and country rock, the areas of most intense mineralisation being related to positive culminations in the batholith roof. It should be noted that similar, although smaller scale, mineralisation and hydrothermal alteration to that seen in SW England characterises many granites of Hercynian age in Brittany and Portugal. It probably follows that the Hercynian lower crust, the probable source through selective fusion for the acidic magmas, was peculiarly rich in elements such as lithium, boron, fluorine, tin and tungsten. It is also a fact that many of the pre-Hercynian granites in the area are also rich in boron, since abundant detrital tourmalines in Mesozoic sediments from the Weald (P. Allen 1972) are derived from the west and give many pre-Hercynian dates.

12c Block–basin tectonics north of the Hercynian thrust front

The Hercynian thrust front is thought to cross S. Ireland, Dyfed and Somerset before it is lost in the subsurface south of the London–Brabant platform, reappearing in Holland, Belgium and N. Germany. However, Mathews' (1974) scepticism about the true lateral persistence of this feature should be noted. He has reasoned, for example, that the thrust front is not present in SW England and that it represents different things in different places to different authors!

In general, north of the supposed thrust front, basinal successions of Old Red Sandstone and Carboniferous are much faulted but deformed into upright folds with only local cleavage development. Block successions, protected and cushioned as it were by shallow rigid granitic basement, are usually only gently tilted, with normal faults being predominant (Fig. 12.2). Many of the previous syndepositional normal faults and hinges at basin margins were rejuvenated during deformation, sometimes acting as reverse faults or thrusts. The major structures in the area – e.g. the Pennines Anticlinorium and S. Wales Synclinorium – were

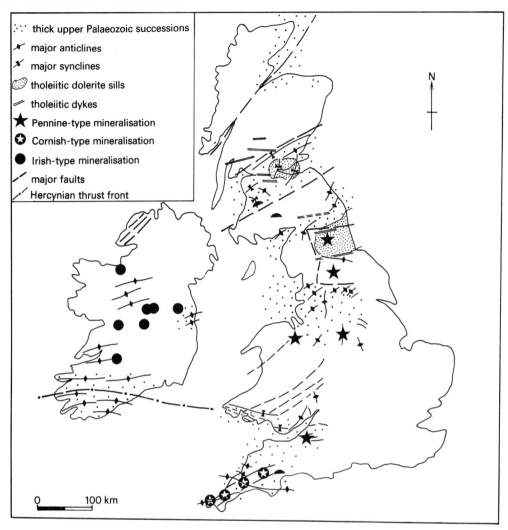

Figure 12.2 Major Hercynian structural features, Permo-Carboniferous mineralisation and tholeiitic intrusions in the British Isles.

produced by the late Carboniferous deformation. Although it may be correct to say that E−W compression produced the major Pennines structure, the local interplay of blocks and basins 'shuffling around' gave rise to many local variations of fold and fault style. It may be noted here that there is no sign of the late Devonian deformation phase in the area, although the middle Devonian (Acadian) event, which is so prominent in eastern N. America, produced major strike-slip movements along the Great Glen Fault and its correlatives in Ireland and the Shetlands and caused minor deformation in the Midland Valley.

Lead−zinc−fluorite−barite mineralisation in veins crosscutting through Dinantian and Namurian strata is a prominent feature of the Pennines area. Such mineralisation, which is mostly post-Westphalian to pre-Permian in age, differs from the Irish-type deposits discussed previously, since (a) it is significantly younger, (b) it is demonstrably epigenetic, and (c) fluorite is abundant. The deposits were formerly regarded as magmatic in origin, with subsurface gravity highs postulated as due to the feeder plutons of Hercynian age. However, the historic Rookhope deep borehole proved the pre-Carboniferous age of the Weardale Granite and made such ideas untenable. More recent fluid-inclusion and isotope studies indicate a lower temperature (<150°C) origin for the mineralisation, with evidence for metal origin in leaching processes due to circulating connate brines heated by geothermal systems.

The tholeiitic magma made available in late Carboniferous times was intruded to form the Whin and Firth of Forth Sills and a prominent suite of E−W-trending dykes (Fig. 12.2). They have been dated by K−Ar methods at *c.* 295 Ma and signify a major change in magma chemistry and crustal stresses at this time (cf. Ch. 11). Francis (1978) has recently drawn attention to a swarm of similar-aged tholeiitic dykes trending ESE−WNW and SE−NW in S. Norway and S. Sweden. He has suggested that these may be continuous with the N. British examples under the North Sea.

12d The Hercynian Orogeny in Europe and N. America

The highly deformed SW England area forms part of a linear WSW−ENE fold belt stretching from Texas to beyond Czechoslovakia, when the N. Atlantic Ocean is closed up to allow for Mesozoic sea-floor spreading (Fig. 12.3). This fold belt comprises the Ouachita belt in Texas, the Alleghenian belt in the Appalachians, the Mauritanian belt in NE Africa and the Hercynian belt in Europe (Fig. 12.4). Syntheses involving the whole of this orogenic zone are made difficult by later Mesozoic to Cainozoic cover in Texas, by the superimposition of Alleghenian events upon earlier orogenic phases in the Appalachians, by the lack of data for the Mauritanides and by Alpine overprinting of Hercynian events in S. Europe.

In Europe, Stille originally recognised three major periods of earth movements that affected the Upper Palaeozoic successions discussed in the last three chapters: the Bretonic (end-Devonian), Sudetic (end-Dinantian) and Asturic (end-Westphalian) phases. Over much of Europe south of the Hercynian thrust front, the Bretonic phase, where it can be identified, seems to have been dominantly an uplift phase, although deformation of this age is well documented in the southern part of the Rhenish Schiefergebirge and in the Bohemian Massif. Similarly, there are some areas, such as Armorica, where a Sudetic phase cannot be readily identified; but it seems that extensive basement regeneration, granite intrusion and low-pressure high-temperature metamorphism occurred at about this time over much of the Moldanubian, Saxo-Thuringian and Pyrenean−Iberian areas. In addition, as already noted, there is widespread evidence for vertical crustal uplift in Europe and N. America in late Dinantian to Namurian times. Gravity nappes are supposed to have slid southwards from a rising dome of crystalline rocks in the Montagne Noire at this time. In the Pyrenees, elongate domes with metamorphic rocks are overlain by a non-metamorphic cover of deformed Devonian to early Carboniferous rocks. Stephanian sediments rest unconformably upon these folded sequences. Zwart

Figure 12.4 The traditional subdivisions of the Hercynian orogenic belt in N. Europe (partly after Read & Watson 1976).
For sections along A−B, see Figure 12.7.

Figure 12.3 The extent of the Hercynian orogenic belt, on a pre-Mesozoic reconstruction of the N. Atlantic area (after Riding 1974).

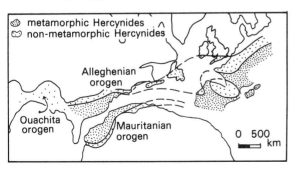

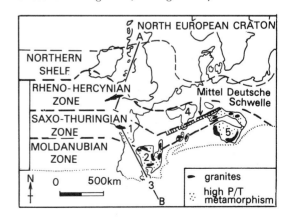

(1967) has concluded that Hercynian metamorphism is dominantly of low-pressure high-temperature type, enacted under a small overburden but with high geothermal gradients, possibly of $150-200°C$ km^{-1}. The later Asturic phase was the terminal end-Carboniferous deformation phase over much of Europe. Metamorphic effects at this time appear to have been slight, although very extensive post-kinematic plutonism occurred over SW England, Armorica and Iberia. The discontinuous northward thrusting of the Hercynian front occurred at this time, with the Ruhr−Ardennes coalfields being tightly folded along ENE−WSW lines. In the Cantabrian mountains of N. Spain, it has been generally accepted that in the upper Carboniferous there were four periods of sedimentation separated by three phases of deformation. However, Reading (1975) has recently proposed that sedimentation and deformation were probably synchronous in adjacent basins, the sedimentary evidence pointing to an origin for the small basins as pull-apart grabens in a strike-slip fault area. Recent strain measurements in Armorica and Iberia have indicated to Ries and Shackleton (1976) that the prominent Iberian-−Armorican 'arc' (Fig. 12.3) is a secondary arc produced by a counterclockwise rotation of the Iberian Peninsula relative to Brittany. In Brittany, Sibuet (1973) has recently identified the S. Armorican shear zone as a major late Hercynian fracture, with a sinistral displacement of possibly hundreds of kilometres. This is roughly parallel to the dextral Biscay wrench and associated faults proposed by Arthaud and Matte (1977*). Such strike-slip faults have a bearing upon certain plate-tectonic schemes for the Hercynides (see below).

In N. America there was extensive reactivation of Lower Palaeozoic basement during the Alleghenian event, with deformation of the Upper Palaeozoic pile and NW-directed thrusting. The Alleghenian front can be traced northeastwards along the western side of the S. Appalachians into Pennsylvania. The front, poorly defined, however, can be traced into New Brunswick and hence into Newfoundland, where recent whole-rock Rb−Sr isochrons indicate a number of Alleghenian−Hercynian granites (325−300 Ma) in the eastern part of the island. The Ouachita belt strikes E−W parallel to the present Gulf of Mexico shoreline and is thought to turn sharply northwards to join up with the Appalachian belt. In the Ouachita Mountains of Oklahoma and Arkansas the sedimentary basins have deposits stretching back to the Cambrian. Cambrian−Devonian sequences of mudstones and cherts (2 km) are followed by thick Carboniferous turbidites (up to 10 km). The late Pennsylvanian deformation caused arcuate folds and N-directed thrusts with throws of up to 80 km.

12e   Plate-tectonic models for the Hercynides

The notion that the Hercynides and their western correlatives resulted from plate-tectonic causes is now a popular one. Since 1971 more than thirteen models have been proposed for the orogen. Let us briefly examine the major reasons why plate-tectonic schemes have been applied and try to rationalise the current models in the literature.

First, palaeomagnetic evidence suggests wide separation of Gondwanaland from Pangaea (Euro-America) in the Lower Palaeozoic (Fig. 12.5). The large ocean, termed the 'Rheic Ocean', separating these two continental agglomerations gradually closed up during the Upper Palaeozoic by the rotation of Gondwanaland and Pangaea along a broad zone roughly coincident with the Alleghenian−Hercynian orogenic belt (A. G. Smith *et al.* 1973). The palaeomagnetic data thus provide a *prima facie* case for ocean closure, but the almost complete absence of data from critical areas such as SW England, Brittany, Spain, NW Europe, N. Africa and the Alpine area leaves room for considerable geological speculation as to the exact nature of the closure.

Secondly, the distribution of faunas in Lower Palaeozoic rocks (e.g. Burrett 1973) has provided evidence to some authors for a mid-European ocean (Fig. 12.6) between S. Europe (Spain, France, S. Germany, Czechoslovakia) and N. Europe (Britain, N. Germany, Scandinavia). However, there are now some doubts about the existence of such faunal provinces. Ager (1975) has pointed out contradictions in the faunal data and suggested that a single broad biotic province extended from Europe to NW Africa during Upper Palaeozoic times. Similar conclusions have been deduced by A. Williams (1976) from Ordovician brachiopod distributions.

Various authors have used the above points in

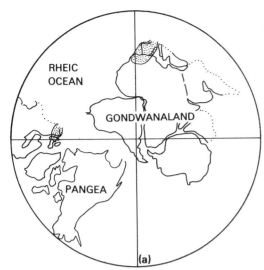

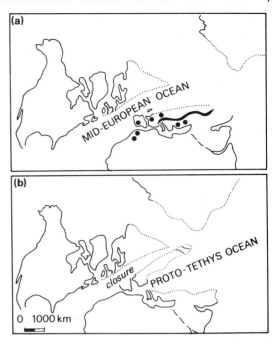

Figure 12.6 G. A. L. Johnson's (1973*) concept of a mid-European ocean during (a) Devonian and (b) Carboniferous times.

combination with their interpretations of the geological evolution of the Hercynides to provide plate-tectonic models, which may be broadly classified as: (a) mid-European Ocean models (Fig. 12.6); (G. A. L. Johnson 1973*, Burrett 1972, Burne 1973, Lorenz 1976); (b) proto-Tethyan Ocean margin models (Nicolas 1972, Floyd 1972*) with marginal back-arc basins (Reading 1973, Leeder 1976b, A. V. Bromley 1976); (c) combination (a) and (b) models (Laurent 1972, Dewey & Burke 1973*); and (d) combination models with microplates (Riding 1974) and oblique collisions (Badham & Halls 1975, Badham 1976, Arthand & Matte 1977*).

Many of these models locate part of the suture zone between the Pangaean and Gondwanaland plates somewhere between Brittany and the Culm trench of SW England, the Lizard Complex assuming great importance as a possible thrust ophiolite sheet. Let us now see how the geological evidence from these areas can help to locate a possible suture zone.

There is gathering evidence for a common metamorphic and structural history for the late Precambrian of S. Britain and of the Armorican area, with the Cadomian event being particularly wide-

Figure 12.5 Southern hemisphere distribution of continental masses during (a) Devonian and (b) Carboniferous times (after A. G. Smith *et al.* 1973). Note: the distribution of continental areas immediately adjacent to the Rheic Ocean (e.g. Iberia, N. Africa, Florida) is not known in detail.

spread (Ch. 2d). That the area was 'intraplate' in
Lower Palaeozoic times is supported by the similarity of lower Cambrian trilobites and of Ordovician
brachiopods throughout Europe, and by the continuity of the Arenigian transgressive quartzites of
Brittany, N. Spain and possibly N. Africa and of the
Llandeilan quartzites and their trilobite faunas
from S. Cornwall (Ch. 10d) to Morocco. More prob-
lematic is the question of whether the Arenigian
quartzites in Europe are the true lateral equivalents
of such units as the transgressive Stiperstones
Quartzite of the Welsh Borders and the Garth Grit
of N. Wales. Such correlations would tend to restrict
a possible suture zone well north of SW England in a
geologically unreasonable position. The influx of
fluviatile facies of early Devonian age into the S.
Devon area (Ch. 10c) clearly restricts the location of
a suture zone to the northern margin of the Gramscatho flysch trough. Since the Devonian volcanics of
Devon and Cornwall and their correlatives in Germany are of alkaline affinities, it seems that they can
have no possible origin as ophiolites related to the
closure of a mid-European ocean, a point supported
by stratigraphic and field relationships. The Lizard
Complex cannot simply represent a slab of
obducted ocean floor, since most of the associated
lavas have a 'within plate' chemical aspect. A simple ocean-floor origin also cannot explain the older
dates recorded from the Landewednack schists and
contact granulites, which are the very rocks with an
'ocean floor basalt' chemical aspect. Finally, severe
problems are raised by the rarity of Hercynian
calc-alkaline plutons and andesites in Britain and
NW Europe, if massive subduction to close a mid-
European ocean is postulated to have occurred dipping northwards from the English Channel area.
Similar problems arise if a S-dipping subduction
zone is postulated. It thus seems that the concept of
a mid-European ocean is partly discredited by the
above conclusions.

Let us now attempt to construct a reasonably
rational plate-tectonic structure for the Hercynian
of NW Europe, a model that can satisfy as many as
possible of the above arguments and also shed light
on the broader palaeogeographic trends deduced in
Chapters 9 to 11. The schematic sections in Figure
12.7 show how a model may be deduced for Europe;
further comment on the fold belt as a whole will be
made below.

The favoured model is based upon the presence of
a marginal basin in SW England, Brittany and other
parts of the Rheno-Hercynian zone, with subduction of lithosphere to close the Rheic Ocean to the
south of the Moldanubian zone of the Hercynides
(Fig. 12.4). The model is a variant of those proposed
by Floyd (1972*), Reading (1973), Leeder (1976b)
and A. V. Bromley (1976).

In early Devonian times the Old Red Sandstone
continent was the site of massive uplift, following
the intrusion of the calc-alkaline magmas of the
'Newer Granite' swarm. Internal and external
molasse basins formed, with Old Red Sandstone
rivers advancing into S. Devon to form the Dartmouth Slates. Far to the south, subduction of Rheic
Ocean lithosphere had begun. This caused uplift in
the English Channel−N. Brittany area and the
beginnings of partial melting. Yoked subsidence in
SW England was followed by deep water sedimentation of early Devonian slates and by volcanism in
the Roseland area of S. Cornwall. Subsequent
introduction of immature clastic debris in turbidity
currents from rising southern hinterlands caused
the Gramscatho flysch phase. Attenuation of the
continental lithosphere by marginal-basin sea-floor
spreading was accentuated in the middle and upper
Devonian, with extensive subsidence forming the
schwellen and becken province and its alkaline
basalt magmatism in the Rheno-Hercynian back-
arc basin. A phase of strike-slip faulting in N. Britain was coincident with the final closure of the
Caledonian Iapetus ocean in N. America in the
middle Devonian. Continued subduction in the
south during the upper Devonian caused uplift,
deformation and granite intrusion in the Moldanubian zone, the deformation front spreading northwards carrying a flysch belt with it, as recorded in
the Rheno-Hercynian successions of Germany. This
was the phase of Bretonic uplift, which extended as
far north as S. Cornwall. A northward marine
transgression over the southern margin of the Old
Red Sandstone continent occurred in late Devonian
to early Carboniferous times. Many of the Carboniferous sedimentary basins formed at this time
as the effects of lower-crustal hot creep towards the
marginal basin caused tensional stresses to exceed
the strength of the upper crust in areas where the
still buoyant 'Newer Granite' plutons were absent
(Leeder 1976b).

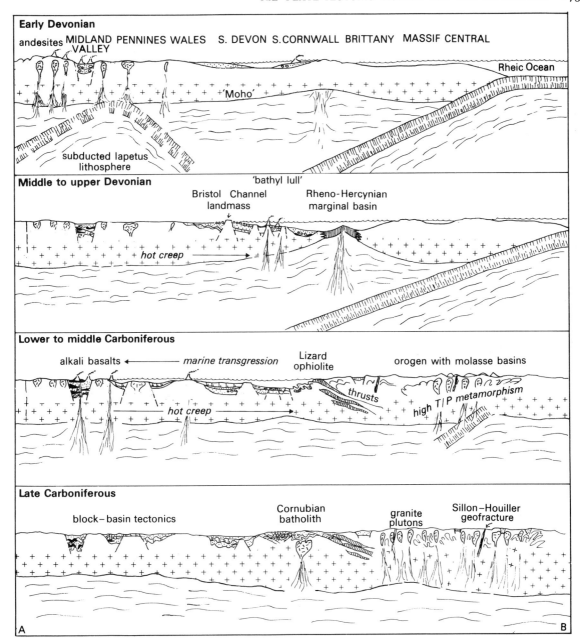

Figure 12.7 A possible plate-tectonic scheme for the European Hercynides along section A–B of Figure 12.4. Such a series of sections cannot show possible lateral or strike-slip motions in the area (see Badham & Halls 1975, Badham 1976 and Arthaud & Matte 1977* for speculations on such features).

Early Carboniferous molasse in Brittany testifies to the rising deformation front produced during the Bretonic phase. Continued pelagic deposition in the Culm trench was followed by the introduction of turbiditic flysch in early Namurian times from the south and north. The Sudetic phase of deformation, plutonism and metamorphism caused further widespread uplift of the southern orogenic belt and further northward spread of flysch in Germany, but it also caused extensive uplift and hinterland rejuvenation over much of Britain and, indeed, N. America. Very large drainage systems were established in the rejuvenated hinterlands, with the fluvio-deltaic Namurian facies being widely established. Basin-fill successions ended with the humid Coal Measures alluvial plains covering much of NW Europe. The final contractions of the Hercynides in late Carboniferous times caused extensive folding and thrusting, with a broad belt of lower crustal fusion producing the late Carboniferous granites of SW England and other areas.

In the above model the Lizard Complex emerges as a piece of thrusted, partly oceanised, Precambrian crust that formed in the marginal basin area. Remnant areas giving 'old' dates are thus explained, as is the chemical composition of the basic lavas, since these are not expected to represent pure ocean crust but marginal basin magmas due to continental crustal separation. The northern margin of the Gramscatho flysch trough is best explained as due to the tectonic juxtaposition of facies belts within the same original marginal basin. It should not be imagined that the above model is capable of explaining all the features of the Hercynides in Europe.

Turning to the whole Alleghenian–Hercynian belt, the hypothesis of Badham and Halls (1975), Badham (1976) and Arthaud and Matte (1977*) should be noted. They have maintained that lateral tectonics, involving oblique collisions and transform/strike-slip faults, is the best mechanism to explain the absence of subduction features such as andesites and the lateral inhomogeneities in the Hercynides. Their hypothesis is an interesting one and must have some degree of truth, for, as previously indicated, large strike-slip faults are a prominent feature of the Armorican–N. Spain areas of the Hercynides. However, the difficulties of such a model for both a Tethyan or Rheic subduction zone *and* subduction zones to close the Alleghenian–Mauritanian ocean between N. America and Africa should not be overlooked. The same strictures concerning the paucity of andesites applies to their model just as it does to the models that they have criticised. Also, it is by no means clear how the Lizard Complex could have been obducted in a

strike-slip fault area such as Badham (1976) has proposed, nor how the Bretonic and Sudetic regional orogenic phases were related to purely strike-slip movements.

In conclusion, we may say that at present there is too much latitude for geological speculation concerning the evolution of the Hercynides. Palaeomagnetic constraints upon plate-tectonic schemes are urgently needed. It seems, however, that the available evidence points to some sort of plate-tectonic mechanism. It surely is insufficient to account for the Hercynides due to 'rising heat fronts that have been induced by physico-chemical phase transformations in the upper mantle' (Krebs & Wackendorf 1973). What caused the phase transformations?

12f Summary

The structural style of SW England, located within the Hercynides proper, reflects both the original variation of primary structures and subsequent refolding events. Radiometric dating confirms the polyphase nature of the deformation and indicates late Devonian (Bretonic), middle Carboniferous (Sudetic) and late Carboniferous (Asturic) events. Radiometric and chemical evidence casts doubt upon the origin of the Lizard Complex as simply an obducted slab of Hercynian ocean floor. North of the discontinuous Hercynian thrust front, a distinctive phase of block–basin tectonics occurred in late Carboniferous times, with the development of upright folds and only local cleavage development. A phase of tholeiitic magmatism and lead–zinc–fluorite mineralisation was associated with this deformation. Current plate-tectonic models for the Hercynides are based upon insufficient palaeomagnetic evidence, but, nevertheless, a model based upon marginal ocean basins north of the Rheic Ocean proper is capable of explaining some features of Upper Palaeozoic evolution. It is also probable that lateral movements occurred during the Hercynian deformation, since important strike-slip faults may be recognised in Maritime Canada, N. Spain and France.

Plate 4 (top) shows a view of East Cliff, Whitby, Yorkshire. The cliff is some 50m high with Whitby Abbey visible at the top. 15m of U. Lias Alum Shales (AS) at the cliff base are followed by a thin, transgressive intertidal ferruginous sandstone known as the Dogger (D). When traced over a wide area the Lias succession seems to have been very gently folded, uplifted and eroded prior to Dogger deposition. The fluviatile Saltwick Formation (SF), formerly the Lower Deltaic Series, forms much of the remaining succession, being composed of 30m of floodplain mudstones and thin flood-deposited sandstones. A prominent channel sandstone deposit (arrow) may be seen downcutting into the Alum Shales on the lower left. The topmost cliff shows the two prominent horizons of the Marine Eller Beck Formation (EBF) overlain by fluvial sandstones of the Cloughton Formation (CF), formerly the middle Deltaic Series. (Photo by Eric Daniels)

Plate 4 (bottom) shows a view of the cliffs south of Hundale Point, Yorkshire. The cliff is some 30m high and shows the outcrop of the fluviatile M. Jurassic, Scalby Formation (formerly U. Deltaic Series). The basal third of the cliff shows the Moor Grit Member, a regional sheet sandstone thought to be of braided stream origin. Sandstones similar and contemporaneous to this form the main reservoirs in many of the Middle Jurassic oilfields of the northern North Sea. (Photo by Eric Daniels)

Theme

During the Permian and Triassic, following the Hercynian continental reorganisation and orogenesis, N. Europe lay within the arid hinterland of the newly formed Pangaea supercontinent. Subsident areas, including the newly formed North Sea Basin, became the sites of deposition of continental facies known as the New Red Sandstone. In the upper Permian, marine transgression from the north established a shallow gulf over N. Germany and the North Sea Basin, in which the Zechstein carbonates and evaporites formed. The sea returned from the south briefly in the middle Triassic, spreading northwards as far as the English Midlands.

At the opening of the Jurassic the arid basins of N. Europe were transgressed from the south and, for much of the period, marine shales, mudstones and limestones became characteristic deposits. Early Jurassic transgression was roughly coincident with the opening of both the central Atlantic and western Tethys Oceans as Pangaea began to disintegrate. Uplift, associated with central North Sea volcanism and rifting, caused temporary regression in the middle Jurassic, spreading fluvio-deltaic facies into northern areas. Later, renewed transgression and subsidence ultimately caused a return to mudstone deposition (Kimmeridge Clay) in the upper Jurassic.

Towards the end of the Jurassic, uplift restricted marine deposition to the North Sea and Wessex–Weald areas. In the southern basin regression culminated, in the early Cretaceous, with the deposition of fluvial Wealden facies, while marine sands and muds accumulated in the North Sea and parts of Lincolnshire and Norfolk. Regression was synchronous with rifting over the site of the Rockall Trough, where oceanic crust was emplaced later in the Cretaceous.

After early Cretaceous uplift the bulk of the period witnessed a progressive transgression from the east, which eventually submerged most of NW Europe beneath the warm clear waters of the Chalk sea. Late Cretaceous tranquility was ended by uplift movements associated with impending rifting between Greenland and N. Europe. This took place in the early Tertiary and was accompanied by major igneous activity in Greenland, W. Scotland, N. Ireland and the Faroes as ocean floor spreading began in the N. Atlantic.

As the Atlantic opened the Tethys closed, and in S. Europe compression accompanying this process culminated in the main Alpine deformations. Tertiary opening of the N. Atlantic resulted in the collapse of the new continental margins forming the present-day marginal shelves. Within the craton, the North Sea and Celtic Sea underwent major phases of rapid subsidence, with the sea periodically spilling on to the adjacent subtropical land-areas from the North Sea Basin. The British highlands were uplifted during the Miocene, and in S. England the post-Palaeozoic cover rocks were thrown into drape folds as their basement underwent both vertical- and wrench-fault motions. Major climatic changes resulted as the area drifted northwards; and by late Tertiary times, with the northern continents encircling the Arctic, a general cooling set in, giving rise to the late Tertiary–Quaternary successions of glacial and interglacial deposits.

13 Permo-Triassic rivers, deserts and evaporating basins

13a Introduction: the Hercynian structural legacy

The final consolidation of Hercynian N. Europe took place during the late Carboniferous and early Permian. From latest Carboniferous (Stephanian) times depositional basins adopted a new style, being mostly fault-controlled troughs separated from each other by discrete horsts or massif blocks. These blocks and basins resulted from differential movements within the basement. Movements occurred along old structural lines inherited from the Caledonian and earlier phases (e.g. the Great Glen Fault) and also from structures newly formed during the Hercynian Orogeny itself. Thus, before the beginning of postorogenic deposition, the Hercynides and much of their foreland were crisscrossed by an 'irregular grid of faults' defining numerous blocks and basins (Read & Watson 1975).

In contrast to the largely compressional stresses of the Hercynian, the NW European Craton was now influenced by tensional stresses. Under these conditions the main palaeogeographic features of the area were established, and many of these features were to persist until the close of the Mesozoic era. First, two large intracratonic basins formed in the Hercynian foreland: the Northern and Southern North Sea Basins. They were separated from each other by a horst comprising the E—W-oriented Mid-North Sea High and Ringkøbing Fyn High (Fig. 13.1). This horst might also have incorporated the N—S-trending Pennines Arch. To the south, St George's Land was still recognisable as the London—Brabant Massif, while the newly created Hercynides provided the cores of massifs such as Cornubia and Armorica (Fig. 13.1; Table 13.1).

Substantial uplift accompanied the final phases of the Hercynian Orogeny but this uplift, particularly in the Westphalian, was not confined to the orogenic belt itself, since it seems also to have affected much of the northern foreland. Thus the Permian sequence is normally separated from the Carboniferous by a substantial unconformity.

The development of positive massifs and the collapse of grabens such as the Oslo Graben were all Permian events that preceded deposition of the Mesozoic cover. The Oslo Graben is a NNE—SSW-trending rift structure about 50 km wide. The boundary faults have throws of 2—3 km, with an infill sequence that commenced with Permian red-beds and associated basalts, alkaline lavas and ignimbrites. Some of the igneous activity generated major ring complexes.

Elsewhere, as in SW England, the Midland Valley of Scotland and the North Sea, volcanism marking the last stages of the Hercynian Orogeny occurred at about the Carboniferous/Permian boundary and continued into the early Permian. The igneous suites comprise both intrusives and extrusives. In SW England a series of potassium-rich vesicular lavas with subordinate olivine basalts form the Exeter Volcanic Series. Most of these lavas have been extensively altered, and the originally amygdaloidal lavas now have their amygdales filled with zeolites and clay minerals. The original magma from which they were derived could have been basaltic, but it was almost certainly contaminated by both water and granitic components possibly associated with the Dartmoor batholith. In the Midland Valley, where many igneous centres are recognised, the igneous suites consist of agglomerates, tuffs, microporphyritic olivine-rich basalts, and undersaturated analcime- and nepheline-bearing basanites. These suites comprise fairly typical sequences accompanying late- and postorogenic block-faulting events throughout the Hercynides. The volcanics are sometimes interbedded with molasse sequences derived from the newly risen highlands.

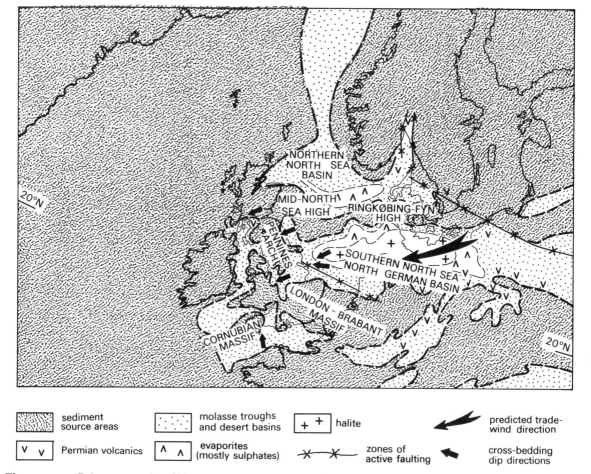

Figure 13.1 Palaeogeography of NW Europe during the deposition of the Rotliegendes, showing palaeolatitude and palaeowind directions.

13b Hercynian molasse, Permian deserts, and small evaporating basins

The lower Permian in the British area, as over much of Europe, was mostly a time of subaerial erosion, with the newly uplifted regions being deeply dissected. By the end of lower Permian time, most of the region had been reduced to a gently rolling peneplain (D. B. Smith et al. 1974*). During this erosional phase, reddening of the surface of the eroded rocks took place under semi-arid conditions, and weathered profiles can be seen immediately below early Permian deposits in many parts of Britain.

Early sediments consist of coarse red water-lain breccias and conglomerates composed of locally derived materials. In SW England, Devonian limestones comprise the dominant clast components;

and in the Midlands, locally outcropping Carboniferous and Silurian limestones, Cambrian quartzites and Precambrian igneous rocks all provided important source materials.

These coarse-grained early Permian sediments constitute a molasse facies, comparable with parts of the Old Red Sandstone (Ch. 9), derived from the uplifted horsts. In all the British sedimentary basins (Fig. 13.2) they appear to represent the deposits of marginal scree tracts that were generated as subaerial sediment fans. Sometimes these gravelly beds provide the fills to mappable valleys that may have been wadi systems. Away from the local massifs, breccias and conglomerates often pass laterally into red sandstones, which may themselves pass both laterally and vertically into red mudstones containing evaporite salts (Fig. 13.2).

Table 13.1 Table of Permian and Triassic stages, time and summary of major geological events

System			Stage	Radiometric age (Ma)	Major Geological Events
TRIASSIC	UPPER		Rhaetian	200	Major transgression begins from the south, producing a complex of quasimarine lagoons over Britain. Volcanism in S. Europe, Aquitaine and New England marks a preliminary phase of central Atlantic rifting, accompanied by the deposition of thick evaporite sequences. Rise of the dinosaurs begins.
			Norian		
	MIDDLE		Karnian	220	Dykes are emplaced in W. Norway. Arid environments return over much of N. Europe.
			Ladinian		Transgression from the south (Tethys) produces the Muschelkalk sequence of Europe and the North Sea and the quasimarine Waterstones succession in the English Midlands. Raised humidities temporarily allow the adjacent desert areas to bloom.
	LOWER		Anisian		Rejuvenation of the Palaeozoic massifs allows the generation of fluvial conglomerates (Bunter Pebble Beds).
			Scythian		
PERMIAN	UPPER		Tatarian	— 235 —	Many typical Palaeozoic taxa become extinct, including trilobites, tabulate and rugose corals, productid brachiopods, goniatite cephalopods and many stalked echinoderms.
			Kazanian		Successive evaporation episodes provide the Zechstein salt sequences of Germany and the North Sea.
			Ufimian		The Zechstein Sea forms as a shallow gulf over Germany and the North Sea. The sea enters the region from the north. Many parts of the world suffer transgression at this time, possibly because the Gondwanaland ice caps have melted. Continental facies persist in SW England and Scotland.
	LOWER		Kungurian	— 240 —	Desert environments established over much of Europe. Uplifted horsts supply coarse sediments via wadi systems into playa basins fringed by dune fields. (The coarser facies provide reservoirs for North Sea gas, derived from the underlying Westphalian coal measures during post-Permian burial.)
			Artinskian	255	
			Sakmarian	265	The Southern and Northern North Sea Basins develop as recognisable entities. Final Hercynian events in NW Europe include the eruption of volcanics in SW England, the Midland Valley of Scotland and the North Sea, and the uplift of older Palaeozoic massifs. Rifting and vulcanism commence in the Oslo Graben.
			Asselian		Shelf seas withdraw from many parts of the world, following the construction of Pangaea.
				280	

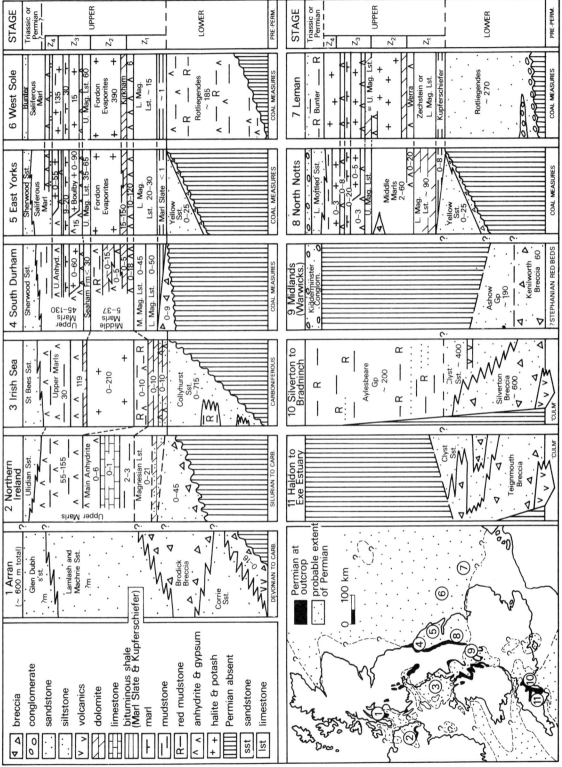

Figure 13.2 Stratigraphic correlation of the Permian in Britain and the location of the major British Permian basins (after D. B. Smith *et al.* 1974*).

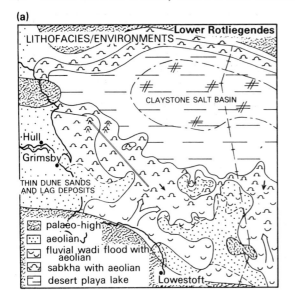

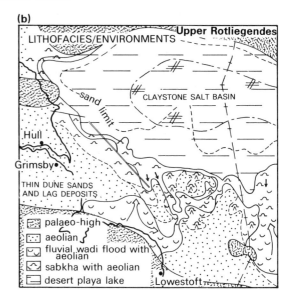

Figure 13.3 Facies and depositional environments in the lower and upper Rotliegendes of the southern North Sea (after Marie 1975*).

From recent work in the southern North Sea a striking model of depositional environments has been proposed, based upon modern desert-facies associations. Here, just as onshore, the Permian commences in unfossiliferous red-beds. Glennie (1972*) and Marie (1975*) have recognised fluvial, aeolian, lake-sabkha and desert-lake environments within these sediments. The sediments were deposited in a westward extension of the German−Dutch Basin (Fig. 13.3) and are known as the 'Rotliegendes'.

The southern limit to the deposition of these sediments seems to have been a fault-controlled escarpment comprising the northern margin of the London−Brabant Massif, the Rotliegendes sediment being banked up against the scarp in some places. Although the present western margin is largely erosional, the original limits to Rotliegendes deposition were probably formed by the Pennines block, and in the north the succession wedges out on to the Mid-North Sea High.

Glennie (1972*) has recognised two main facies associations within the Rotliegendes, each having a clearly defined geographical distribution (Fig. 13.3). The first consists mainly of sandstones with some conglomerates and breccias. It is confined to the southern margins of the basin, has a NW−SE trend and abuts against the London−Brabant Mas-

sif. The conglomerates and breccias are water lain and contain locally derived clasts from the underlying Westphalian coal-measures. Sequences (Fig. 13.4) frequently comprise interbedded conglomerates and sandstones. The latter consist of very well-sorted cross-stratified sandstones with highly rounded grains and less well-sorted muddier sandstones. The well-sorted sands may represent aeolian dunes, but the more muddy sandstones and the conglomerates probably resulted from desert flash-flood processes. Clay laminations within the fluvial sands are frequently cut by desiccation cracks, and reworked desiccated clay laminae often seem to have provided claystone intraclasts. Excellent exposures in these facies are available on Arran, and in S. Devon and many other localities along the Permian outcrop.

Foreset laminae within the cleaner aeolian sandstones dip mostly towards the west and southwest and imply a fairly constant wind direction from the east and northeast. Similar trends had been observed onshore from the Permian sandstones of N. England (Shotton 1956). Such regional wind directions are entirely consistent with the idea that the area lay within the northern trade-wind belt. This conclusion is supported by palaeolatitudes derived from palaeomagnetic data (Van der Voo & French 1974; Fig. 13.1).

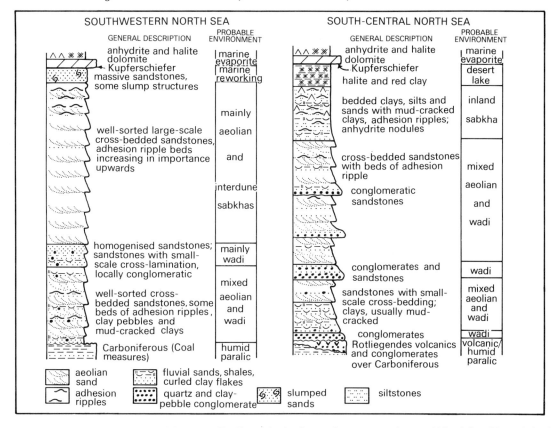

Figure 13.4   Facies development of the upper Rotliegendes in the southwestern and central North Sea. No scale is given because total thickness and relative proportions vary from place to place (after Glennie 1972).

In the northern and central parts of the Rotliegendes basin (Fig. 13.3) the coarser facies pass laterally into the second major facies association, of siltstones and claystones with evaporites. Vertical sequences in the southeastern part of the Rotliegendes basin normally show the main aeolian interval (Fig. 13.4) being succeeded by a monotonous succession of red claystones, with minor local dolomitic and anhydritic sandstones. These claystones, with their frequent desiccation cracks and evaporites, probably represent sabkha sequences generated on the fluctuating shoreline of an inland lake. The absence of any marine flora or fauna seems to preclude the possibility of a marine environment.

The lake shoreline sediments grade both laterally and vertically (Fig. 13.5) into a muddy sequence containing up to 30% halite. Glennie (1972*) has suggested that these represent the deposits of a desert lake that occupied the centre of the Rot-

liegendes basin. Periodic fluctuations in lake level over the marginal zones might have been caused by seasonal variations in rainfall in the Hercynian highlands to the south (Fig. 13.1).

The Rotliegendes overlie Carboniferous coal measures, and during later burial the coals devolatilised, emitting gases that became trapped in the more porous sandstones. The aeolian sandstones of the Rotliegendes with their high porosities provide the prime gas reservoirs, whereas the more muddy sediments are less attractive or have no prospects. Thus the environmental model deduced above provides a basis for predicting the distribution of sediments with the best potential reservoir characteristics.

The lack of chronologically significant fossils makes dating and correlation of these early Permian deposits very difficult (D. B. Smith *et al.* 1974*). Reptilian footprints are sometimes seen and plant beds occur, particularly low down in the sequence,

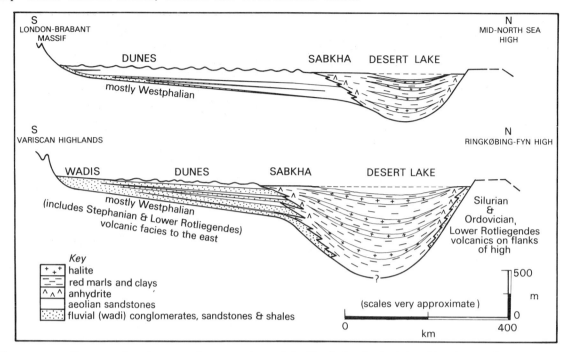

Figure 13.5 Conceptual environmental cross-sections of the North Sea Basin (after Glennie 1972*).

but these are of limited stratigraphic value. The various facies are almost certainly diachronous; so lithological correlations need not imply a time equivalence. Beds with plant debris are limited to the earliest deposits, which may have been produced in initially humid post-Hercynian times. Later, aeolian deposits and evaporites become dominant, indicating an arid climate.

At the close of the early Permian the British area thus comprised a shimmering desert underlain by a block-faulted basement. The North Sea area and other basins, such as those of N. Ireland and NW England, had been subsiding more rapidly than adjacent areas, and their land or lake surfaces might well have lain below sea-level before the beginning of the upper Permian. The Rotliegendes basin compares both in area and in general environmental context with certain large land-locked desert basins of the present time, e.g. the Caspian Basin of central Asia.

13c The Zechstein Sea

The upper Permian began with a major marine transgression which spread an epicontinental sea over the Irish and North Sea Basins. It extended southwards over parts of Holland and N. Germany (Fig. 13.6) and led to the deposition of the Zechstein sequences. The sea probably entered the area from the north (Fig. 13.6), because marine sequences found in E. Greenland and Spitsbergen contain faunas closely comparable with those from the British and German Zechstein.

The upper Permian roughly corresponds with the time of deglaciation in Gondwanaland, and it has been suggested (e.g. D. B. Smith *et al.* 1974*) that a glacio-eustatic change was responsible for the late Permian transgression. The initial transgression might have been fairly rapid, because basal conglomerate beds exist in places and, locally, dune sands below the first marine deposits have been reworked and exhibit slumping compatible with sudden drowning (Glennie 1972*). Later the transgression seems to have progressed in two or three stages, with dune re-establishment between each marine advance. Transgression seems to have significantly reduced the size of sediment source regions in the British area, thus restricting the availability of terrigenous clastic material. Plant cover was locally re-established and provided debris in earliest upper Permian deposits. This probably

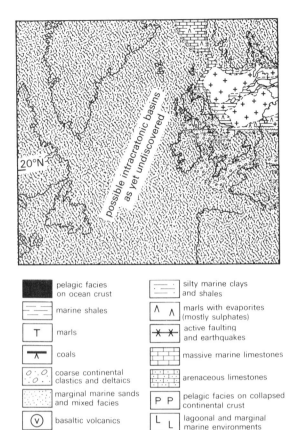

Figure 13.6 Generalised Zechstein (upper Permian) facies and palaeogeography of the British area.

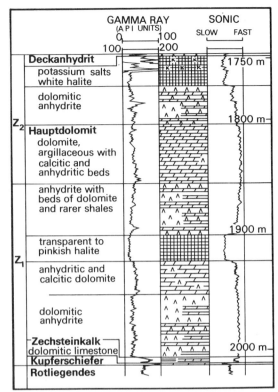

Figure 13.7 Typical sequences in two Zechstein cycles from the southern North Sea (after Rhys 1974).

reflects a temporary increase in the humidity of areas adjacent to the newly formed Zechstein seaway. However, this trend was short lived, because later thick sequences of evaporites once again began to form within the area.

Beneath NE England, in the North Sea and in Germany, Zechstein deposits comprise five major sedimentary cycles, each commencing in shelf carbonates and grading up into evaporites (Fig. 13.7). The carbonates are best developed around the margins of the basin (Fig. 13.8), and the thickest evaporites (c. 1000 m) occur within the basin itself. Later salt diapirism (Chs 13j and 14) has complicated the thickness patterns considerably.

The enormous extent of the Zechstein evaporites and the presence within the cycles of potassium salts provides a striking contrast between these deposits and modern evaporite associations (Hsü 1972). The base of each cycle is sometimes marked by a widespread bituminous shale which passes up into the main carbonate sequence. The cycles form individual progradational wedges of sediment and, although each cycle has its own characteristics, we can generalise some of the main features.

The basal shaley member is typified by the Kupferschiefer in Germany, and its correlative is the Marl Slate in NE England. These are finely laminated shales (with fish-beds attributed to mass mortality) and benthic faunas are either absent or very restricted. Such shales reflect the low energy of the epicontinental basin and probably mark the commencement of brine-induced density stratification of the water column. Bottom waters were not circulated and became oxygen deficient, leading to the formation of organic-rich sapropelic shales.

At first, around the basin margins, life flourished. Bryozoans and algae constructed reefal build-ups (Fig. 13.9) around which other organisms also lived. Productid and other brachiopods were associated with foraminiferans, bivalves, gastropods,

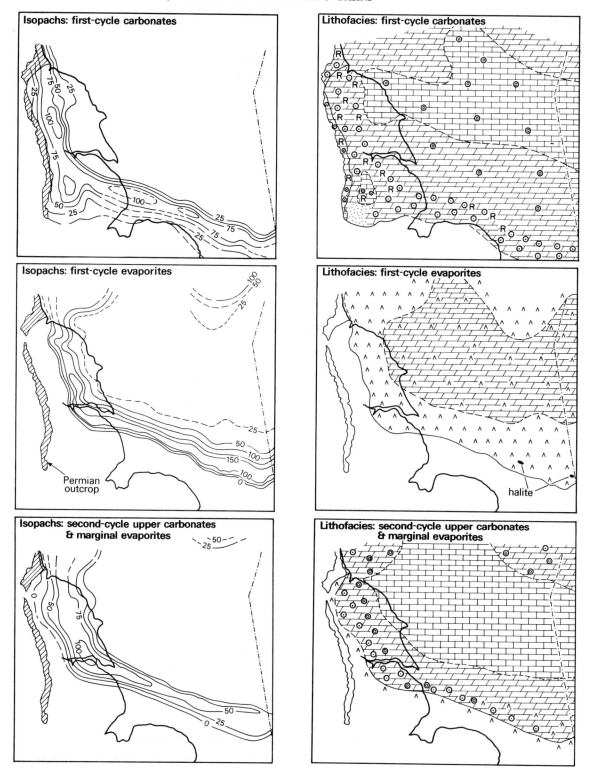

Figure 13.8 Generalised facies and isopachs for individual Zechstein cycles in the southern North Sea (after Taylor & Colter 1975*).

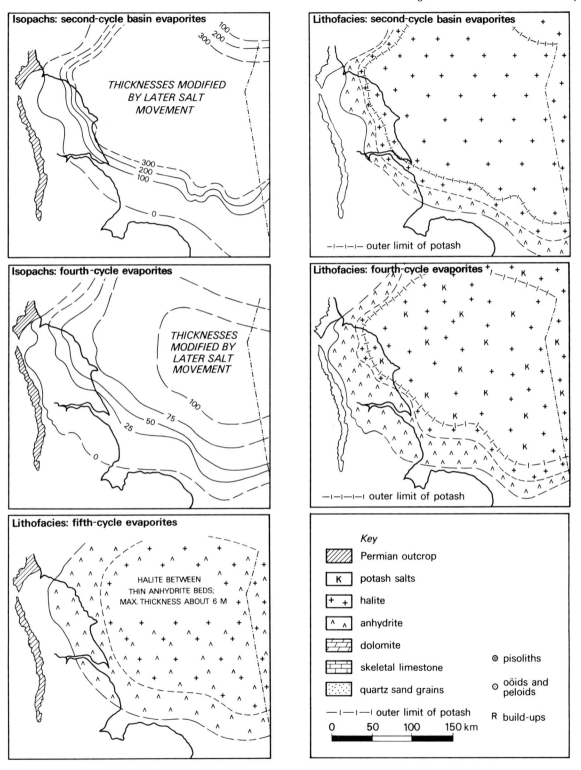

Figure 13.8 (cont.)

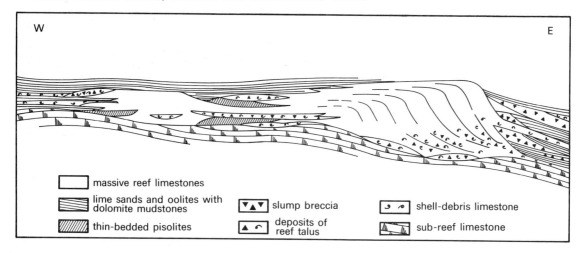

Figure 13.9 Sketch section to show build-up development in the upper Permian of NE England (after Smith 1958).

ostracods, corals, echinoderms and nautiloids (Pattison 1970; Ramsbottom 1978). In NE England, the dolomitic carbonates at the base of each cycle are known as a Magnesian Limestone (Fig. 13.10). In general, the most diverse faunas are associated with the carbonates of the first cycle; but even here ammonoids and fusulinid foraminiferans are absent, so that precise stratigraphic correlations with the world's type sections are impossible.

In each cycle the earliest carbonates of the basin margins are often oölitic dolomites. Locally, the nearshore facies are micritic with a fauna of bivalves, e.g. *Bakevillia*. In these facies, abundant but low-diversity faunas reflect restricted, possibly hypersaline, environmental conditions. Carbonate grains have mostly been micritised, coated with oölitic laminae and subsequently dolomitised, so that diagenetic modification of the original sediment has been considerable. The carbonates also include peloids, pisoliths and algal stromatolites.

Basinwards, the carbonates normally give way to anhydrite (Fig. 13.10). In cycle 1 this is an abrupt change. The anhydritic units thicken into the basin, reach a maximum and then thin towards the centre of the basin. Internally, these anhydritic deposits show many features, such as 'chicken wire' texture and enterolithic folding, that are considered to be typical of marine coastal sabkha deposition (Taylor & Colter 1975*).

If the thick anhydrite sequences are comparable with those from modern sabkhas along the shores of the Persian Gulf, then the geometry of the Zechstein basin must have been totally altered from that at the time of carbonate formation. On the modern Trucial Coast of the Persian Gulf, carbonates (including oölites and reef limestones) form offshore while dolomitic and anhydritic sediments accumulate in the supratidal sabkha (see Purser 1973 for reviews). However, in the Zechstein basin evaporites formed towards the basinal areas while carbonates occurred as platform facies at the basin margin. Clearly, if the anhydritic wedges accreted as sabkha deposits, they must have formed at a time when the basinal water level had dropped, leaving the marginal carbonates exposed to subaerial processes. Rapid sabkha progradation providing the anhydritic wedges then presumably followed.

Cycles 2 to 5 continue with thick halite sequences. In order to produce the observed thickness of both anhydrite and halite (Fig. 13.10), some process allowing the continual replenishment of the basin waters must have occurred; otherwise only thin evaporite layers would have resulted. For example, if the present Mediterranean Basin evaporated to dryness, only *c.* 25 m of salts (mostly halite) would be generated. The halite phases in each cycle reflect the final phases of evaporation of water that was already deficient in calcium ions. Total desiccation in cycles 2 to 5 resulted in the deposition of the most soluble potassium salts (carnallite and sylvite) in the more central parts of the basin (Fig. 13.8).

Each of the five cycles thus represents a phase of basin filling and progressive evaporation. In order to allow the replenishment of marine waters in the

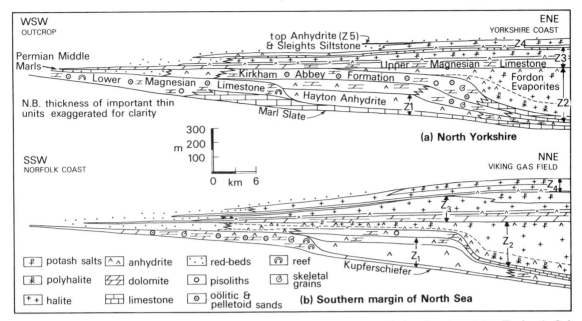

Figure 13.10 Generalised sections to show Zechstein sequences as progradational wedges (after Taylor & Colter 1975*).

basin, a barrier or partial barrier system must have been present. It has already been shown that the marine entrance to the basin probably lay to the north (Fig. 13.6). However, within the basin itself secondary barriers also existed. Sequences are incomplete on the Mid-North Sea and Ringkøbing Fyn Highs, as revealed by hydrocarbon exploration drilling. To the west, from Cumberland to N. Ireland, there periodically existed a sub-basin of the main North Sea Basin. This sometimes received marine waters which subsequently were desiccated to provide evaporites.

Elsewhere in Britain (e.g. SW England and Arran) the individual fault-controlled basins of the early Permian persisted, mostly as desert lakes and playas on similar lines to those of the Rotliegendes basin considered earlier (Ch. 13b).

13d Permian extinctions: the end of an era

Although impoverished by comparison with their Carboniferous ancestors, the faunas of Zechstein carbonates consist of groups wholly typical of the late Palaeozoic. This is the case for all known Per-mian marine assemblages. The close of the Permian was marked by the extinction of many of these typically Palaeozoic organisms and has been considered by some palaeontologists to reflect a major crisis in the history of life (reviewed in Raup & Stanley 1971 and Valentine 1973). As far as can be judged from the fossil record, this phase of extinction occurred within a relatively short period of time. It affected marine groups as diverse as trilobites, eurypterids, tabulate and rugose corals, goniatite cephalopods, productid brachipods, orthid brachiopods, fusulinid foraminiferans and many groups of stalked echinoderms. On land, many synapsid reptiles (mammal ancestors) also became extinct.

The demise of these groups was followed, in the Triassic, by the appearance of many new groups. Some filled ecological niches left vacant by the removal of the old faunas; others moulded new niches that had not existed before. The Bivalvia underwent tremendous diversification in the Triassic, unionids invading fresh waters while many other new groups adopted a burrowing mode of life (Stanley 1968). Many new families of gastropods arose; and a new ammonoid family (the phylloceratids), ancestral to the Mesozoic ammonites, replaced the Palaeozoic goniatites.

Archosaur reptiles (including dinosaurs, crocodiles and pterosaurs) replaced the paramammals as the dominant terrestrial organisms. Other new reptile groups also developed, including ichthyosaurs, plesiosaurs, squamata (snakes) and rhynchosaurs.

These great faunal change-overs obviously require some explanation. Extinction hypotheses have included many types of catastrophe – e.g. cosmic radiation, volcanic pollution, worldwide salinity crises in the oceans, wild climatic fluctuations – but none of these explanations are entirely satisfactory because of the selective nature of the extinctions.

Oceanic salinity changes might have occurred because of the removal of Permian evaporites. Such changes, however, should have had a more general effect and would not have affected terrestrial faunas. Temperature or climatic deteriorations do not seem to have caused mass extinctions during the Pleistocene, and there is no evidence to suggest that Permian climatic fluctuations were any more severe than those of the last few million years. Thus it is unlikely that such events could have caused major extinction phases.

However, the generation of the Pangaea supercontinent from originally dispersed continents, and the concomitant construction of the Hercynian ranges, had totally changed the world's aspect. Continental shelf areas were progressively restricted to the marginal zones of Pangaea as most of the late Palaeozoic epicontinental seaways disappeared. One consequence of the decreasing shelf area was increased competition for the available niches, favouring the survival of the most tolerant groups. On the land, originally isolated terrestrial populations were brought into direct competition, and habitats on the continents themselves severely deteriorated with increasing continentality.

Valentine (1973), reviewing mass extinction phases, has explained how increasing continentality produces raised environmental stresses both on land and on adjacent shelves. This leads to the survival of only the most tolerant groups and to the extinction of the organisms most highly specialised to the conditions of the previous regime. Such conclusions seem adequate to explain the extinctions of the late Permian, and cosmically induced catastrophes appear to be unnecessary.

13e The Triassic stratigraphic framework

In Britain the Triassic period is almost wholly represented by continental red-beds. The standard European stages (Table 13.1) are, however, defined on the basis of ammonite faunas in the marine sequences of the Alps. Thus precise zonal correlations in Britain and between Britain and S. Europe are not yet possible. In Germany the lower and upper divisions are of continental facies, the Buntsandstein ('mottled sandstone') and Keuper ('red marl'), and the middle division is represented by the marine Muschelkalk ('mussel limestone').

Although the Muschelkalk is absent in Britain, the terms 'Bunter' and 'Keuper' were imported by Sedgwick (1835) as names for the generally sandy and muddy phases of the British Triassic, and these terms have been haphazardly used ever since. They have been employed lithostratigraphically, chronostratigraphically and even chronologically (Ager 1970).

Although macrofossils of precise biostratigraphic significance are absent from the red-bed facies, recent work on plant spores has established a stage-by-stage correlation from Germany into Britain via the North Sea (e.g. Geiger & Hopping 1968*, Warrington 1970, Smith & Warrington 1971). The diachronous nature of some facies has been clearly demonstrated. With these improving correlations it is also becoming increasingly clear that certain sandstone units were deposited very rapidly over large areas and so provide 'event horizons' for correlation. The middle Triassic marine transgression that produced the Muschelkalk also seems to have been approximately isochronous, affecting areas as far afield as NW Africa, S. Europe, Germany, Poland, France and the southern North Sea. The English Midlands also experienced their first Triassic 'whiff of the sea' during this phase. Such a widespread transgression might have been due to simultaneous subsidence in all areas, a eustatic sea level change or a combination of both factors. As will be shown, this widespread incursion of marine waters seems to reflect the extraordinary flatness of the later Triassic topography, which allowed short-term climatic, eustatic or tectonic fluctuations to have very widespread effects.

Where Triassic rocks overlie the Permian there is mostly little evidence of unconformity or major hiatus, but locally Triassic beds lie unconformably upon older rocks (Ch. 13f). There is still no direct palaeontological evidence for the definition of the base of the Triassic in Britain. It is taken at the base of the so-called Bunter Pebble Beds (Audley-Charles 1970*), which correlate in a general way with the earliest Triassic beds on the continent of Europe. The top is taken at the top of the Penarth Group, formerly named the Rhaetic Beds.

13f   Triassic flash floods

The hot and mainly arid climate that typified N. Europe during the Permian continued into Triassic times. During the late Permian much of the British area constituted a virtually featureless peneplain. At the opening of the Triassic some of the Hercynian massifs were strongly rejuvenated (e.g. Wales, Scotland, parts of Ireland, Armorica and the Massif Central) and once more began to supply coarse sediments to adjacent basins. Other massifs (e.g. London–Brabant, Pennines, Lake District and Cornubia) were only moderately elevated and, at first, provided little coarse clastic detritus. Close to massif areas, this phase of uplift is recorded by the presence of an unconformity (Fig. 13.11).

Onshore in Britain, Triassic rocks are mostly represented by three interdigitating facies associations: (a) coarse-grained breccias, conglomerates and sandstones; (b) siltstones and mudstones; and (c) evaporites. The coarse-grained sediments (Fig. 13.12) include matrix-supported paraconglomerates and poorly sorted breccias which, in S. Wales, occur against fossil clifflines composed of Carboniferous limestone (Fig. 13.13). The breccias are interpreted as fossil scree deposits, whereas the matrix-supported conglomerates bear all the hallmarks of mudflow emplacement (Tucker 1977*). Similar mudflow deposits have been described from NW Scotland (Steel 1974, Steel et al. 1975*).

The great bulk of the coarse-grained facies consists of sorted conglomerates and cross-bedded sandstones arranged in fining-upwards cyclothems a few metres thick (Fig. 13.12), representing streamflood and braided stream deposits in which the pebbles were derived from locally outcropping sources. In S. Wales, for example, the clasts consist of rounded Carboniferous limestone, Coal Measure

sandstone and Old Red Sandstone, whereas in W. Scotland schists and quartzites of possible Moinian and Cambrian origin are common components. Thus these, and most other, conglomeratic facies seem to have been produced as local fringes of alluvial fan origin adjacent to the uplifted massifs. However, from Devon (Budleigh Salterton Pebble Beds) to the N. Midlands (Bunter Pebble Beds) the earliest Triassic conglomerates also contain more exotic clasts. On the basis of pebble petrography, palaeocurrents and the isotopic composition of the detrital micas, these conglomerates seem to have been derived, in part, from an Armorican (Brittany–English Channel) source area (Fig. 13.14).

Red sandstones and siltstones are found capping fining-upwards sequences within the conglomeratic facies (Fig. 13.12). These siltstones are mostly structureless, except for the presence of occasional ripplemarks and extensive surfaces traversed by polygonal desiccation cracks. Bands of nodular carbonate are frequently found in association with these red siltstones, and they exhibit concretionary pisolites, microlamination and microbrecciation typical of calcrete soil horizons (Tucker 1975, Steel 1974, Steel et al. 1975*). Comparable horizons (cornstones) are found in Old Red Sandstone deposits (Ch. 9).

Red silts constituted the 'background' sediment to the streamflood and sheetflood deposits represented by the coarser material. As well as containing calcretes and desiccation cracks, this facies is sometimes associated with evaporites or contains petrographic features indicating the former presence of gypsum or anhydrite. Such siltstones and their associated mudstones probably accumulated as the overbank deposits of established fluvial systems or in larger-scale playa lakes. Steel (1974) has shown how some of these siltstones comprise the finer portions of fining-upwards sequences that are comparable with those from the Old Red Sandstone (Ch. 9).

The coarser-grained facies often show an evolution, both upwards and outwards from the massif areas, from scree breccias and mudflow deposits through streamflood sequences into braided stream material. Such sequences conform well with those from modern alluvial fans (Tucker 1978, Steel et al. 1975*). Indeed, Tucker has even mapped out the location of individual fans and suggested that such

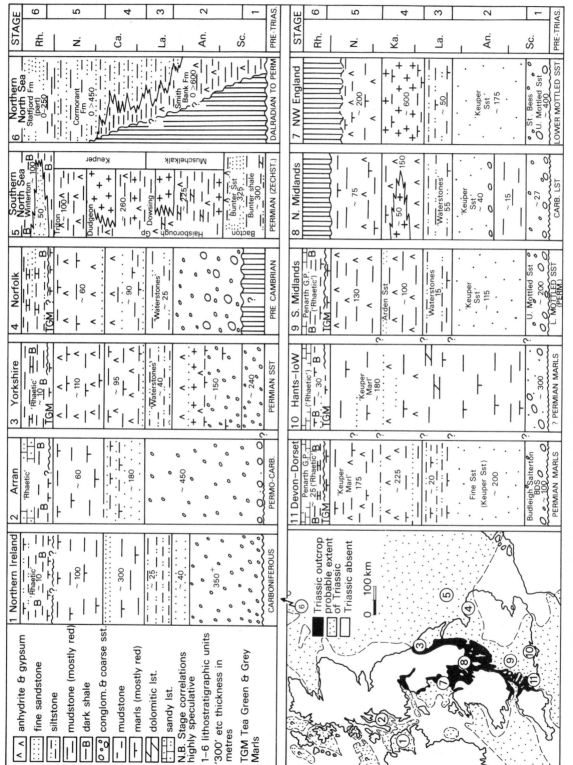

Figure 13.11 Stratigraphic correlation of Triassic sediments in Britain (after Audley-Charles 1970*).

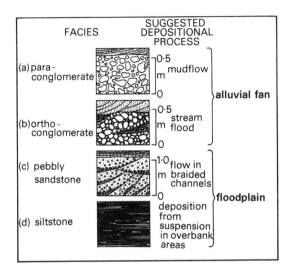

Figure 13.12 Four major facies associations found in the British Triassic. The siltstone facies may also be associated with evaporites (after Steel *et al.* 1975*).

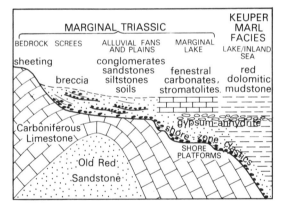

Figure 13.13 Facies relationships in the marginal Triassic deposits of S. Wales (after Tucker 1977*).

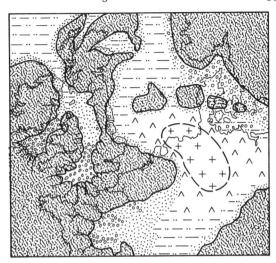

Figure 13.14 Early Triassic facies and palaeogeography of the British area (after Warrington 1970). For key see Figure 13.6.

rivers as the modern Taff are currently flowing down exhumed Triassic canyons (Fig. 13.15).

By comparison with Recent calcretes, those of the Triassic probably required many thousands of years for their development. In some cases the nodular carbonates are associated with authigenic clay minerals such as palygorskite, which also forms under calcrete soil-forming conditions. Indeed, pedogenesis of this type is an indication that the climate was semi-arid, as soils forming, and preserving, palygorskite today receive an average rainfall of less than 300 mm a⁻¹ (Watts 1976). Confirmation of such a climatic regime is also supported by the presence locally of exfoliated pebbles and sheeted bedrock, features that form by insolation processes in modern deserts (Tucker 1975).

13g Triassic playa lakes

The red Keuper facies consists of homogeneous dolomitic mudstones whose monotony is only relieved by greenish bands, desiccation cracks, laminae with rainprints, and horizons of gypsum and anhydrite nodules. The clay minerals consist mostly of illite and chlorite. Kaolinite is virtually absent, and this fact, coupled with the unweathered nature of included feldspars, is a further indication that chemical weathering was unimportant in the hinterlands.

The lack both of diagnostic sedimentary structures within these red marls and of macrofossils makes the interpretation of the depositional environment very difficult. Two contrasting views exist: (a) that the sediments represent subaqueous deposition in an inland hypersaline lake or inland sea, or (b) that they comprise giant playa or desert plain sediments.

Important evidence in favour of the latter interpretation is again provided by the sequence in S. Wales (Tucker 1977*, 1978), where coarse marginal facies rest upon wave-cut platforms with fossil cliff notches and pass laterally into red marls with sabkha-type evaporites. The gravelly sediments are confined to narrow belts adjacent to the fossil cliff-lines and are interpreted as shore zone deposits. Their restricted distribution suggests that the basinal waters had little turbulence. The shore zone sediments include carbonates with dripstone, crypt-algal and fenestrate fabrics. These features suggest that deposition took place under shallow subaqueous conditions, punctuated by phases of subaerial exposure. After occasional but torrential rains a lake is presumed to have developed, which was subjected to a long period of evaporation that transformed it first into an extensive mudflat and finally into a desiccated evaporite-encrusted pan (Tucker 1978). In this way basinal areas such as the Worcester Graben and Cheshire Basin were able to accumulate thick sequences of evaporitic sulphate salts with associated halite.

If the red mudstones represent distal lacustrine sediments, why are they not laminated as are those of the Old Red Sandstone Orcadian lake (Ch. 9)? Glennie (1970), for example, has recorded well-developed laminations from modern desert lakes and playas where sediments are introduced by both wind action and water flow. It is possible that soil-forming processes operating during the protracted phases of desiccation could have destroyed any original bedding, leaving the typically massive red mudstones now seen. Some of the mudstones may also represent wind-borne loess trapped on mudflat surfaces while they were still damp.

Although the above discussion has concentrated on certain closely studied areas, facies comparable with those discussed were established over much of N. Europe in the Triassic, with alluvial fans adjacent to massifs passing laterally into distal desert playas.

13h A middle-Triassic 'whiff of the sea'

In the early Triassic, marine conditions were restricted to Alpine Europe in the south, and to E. Greenland, Spitsbergen and the Central North Sea Graben in the north (Fig. 13.16). The Greenland seaway seems to have opened northwards into the Arctic (Birkelund & Perch-Nielsen 1976), while the southern margin of Europe comprised a carbonate shelf with reefs (Zankl 1971) fringing the Tethys Ocean.

During the middle Triassic, however, the sea spread from S. Europe, across Germany and the southern North Sea, and flooded large parts of the British Isles (P. A. Ziegler 1975*, Audley-Charles 1970*, Warrington 1970). Transgression in Germany produced the Muschelkalk sequence.

In Saxony, the lower Muschelkalk consists of grey dolomitic and argillaceous limestones which contain a high percentage of open-marine microplanktonic organisms. The middle Muschelkalk contains dolomitic mudstones with sabkha-type anhydrite and halite layers. The biota is greatly restricted. The upper Muschelkalk comprises massive crinoi-

Figure 13.15a General distribution and thickness of Triassic sediments in S. Wales (after Tucker 1977*).

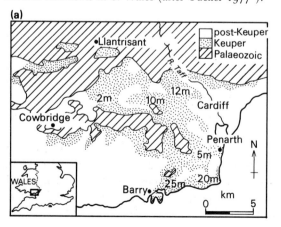

Figure 13.15b Triassic depositional environments in S. Wales (after Tucker 1977*).

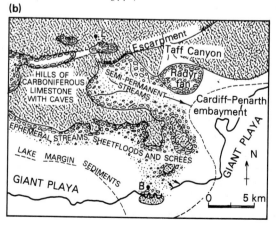

dal limestones overlain by limestones and marls containing a rich bivalve fauna and an open marine microflora (Geiger & Hopping 1968*). The upper Muschelkalk records the most important transgressive phase. The sea entered N. Europe from the south through the Hessian and Moravian depressions (Fig. 13.16), and with the marine waters came Tethyan faunas such as ceratitid ammonites. As yet there are no records of marine Muschelkalk sequences north of the Mid-North Sea High (Brennand 1975*). Thus the direction from which transgression came contrasted with that of the Zechstein (Ch. 13c).

In the southern North Sea the Muschelkalk is developed in a semi-marginal facies, with limestones alternating with anhydritic mudstones. Marine influences spread through the Anglo-Dutch Basin and into the English Midlands. Marine (or quasi-marine) acritarchs have been recorded from the Waterstones Formation (Warrington 1970), and communities of *Lingula* were established locally. An abundant trace fossil assemblage in sediments of marine intertidal aspect has recently been discovered in the Waterstones of Cheshire (Ireland *et al.* 1978).

The approach of the Muschelkalk sea must temporarily have raised humidities in the area, because populations of both aquatic and terrestrial vertebrate animals settled the region (A. D. Walker 1969). This fauna includes vegetarian dinosaurs, carnivorous reptiles and marine fish such as *Gyrolepis*, which is also found in the Muschelkalk of Germany. In Worcestershire the Waterstones consist of an interbedded series of shales, siltstones and sandstones, which interdigitate southwards with alluvial sequences containing lungfish, scorpions and plants (horsetails and conifers). Surprisingly, many of the components in the terrestrial fauna have an almost cosmopolitan distribution with certain groups being found in the Far East and the Americas. Such distributions of terrestrial faunas indicate both climatic homogeneity and land connections between these areas, and lead to the conclusion that the Pangaea supercontinent was still united.

The marine phase discussed above was shortlived, and environments in the British area subsequently reverted to the more arid non-marine ones that had existed previously.

13i The late Triassic marine invasion

The youngest Triassic rocks in Britain record a progressive marine transgression over the area, with continental red-beds finally giving way to shales and limestones containing marine and semimarine organisms.

In NW Scotland, increased humidities towards the end of the Triassic seem to have caused the development of meandering streams that lacked calcrete profiles in their floodplains (Steel 1974). There was also a decrease in source area relief during the filling of some Scottish basins, such as the Minch Basin.

In England the red Keuper facies is replaced upwards by the greyish and greenish dolomitic mudstones of the Tea Green Marls and Grey Marls, which contain latest Triassic (Rhaetian) microplankton. This change in sediment colour occurs from SW England to Yorkshire and probably reflects the higher initial organic contents of the sediment. Such a situation would have promoted anaerobic diagenetic conditions, in contrast to the oxidising burial conditions that are thought to have characterised the abiotic red playa sediments.

These green and grey sediments consist of rhythmically interbedded laminated mudstones and unlaminated fine-grained dolomites which locally contain nodular gypsum. Sequences are comparable with those from modern sabkhas, with the associated acritarchs and rare U-shaped burrows signifying marine, rather than lake, shoreline conditions (Sellwood *et al.* 1970).

Above the green marls there is normally a dramatic change of facies. The top of the marls is often intensely burrowed, with the burrows filled with mud clasts. Sometimes bone-rich debris is also present. The surface marks the base of the Rhaetian sequence, now named the Penarth Group, and is overlain by beds containing marine fossils, including oysters, pectinids and echinoids (Ivimey-Cook 1974), which indicate a major marine incursion into the British area. The bone-beds that recur within the basal part of the succeeding Penarth Group beds, probably reflect successive reworked strandline deposits. Teeth, scales, spines and other bones would have been the most durable materials available as the sea spread over and reworked the broad and almost featureless playa and sabkha plains. The bones of both marine and freshwater animals are represented in the bone-beds, with teeth of the lungfish *(Ceratodus)* being very characteristic. Although palaeontological evidence is not adequate to give precise dating, it is probable that the marine flooding phase occurred very rapidly. Kent (1970), for example, has suggested that the whole of the Midlands was inundated almost simultaneously.

After this initial flooding, whose effects can also be seen in N. Ireland and W. Scotland, dark shales of the Westbury Formation accumulated. This facies characterises the basal Rhaetian beds from the Alps to the Baltic (Kent 1970). Their fauna of specialised epifaunal bivalves *(Rhaetavicula contorta)* probably reflects an unfavourable habitat, with 'soupy' substrates and oxygen-deficient bottom waters extending uniformly over this wide area.

Subsequently there must have been a widespread, but temporary, retreat of the sea, because the succeeding beds, the Cotham Member of the Lilstock Formation, are non-marine green and grey marls. At most localities these contain abundant fossil water-fleas *(Euestheria)* which are comparable with those from modern African lakes. Beneath Lincolnshire the Cotham facies passes into red Keuper-type marls, which persist into the North Sea. In SW England, however, the Cotham facies contains quasimarine populations, including oysters. These features suggest that marine influences were now coming from the south and west and no longer from the east (Fig. 13.16).

The Cotham Member locally contains well-developed domal stromatolites which often reach a height of 200 mm (Hamilton 1961). Desiccation cracks frequently cut the stromatolites, whose morphology compares well with modern forms that develop intertidally in very restricted embayments, such as Shark Bay, W. Australia. In the English Midlands certain limestone beds within the Cotham facies exhibit well-developed slump rolls, which may be expressions of contemporary earthquake activity.

The youngest Penarth Group beds, the Langport Member, consist mostly of limestones with dark shales and mark a renewal of marine advance. The clay minerals show dramatic differences from underlying beds, with chlorite being almost wholly supplanted by kaolinite, a feature that may denote more intense hinterland weathering and hence increased humidity. Terrestrial plant debris (e.g. from conifers) is more abundant; and the beds often contain a rich fauna, which is dominated by oysters and pectinid bivalves but also includes echinoids, solitary corals and trace fossils. Cephalopods are, however, absent in all of the Rhaetian beds in Britain, possibly due to moderate hypersalinity, as indicated locally by the occurence of evaporite pseudomorphs.

In Devon the limestones of the Langport Member (formerly known as the White Lias) exhibit major slump sequences up to 2 m thick (Hallam 1960) and contain micrite-supported intraclast conglomerates attributable to mudflows. Similar structures occur at about the same stratigraphic level in N. Somerset and elsewhere and may indicate a further phase of late Triassic earthquake activity.

The late Triassic in Britain thus witnessed an important phase of marine transgression from the south and southwest. This produced a succession of marine sabkha shorelines represented by the Tea Green and Grey Marls, and later a complex of shallow muddy lagoons as deduced from the Westbury Beds. A minor phase of marine retreat, possibly promoted by earth movements, caused a reversion

Figure 13.16 Middle Triassic facies and palaeogeography of the British area. Transgression (arrowed) via the Moravian Depression (MD) and Hessian Depression (HD) produced the Muschelkalk sea in central Europe, and marine waters spilling round the northern margin of the London–Brabant Massif produced the Waterstones transgression in the English Midlands (after Warrington 1970 and P. A. Ziegler 1975*). For key see Figure 13.17.

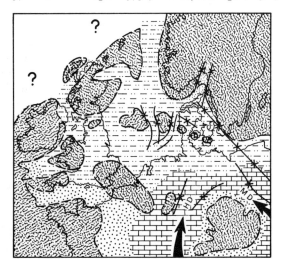

to lacustrine conditions over the Midlands and N. Britain, while the southern and western parts of the country became an extensive inhospitable tidal-flat. Subsequently there was renewed advance of the sea, which converted the lakes and lagoons into a shallow and hypersaline embayment. Transgression continued, and later the whole region, with the exception of the ancient massif areas, became submerged by a broad seaway in early Jurassic times.

13j Climatic and tectonic controls

As shown above, Triassic deposition in the British area mostly took place under semi-arid and arid conditions that may be considered typical for a low-latitude continental interior. The presence of substantial evaporite sequences, the existence of exfoliated pebbles, calcrete soil profiles and associated clay mineral suites all support this inference. The striking red colour of most beds probably developed during diagenesis as the unstable original minerals oxidised during burial and provided iron-oxide grain coatings (T. R. Walker 1976). Oxidising burial phases and immature mineralogies would have been promoted by the lack of intense weathering in the sediment source areas and by the absence of buried organic matter.

Periodically, however, the climate became more humid, plant cover developed temporarily, and green, rather than red, diagenetic colours resulted. Some of the green colours developed long after burial and are parallel to joints and faults. Terrestrial and aquatic organisms migrated into the area, and permanent meandering streams were established in place of the ephemeral alluvial systems of the drier periods. In a general way the phases of increased precipitation correlate with marine transgressions, and it is tempting to suppose that the establishment of an inland seaway created its own more humid microclimates.

Triassic rocks of all stages exhibit substantial changes of facies and thickness, especially approaching the ancient massif areas that were supplying the sediment. The most dramatic changes can sometimes be related to currently exposed faults (Tucker 1977*). In other cases, around the Mendips for example, they occur over buried faults that are identifiable on geophysical

grounds. Audley-Charles (1970*), Tucker (1977*), P. A. Ziegler (1975*), Brennand (1975*) and many others have recorded the influence of contemporary tectonism on Triassic facies distributions. It is also well established that the movement of Zechstein salt plugs and domes commenced during Triassic time and that these motions were probably initiated as the result of earth movements (Heybroek 1975).

In the North Sea, the subsidence that had begun in a 'dignified manner' during the Permian (Selley 1977) rapidly accelerated during the Triassic. Block faulting led to the establishment of horsts and grabens with a new N−S trend which, as will be shown, became strongly accentuated during the subsequent Jurassic period.

Expressions of earth movements are documented onshore by the presence of slumps and mass flow

Figure 13.17 Generalised late Triassic facies and palaeogeography of the British area (after Hallam & Sellwood 1976*).

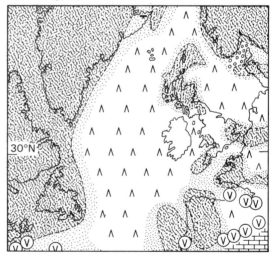

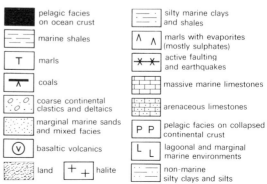

deposits which continued into the latest Triassic. But more dramatic evidence of very late Triassic earthquakes is provided by fissure deposits in the Mendips and S. Wales. Here, actually within ancient massif rocks (particularly Carboniferous limestone), E—W-trending fracture systems contain fillings that include Rhaetian terrestrial floras and faunas. Tensional stresses probably led to their initiation in the late Triassic and, once formed, such fissures were intermittently infilled throughout the Mesozoic.

In S. Europe, the Aquitaine and eastern N. America (Fig. 13.17) an important phase of alkaline volcanism occurred during the late Triassic (Hallam & Sellwood 1976*), and the European area was subjected to tensional stresses causing basement faulting. As will be shown later, volcanism in the W. Mediterranean seems also to have marked the preliminary phases of the westwards extension of the Tethyan Ocean during late Triassic and early Jurassic times.

13k Summary

After the Hercynian consolidation of N. Europe, the area became influenced by predominantly tensional stresses, and a well-defined series of basins and massifs was established. Throughout Permian times the area lay in the trade wind belt, having a hot and arid climate. Ephemeral streams cut wadis in the highland massifs, which became fringed with talus.

The basins became the sites of aeolian dunefields and playa lakes in which evaporites formed. In the upper Permian the North Sea area was flooded by the sea which entered from the north and spread over some adjacent basinal areas in Britain and northern continental Europe. Marine faunas flourished initially but became progressively more restricted as the sea became increasingly saline. Subsequently, major evaporite sequences were produced during several repeated phases of flooding and desiccation.

At the end of the Permian many groups of Palaeozoic organisms became extinct both on land and in the sea. This crisis in the history of life probably resulted from the increased environmental stresses, affecting many habitats, that followed the construction of the Pangaea supercontinent.

The Triassic in N. Europe opened with a phase of uplift which affected many of the ancient massif areas. The remaining Triassic recorded the peneplanation, under arid conditions, of the uplifted regions and the establishment of widespread low-lying continental-interior basins which underwent strong subsidence. Periodically, transgressions occurred from the south, affecting large portions of the continental interior. These events raised humidities locally and permitted the temporary establishment of both rich terrestrial and lagoonal communities. Finally, at the end of Triassic time much of the continental hinterland was flooded from the south to produce a seaway that persisted over the region throughout the early Jurassic.

14 The epeiric sea of Jurassic Europe

14a Introduction and chronostratigraphic framework

Marine transgression transformed much of Jurassic Europe into an enormous, generally shallow, epicontinental (or epeiric) sea (Fig. 14.1) whose rich fauna included ammonites which now provide the key to precise Jurassic correlation. Within the lower Jurassic alone no less then twenty ammonite zones are recognised, each one being divided into a number of subzones. Where ammonites are scarce or lacking, correlations have been attempted using other fossils (foraminiferans, ostracods, dinoflagellates, pollen and spores), but these are generally less precise.

Ammonite zones consist of subzones defined by the range of an individual ammonite genus or species. The stages (Table 14.1) are defined as groups of zones. Zonal correlations can be achieved throughout Europe for much of the Jurassic, and for the early Jurassic the same ammonite genera can be found as far afield as N. and S. America and the Far East. However, ammonite provinciality began to occur before the end of the lower Jurassic, and by the late Jurassic two major provinces or realms, the Tethyan and Boreal, had become established (Ch. 14m). A consequence of this provincialism is the recognition of the Volgian Stage in Boreal Europe and the equivalent Tithonian Stage in S. (Tethyan) Europe. Both stages are partially equivalent to the

Figure 14.1 Generalised palaeogeography of the British area during the lower Jurassic, showing the distribution of the major positive regions and the area known to have been affected by the movement of Permian salt (after Sellwood & Jenkyns 1975*). 1 = Meseta; 2 = Massif Central; 3 = Armorican Massif; 4 = London−Brabant Massif; 5 =Mendip−Glamorgan archipelago; 7 = Grampian−Shetland platform; 8 = Ringkøbing−Fyn High.

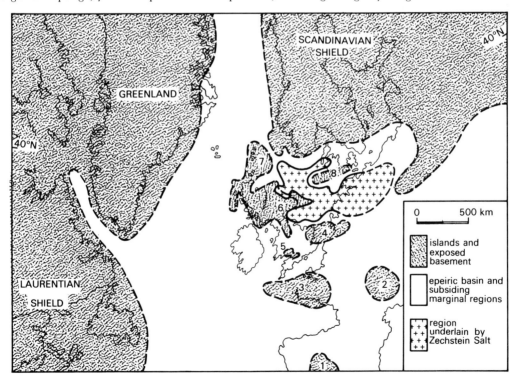

Table 14.1 Jurassic stages, time and summary of major geological events

System			Stage		Radiometric Age (Ma)	Major Geological Events
			Britain	Elsewhere		
JURASSIC	UPPER		Volgian	Kimmeridgian / Volgian / Tithonian / Kimm.	c. 135	Regression results from both eustatic lowering of sea level and the reactivation of local horsts. Large-scale differential subsidence is accompanied by the deposition of the thick organic-rich muds, locally interbedded with mass flow deposits derived from adjacent rising horsts. Volcanism occurs west of Britain.
			Oxfordian			
	MIDDLE		Callovian		c. 139	Downwarp stage begins. Crustal collapse allows resumed transgression.
			Bathonian		c. 163	Uplift and rifting focused on the central North Sea occur, accompanied by extrusion of alkali basalts. Fluvio-deltaic sediments generated from the uplifted rift margins are deposited both axially in the graben and laterally in NE England. Lagoonal and carbonate shelf facies accumulate in areas peripheral to the main locus of uplift.
			Bajocian			
			Aalenian		c. 170	
	LOWER	UPPER	Toarcian			Uplift begins as a prelude to rifting and results in the deposition of marine sands in both Yorkshire and Dorset.
		MIDDLE	Pliensbachian			Opening of the central N. Atlantic begins, accompanied by crustal tension in NW Europe. In this area the Palaeozoic floor already possesses its own structural grain, and movements along these ancient lines are triggered by tensional stresses. Salt diapirs from the Zechstein are also initiated, and their orientations often parallel ancient structural trends. Mostly marine mud deposition occurs, with some sand and condensed limestones and ironstones.
		LOWER	Sinemurian			
			Hettangian			Major transgression continues. Limestone–marl alternations are deposited in S. Britain while fluvial sequences accumulate in the northern North Sea and Moray Firth areas. Evaporites form to the south in Aquitaine and the central Atlantic.
TRIASSIC			Rhaetian		c. 194	Transgression commences on a world scale.

topmost Kimmeridgian Stage recognised in Britain (Table 14.1).

14b Eustatic changes

When the area of continents covered by marine sediments is plotted for the Jurassic (Hallam 1975*), a picture emerges of progressive overstep by later formations upon earlier ones. With the exception of the early Bathonian Stage, this pattern holds good until the Oxfordian–Kimmeridgian (Figs 14.2, 14.3 and 14.4). However, towards the end of the Jurassic, the area of the epeiric seas became significantly reduced. Synchronous transgression and regression affecting so many continental areas can hardly have been produced by merely local subsidence and uplift. The actual amount of increase or decrease in sea level reflected by the curve (Fig. 14.2) is difficult to assess. In many areas, the replacement of low-lying 'deltaic' or coastal evaporitic environments by more open marine environments might have required only a few metres of actual sea-level rise. Certainly, eustatic changes of hundreds of metres would have been unnecessary.

Eustatic changes of sea level result either from the freezing and melting of ice caps or from changes in the volume of the ocean basins. There is no geological evidence for Jurassic glaciation and, from palaeomagnetic evidence (Figs 14.3 and 14.4), no continent is thought to have been polar in location. Thus we can be fairly certain that ice caps on the scale of those of the Quaternary did not exist (Ch. 14m). The construction and subsequent collapse of oceanic ridges produce volumetric changes within oceans, and Pitman and Talwani (1972) have suggested that central Atlantic spreading occurred

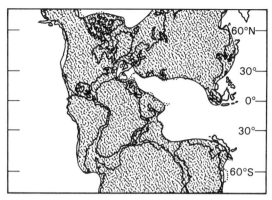

Figure 14.3 Generalised world palaeogeography for the early Jurassic (Hettangian). Land areas shaded (after Hallam 1975*).

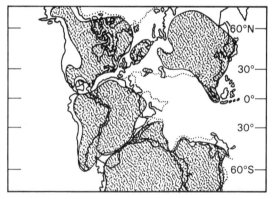

Figure 14.4 Generalised world palaeogeography for the late Jurassic (Oxfordian). Land areas shaded (after Hallam 1975*).

from the early Jurassic onwards. Thus ridge construction would correlate well with the onset of transgression (Hallam 1975*). A slow-down in spreading rates towards the end of the Jurassic could have led to the subsidence of ridges and the retreat of seas from continental areas. Local epeirogenesis, however, also complicated this simplified picture, as will be discussed next.

14c Epeirogenic movements and rifting

Stratigraphic, sedimentological and igneous evidence suggests that periodically sectors of continental lithosphere were locally upwarped, e.g. in the North Sea (Hallam & Sellwood 1976*). These movements would have been accompanied by tension and the downward propagation of normal

Figure 14.2 Approximate percentage of the continental area covered by sea during the Jurassic (after Hallam 1975*).

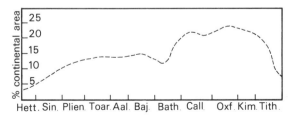

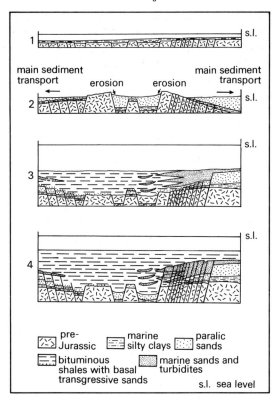

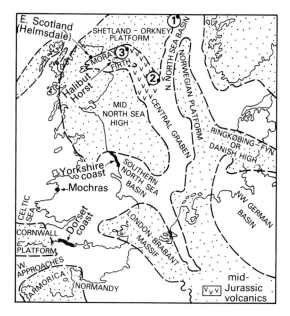

Figure 14.6 Localities of outcrops and features referred to in the text and the sites of the more persistent structures.

1 = Brent oilfield; 2 = Forties oilfield; 3 = Piper oilfield.

Figure 14.5 Rifting model illustrating a possible mechanism for facies development in the North Sea area through the Jurassic.
(1) Early Jurassic marine mudstones are deposited on older rocks already dissected by a grid of faults.
(2) Doming of the crust in the middle Jurassic leads to graben collapse associated with local volcanism. Uplift associated with doming provides both the graben-fill sediment and more regional regression.
(3) Later Jurassic subsidence allows marine transgression to occur, and much of the area is enveloped in basinal shales. Local horsts provide coarse-grained clastics which sometimes enter the basin as tectonically triggered mass flows.
(4) Continued subsidence results in most of the area being enveloped in bituminous shales (Kimmeridgian).

faults, particularly along ancient structural lines (Fig. 14.5). In the British area, uplift of the central North Sea is thought to have commenced in the late Lias. In N. England these movements at first provided detritus for marine sandstones, but subsequently led to folding and erosion of the early Jurassic sediments. Middle Jurassic time witnessed the strongest differential vertical movements with the collapse of the Central North Sea Graben (Fig.

14.5), the modest uplift of marginal horsts and the eruption of alkaline olivine basalts (Howitt *et al.* 1975). Volcanism was centred on the junction of the Central North Sea Graben with the Moray Firth—Viking Graben (Fig. 14.6). Late Palaeozoic rocks were exposed upon numerous horsts which supplied coarse clastic materials to adjacent subsiding regions (Fig. 14.5). Fault-bounded blocks such as the London—Brabant Massif away from the main locus of uplift were also rejuvenated at this time, as indicated by the periodic spread of sandy paralic facies northwards into the Midlands (Sellwood & McKerrow 1974, Horton 1977).

As seen in the late Tertiary evolution of the Red Sea (e.g. Lowell & Genik 1972), events such as those described above accompanied the rifting phases that preceded sea floor spreading (Burke & Dewey 1973, Hoffman *et al.* 1974). However, seismic refraction data show that oceanic crust was never generated in the North Sea; instead, strong subsidence affected the region during the later Jurassic (Fig. 14.5). Subsequently a new cycle of uplift and rifting west of Britain is postulated to have occurred in the early Cretaceous (Ch. 15e).

14d Lower Jurassic facies in Britain

Although the latest Triassic sediments were marine influenced, wholly marine environments did not become established over most of Europe until the beginning of the Jurassic. N. Europe is the only region outside the Pacific margins or parts of the Tethyan zone (the present Alpine–Himalayan belt) where a marine Rhaetian–Hettangian sequence is known. In NW Europe transgression produced a broad epicontinental sea that spread, more-or-less progressively, over the cratonic hinterland (Fig. 14.1).

Modern shelf seas are mostly confined to narrow submergent strips adjacent to the world's oceans, and facies distributions within them are strongly influenced by storm, wave and tidal currents originating in the adjacent oceans. In sharp contrast, the early Jurassic shelf of NW Europe seems to have been of large areal extent, and the facies that developed there included laminated bituminous shales, rhythmic sequences of calcilutite and marl, oölitic ironstones and condensed cephalopod limestones all of which appear to lack direct modern analogues.

In Britain, the lower Jurassic (or Lias) is developed primarily in marine shale facies which are more calcareous in the south and more sandy in the north (Figs 14.7 and 14.8). The two major shale groups of the early and later Lias are separated by the shallower water series of sandy shales, sandstones and oölitic ironstones of the middle Lias (Fig. 14.8).

The Jurassic sequence begins with a series of shale–marl and micritic limestone rhythms termed the Blue Lias Formation (Fig. 14.8), which follows the latest Triassic limestones above a minor disconformity. Much of the fine-grained carbonate of the Blue Lias might have been produced by coccolithophorids that lived in the marine surface waters. The rhythmic alternations of limestones and marls may be interpreted in terms of fluctuating rates of clastic sediment supply followed by selective cementation of the clay-starved limestone units (Fig. 14.9). Winnowed ammonite-shell concentrates suggest that the sea floor was below the limits of normal wave activity but within the reach of storm-driven waves. Phases of slackened clay input seem to have promoted the formation of condensed limestones rich in ammonites, and microplankton blooms in the surface waters might have caused mass mortalities of nektonic animals.

Although the limestone–marl lithologies are typical of the earliest Lias over most of S. and central Britain, in the Mendips and Glamorgan areas (Fig. 14.1) sequences are condensed and in places comprise oölitic calcarenites that rest directly upon Palaeozoic rocks. During the Triassic and early Jurassic, cave and fissure systems developed in the Carboniferous limestone, and within these caverns terrestrial material accumulated including reptilian and terrestrial plant debris (Ch. 13j). Thus the Mendips–Glamorgan area seems to have comprised an archipelago of low islands fringed by carbonate beaches.

In the Mendips area the earliest marine deposits

Figure 14.7 Generalised late Lower Jurassic facies and palaeogeography for the British area (after Hallam & Sellwood 1976*).

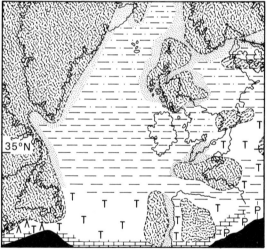

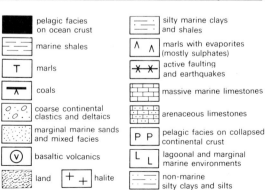

■ pelagic facies on ocean crust		silty marine clays and shales
marine shales		∧ ∧ marls with evaporites (mostly sulphates)
T marls		✳—✳ active faulting and earthquakes
⊼ coals		massive marine limestones
o·.o· coarse continental clastics and deltaics		arenaceous limestones
marginal marine sands and mixed facies		P P pelagic facies on collapsed continental crust
Ⓥ basaltic volcanics		L L lagoonal and marginal marine environments
land + + halite		non-marine silty clays and silts

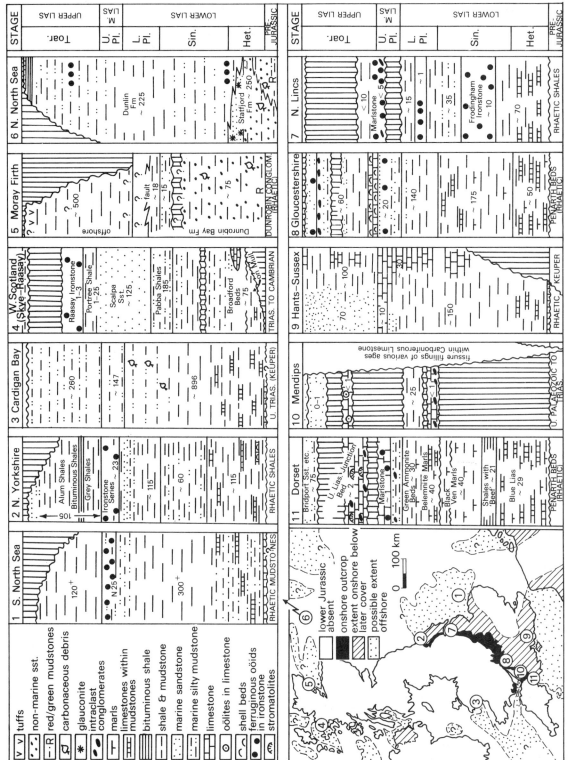

Figure 14.8 Stratigraphic correlation chart for the lower Jurassic.

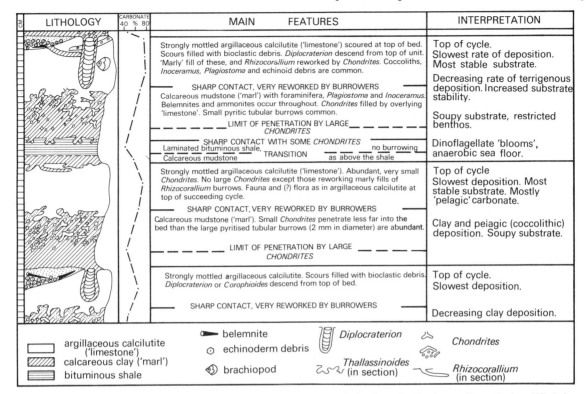

CM	LITHOLOGY	CARBONATE 40 % 80	MAIN FEATURES	INTERPRETATION
			Strongly mottled argillaceous calcilutite ('limestone') scoured at top of bed. Scours filled with bioclastic debris. *Diplocraterion* descend from top of unit. 'Marly' fill of these, and *Rhizocorallium* reworked by *Chondrites*. Coccoliths, *Inoceramus*, *Plagiostoma* and echinoid debris are common.	Top of cycle. Slowest rate of deposition. Most stable substrate.
			SHARP CONTACT, VERY REWORKED BY BURROWERS	Decreasing rate of terrigenous deposition. Increased substrate stability.
			Calcareous mudstone ('marl') with foraminifera, *Plagiostoma* and *Inoceramus*. Belemnites and ammonites occur throughout. *Chondrites* filled by overlying 'limestone'. Small pyritic tubular burrows common.	
			LIMIT OF PENETRATION BY LARGE CHONDRITES	Soupy substrate, restricted benthos.
			SHARP CONTACT WITH SOME *CHONDRITES*	
			Laminated bituminous shale, no burrowing TRANSITION as above the shale Calcareous mudstone	Dinoflagellate 'blooms', anaerobic sea floor.
			Strongly mottled argillaceous calcilutite ('limestone'). Abundant, very small *Chondrites*. No large *Chondrites* except those reworking marly fills of *Rhizocorallium* burrows. Fauna and (?) flora as in argillaceous calcilutite at top of succeeding cycle.	Top of cycle Slowest deposition. Most stable substrate. Mostly 'pelagic' carbonate.
			SHARP CONTACT, VERY REWORKED BY BURROWERS	
			Calcareous mudstone ('marl'). Small *Chondrites* penetrate less far into the bed than the large pyritised tubular burrows (2 mm in diameter) are abundant.	Clay and pelagic (coccolithic) deposition. Soupy substrate.
			LIMIT OF PENETRATION BY LARGE CHONDRITES	
			Strongly mottled argillaceous calcilutite. Scours filled with bioclastic debris. *Diplocraterion* or *Corophioides* descend from top of bed.	Top of cycle. Slowest deposition.
			SHARP CONTACT, VERY REWORKED BY BURROWERS	Decreasing clay deposition.

belemnite   *Diplocraterion*   *Chondrites*

echinoderm debris   *Thallassinoides* (in section)   *Rhizocorallium* (in section)

brachiopod

argillaceous calcilutite ('limestone')
calcareous clay ('marl')
bituminous shale

Figure 14.9 Main features and interpretation of limestone—marl rhythms in the lower Jurassic (modified from Sellwood 1970).

range in age from early to middle Jurassic. In any one place throughout this period both expanded and condensed marine sequences accumulated, suggesting that the Mendips region was undergoing differential vertical movements throughout the early Jurassic. It sometimes acted as an elevated block (as a swell or shoal) and at other times subsided by the same amount as the surrounding basins (Sellwood & Jenkyns 1975*).

The calcarenitic sediments of the swell areas were richer with respect to suspension-feeding faunas than were the more muddy basinal sediments. Those of the Mendips and Glamorgan regions contain abundant infaunal and epifaunal suspension feeders, such as *Astarte*, *Cardinia*, *Pleuromya*, *Pholadomya*, *Gryphaea*, brachiopods, colonial corals and the trace fossils *Thalassinoides* and *Diplocraterion*.

In S. Britain, shallow marine facies of middle to late Lias age are represented by cyclically arranged mudstones, sandstones, limestones and ironstones (Sellwood & Jenkyns 1975*). The mudstones typically contain a fauna of protobranch, lucinoid and thin-shelled pectinid bivalves, ammonites and belemnites. These features all indicate a shelf mud assemblage. The mudstones often pass upwards into wave-rippled quartzitic sandstones containing a more diverse assemblage of infaunal and epifaunal suspension-feeding organisms. *Diplocraterion*, *Skolithos* and *Thalassinoides* burrows are locally abundant. In N. Britain, cyclic sequences of this type also typified much of the early Jurassic with greater and lesser developments of sands occurring in different places at different times (Sellwood 1972).

By comparison with the carbonate and ferruginous facies (see below), these sandy and silty sequences accumulated very rapidly; 30 m or more of sediment often represent little more than one ammonite zone, whereas a mere 1 m or so of calcareous and ferruginous beds may include several zones (Fig. 14.8).

There are several factors controlling the deposition of mud, as opposed to sands, in modern shelf seas. These include the density of suspended matter, the depth of water and the influence of tidal currents and waves (McCave 1971). In these early Jurassic sediments, symmetrical ripplemarks are the most important structures, suggesting that wave activity was the main factor controlling the deposition of mud. The transition from mud to sand in Lias coarsening sequences most probably resulted from enhanced wave activity caused by shallowing.

Ironstones and shelly limestones locally cap these coarsening sequences and, although volumetrically subordinate to other facies, chamositic ironstones developed from time to time during the early and middle Jurassic. Significant examples include the Frodingham, Banbury, Cleveland, Northampton and Raasay Ironstones. Chamosite, siderite and goethite constitute the most important minerals in the ironstones and can be present as either oöids or matrix. Chamositic ironstones are more normally associated with sediments containing slightly coarser, terrigenous clastic components, such as occur in N. and Midland Britain (Fig. 14.8). Chamositic ironstones are often cross-stratified and contain abundant and diverse shallow-marine faunas dominated by suspension feeders (Hallam 1975*). As noted above, by comparison with laterally equivalent shaly and sandy sequences, the ironstones are stratigraphically condensed (Fig. 14.8).

Chamosite and siderite both require reducing environments for their formation (Curtis & Spears 1968), whereas the oölitic structure, cross-bedding and palaeontological evidence suggests that mobile aerobic substrates actually existed. In an attempt to explain this paradox, most workers agree that the mineralogy of these ironstones resulted from authigenic and diagenetic processes (Hallam 1975*). Modern chamosite is known from the Niger and Orinoco deltas (Porrenga 1966) and from partially land-locked basins (Rohrlich *et al.* 1969). With both occurrences there is a close association between faecal pellets and chamosite. This suggests that a high organic content in the sediment is necessary for chamosite formation.

Iron enters modern marine basins either colloidally within organic complexes or as iron oxide coatings upon clay micelles. Over the sites of ironstone formation some mechanism of iron preconcentration probably operated, along with a diminution of normal terrigenous clastic influx. Hallam (1975*) has suggested that ironstones

accumulated on offshore shoals or swells. Although this model satisfactorily explains the stratigraphic relationships, the exact geochemical prerequisites for ironstone formation remain enigmatic. Perhaps it was necessary for iron to be concentrated biologically, by either bacterial or algal processes.

As with the chamositic ironstones, the associated limestones are usually stratigraphically condensed and consist of biomicrites and biosparites that are sometimes ferruginous. They may contain intraformational conglomerates, stromatolitic crusts and oncolites, as in the Lias Junction Bed of the Dorset coast and the Cephalopod Bed of SW England. Ammonites are locally abundant and may sometimes be seen to have suffered corrosion upon the sea floor prior to encrustation by serpulids or algae. Corrosion of aragonitic ammonite shells probably reflects the slow rates of deposition, which allowed long exposure of the shells to sea water undersaturated with respect to $CaCO_3$. The micritic Junction Bed has significantly higher manganese concentrations than most N. European early Jurassic deposits and is comparable with ammonite-rich condensed pelagic limestones of S. Europe (Ch. 14g).

Condensed limestones accumulated in environments with reduced terrigenous clastic deposition. Conglomeratic and stromatolitic structures within the limestones indicate shallow and photic environments of deposition, and the essentially local nature of these beds, at any one time, suggests a shoal or swell type of setting (Sellwood & Jenkyns 1975*).

Shales and mudstones are the sediments most typically associated with the early Jurassic. The normal mudstones with more-or-less silt- and sand-grade quartz have already been discussed. Benthic faunas are normally dominated by deposit feeders. However, at times environmental conditions gradually deteriorated, and even these faunas became progressively more restricted. Ultimately even the tolerant protobranch bivalves were excluded, and unbioturbated sequences of bituminous laminated shales, containing only the remains of swimming or floating organisms, developed. In Yorkshire, such a set of conditions developed during the late Lias (Fig. 14.8) and culminated in the deposition of the Jet Rock, which, in addition to logs of wood (the Jet), also contains well-preserved marine reptiles. At the same time as Jet Rock deposition in Yorkshire, similar conditions extended as far south as central Germany, allowing the deposition of the Posidonienschiefer. Here, apart from rare logs encrusted with bivalves and crinoids, the indigenous faunas consist of ammonites, belemnites, fish, ichthyosaurs and plesiosaurs, often remarkably preserved. Bituminous material is present in the shales as kerogen, i.e. complex hydrocarbon molecules of high molecular weight. These shales are clearly potential hydrocarbon source-rocks, because occasional uncrushed fossils have their internal cavities filled with oil.

Oxygen-deficient conditions affecting the sea floor obviously inhibited the development of a burrowing benthos and allowed the undisturbed laminations to accumulate. Some laminae in the Posidonienschiefer contain abundant coccoliths and may represent 'blooms' of plankton or seasonal increases in coccolithophorid production. Oxygen-deficient conditions are known in modern seas, but they are usually confined to the deeper zones of partially enclosed basins (e.g. the Black Sea, Mediterranean and Gulf of California). These early Jurassic shales are both vertically and laterally associated with shallow marine deposits and clearly did not accumulate at great depths. However, facies like the Jet Rock and Posidonienschiefer represent the most restricted, and perhaps the deepest water, deposits of the N. European epeiric basin. They obviously accumulated below the limits of wave effectiveness and were poorly ventilated.

Most of the sediment deposited over NW Europe during the early Jurassic was of terrigenous clastic origin, and clearly some major source areas were constantly supplying this material. In the North Sea province of the Viking Graben, however, continuous red-bed and fluvial deposition continued across the Triassic/Jurassic boundary as the currently oil-bearing Statfjord Formation. Diagnostic fossils are lacking, but later lower Jurassic beds are represented by the conformably overlying marine shales, siltstones and mudstones of the Dunlin Formation (Jones *et al.* 1975, J. M. Bowen 1975*). The Statfjord Formation represents a basal sequence of braided river sands and is succeeded by a coastal-barrier sand complex. Onshore, early Jurassic coal-bearing sequences are encountered in Scania, on the Danish island of Bornholm (Fig. 14.1), and as carbonaceous fluvial sands in E. Scotland. On Bornholm, coals are associated with both fluvial and marginal marine deposits which include fining-upwards and flaser-bedded successions influenced by weak tidal currents at the time of deposition (Sellwood 1975).

Coarse terrigenous clastic material is more dominant towards the northern and eastern parts of Europe and was mostly derived from the Scandinavian Shield. On Bornholm, early Jurassic sediments rest upon deeply kaolinised basement rocks. The rich flora in these sediments shows that the hinterland was vegetated. In Greenland, source areas lay in the north and west on the Greenland Shield, and both the Greenland and Scandinavian Shield areas were periodically uplifted throughout the Mesozoic to provide most of the muddy sediments that accumulated in the adjacent basins.

14e Early Jurassic islands, horsts and swells

Recognition that both stratigraphically condensed and expanded facies developed simultaneously led first to the concept of axes of uplift and later to that of swells (for reviews see Arkell 1933*, Hallam 1958 and Sellwood & Jenkyns 1975*). These swells produced shoal areas upon which a variety of condensed deposits accumulated. Thicker sequences were deposited in intervening areas, which functioned as active basins. In some areas condensed sequences occurred at discrete intervals and were separated by episodes of more rapid sedimentation. But other areas, although subsiding intermittently, acted as swells for much of lower Jurassic time. Of these latter areas the best known are the London−Brabant, Mendips and Armorican Massifs and the Market Weighton block. At times they were minor producers of terrigenous clastic material and constituted islands. Locally, ferruginous sands and gravels rest unconformably upon weathered Palaeozoic basement (e.g. Normandy and in boreholes under S. England). Restriction of clastic deposits to thin basal gravels suggests that these island areas were topographically subdued, supporting only minor river systems.

The Market Weighton block persisted as a positive feature throughout the whole of the Jurassic (and much of the Cretaceous). However, the

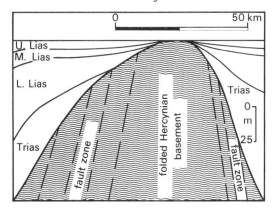

Figure 14.10 Diagrammatic section across the Mendips area, showing the relationship between Mesozoic sediments and the Hercynian basement rocks (after Sellwood & Jenkyns 1975*).

London–Brabant and Mendips Massifs were sometimes islands or swells but at other times were actively subsiding. On geophysical grounds, some of these massifs can be shown to have fault-controlled margins (e.g. the London–Brabant and the Mendips; Fig. 14.10), but for others (e.g. the Market Weighton) the nature of the margins is uncertain.

A likely mechanism for the relative buoyancy of these areas is that they contain light crustal rocks at depth and that the blocks themselves are uncoupled from adjacent crustal units by old fault lines. Such a mechanism would help to explain why their movements were pulsatory and, therefore, comparable in some respects to the movements of salt diapirs. The

Figure 14.11 Distribution and trends of salt structures in the southern North Sea, some of which began to form in the Triassic and Jurassic (after Christian 1969). Tf = Tornquist Line fault belt.

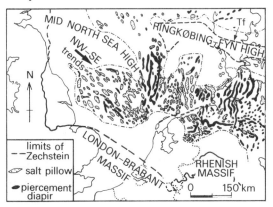

impetus for these differential movements might have come from the tensional stresses acting upon the whole N. European area as a result of the opening of both the Tethys and the central Atlantic (Ch. 14g).

In the North Sea region, the late Triassic witnessed the commencement of salt movements, termed halokinesis. Salt domes and pillows originating in the Zechstein (Fig. 14.11) were propagated over the sites of ancient basement structures (Brunstrom & Walmsley 1969, Christian 1969, Read & Watson 1975). Condensed sequences could have formed over the crests of salt structures, which thus constitute a further potential mechanism for swell development in the North Sea area.

14f Early Jurassic palaeogeography

By employing the stratigraphic framework given by ammonites, it is possible to reconstruct the facies distribution (Fig. 14.12) for individual zones.

Tidally influenced sediments have only been recognised in Scandinavia, and it is probable that the N. European Basin as a whole was wave-dominated. During the early Jurassic, lateral and vertical variations in thickness and facies were considerable (Fig. 14.8), with these variations being related to faulting in the pre-Mesozoic basement. Transgression in NW Europe during the Lias was probably facilitated by a combination of eustatic rise and a major phase of regional tension that slightly thinned the crust and led to a general depression of its upper surface. Local blocks, however, acted independently of the general movements.

In mid-lower Jurassic times the area received more regressive facies, and the first coarsening-upwards sequences, with marine sands, were deposited over SW England. Indeed, at about this time other areas, such as W. France and the Lusitanian Basin of Portugal, started to receive their first Jurassic sands from source areas to the west. These limited facts suggest that sourcelands were uplifted west of Portugal and southwest of Britain during the late lower Jurassic. From palaeomagnetic evidence, Pitman and Talwani (1972) have inferred that rifting occurred between Africa and N. America at about this time (Table 14.1), and such a process (Fig. 14.5) could have been accompanied by

(a)

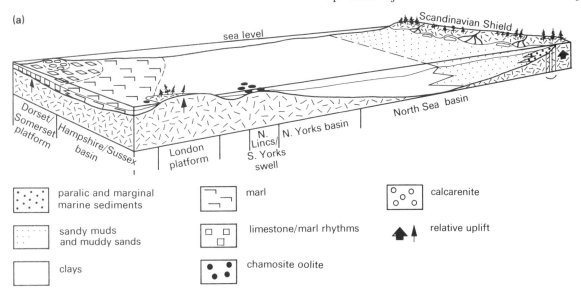

Figure 14.12a Lower Jurassic lithofacies relationships in NW Europe (after Sellwood 1972).

(b)

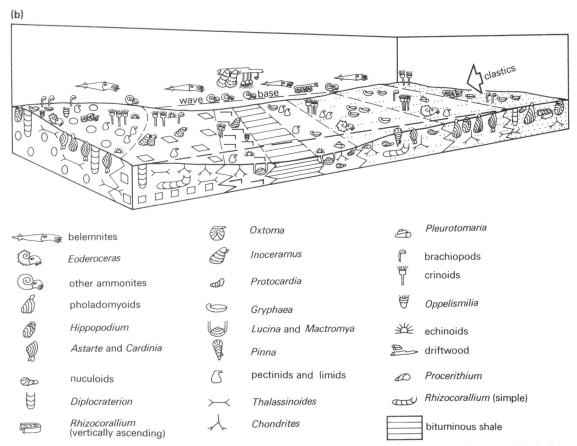

Figure 14.12b Lower Jurassic biofacies relationships in NW Europe (after Sellwood 1972). For key to lithologies see Figure 14.12a.

arching of the lithosphere. Uplift associated with rifting provides a mechanism for the generation of these western sourcelands.

In N. Britain lower Jurassic marine shales pass locally upwards into sands; these were gently folded before renewed deposition occurred in the middle Jurassic (Hemingway 1974). Although these events follow a pattern generally similar to that of events in the south, their cause, as will be shown, may be related to uplift in the North Sea area rather than to Atlantic events (Ch. 14h).

14g   Early Jurassic environments beyond NW Europe

Over much of S. Europe (Fig. 14.13) the Triassic/Jurassic boundary occurs in marine limestones (Bernoulli & Jenkyns 1974*). The carbonates consist dominantly of stromatolitic and peloidal

micrites, with oölitic grainstones and biosparites. Desiccation cracks and 'birdseye' vugs are common, and gastropods and calcareous algae also attest the shallow water origins of most beds. Ammonites occur but are rare; consequently there are great problems in assigning an accurate age to the beds. The limestone facies and structures compare closely with those from modern carbonate platforms, such as the Bahama Banks (Bernoulli & Jenkyns 1974*).

Later in the lower Jurassic, the shallow-water platform limestones suddenly give way to a variety of pelagic sediments including ferromanganese deposits, and biomicrites containing *Bositra* (formerly *Posidonia*), sponge sicules, radiolarians and coccoliths. Ammonites and ammonite aptychi are present, and the sediments often display *Zoophycos* burrows. This pelagic association also includes turbiditic beds of platform-derived grainstones, while, on presumed submarine highs, condensed ammonite-rich red limestones accumulated (Fig. 14.13; Bernoulli & Jenkyns 1974*).

Figure 14.13   Successive stages in the disintegration of carbonate platform environments margining the Tethys Ocean in S. Europe (after Bernoulli & Jenkyns 1974*).

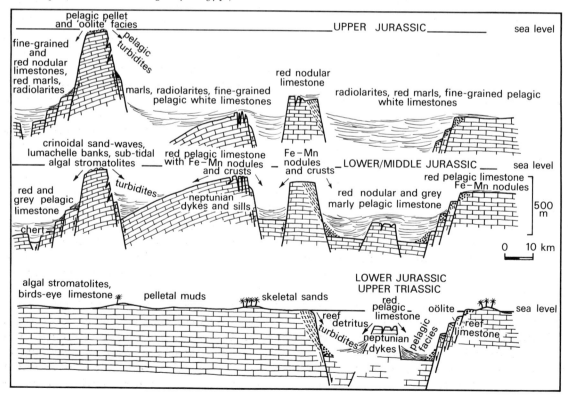

The change from shallow to deep water environments deduced for S. Europe was roughly synchronous with the major change from sandy to more muddy sedimentation in N. Europe (i.e. middle Pliensbachian). Subsidence of the Tethyan carbonate platforms is regarded as due to continental margin collapse following the westward extension of the Tethys Ocean. The opening of the central N. Atlantic Ocean is dated at about this time (c. 180 Ma according to Pitman & Talwani 1972), and it cannot be chance that major facies changes occur concurrently over N. Europe. However, it is as yet not clear whether facies changed in response to a direct tectonic control or to eustatic changes resulting from the subsequent construction of new oceanic ridges.

The central Atlantic rifting phase continued into the lower Jurassic and, on the Nova Scotia shelf and in Aquitaine, Triassic red-beds and older basement rocks are overlain by a thick sequence of early Jurassic evaporites (Hallam & Sellwood 1976*). In E. Greenland, the earliest Jurassic is developed in a coal measure facies with coarse sandstones. Here too, marine transgression provided a major change of facies in the Pliensbachian (Hallam & Sellwood 1976*). The incidence of coals north of Britain and of evaporites in the south provides clear facies evidence of climatic zonation.

14h Middle Jurassic environments in Britain

The middle Jurassic opened with the development of regressive facies. Fluvio-deltaic environments spread southwards into Britain while, at times, central England and W. Scotland became the sites of lagoons. S. England was often an area of shallows starved of terrigenous clastic detritus, and here extensive suites of shoalwater carbonates formed. It was only in SW England and further south that open marine environments were maintained (Figs 14.14 and 14.15). Periodic transgressions spread marine environments northwards into the fluvially dominated regions, and these rare events allowed ammonites to invade the area. Normally, however, cephalopods were excluded from the region, and so a precise stratigraphic base is lacking.

The central and northern North Sea, E. and W. Scotland and NE England were the main areas of fluvio-deltaic deposition, although freshwater environments became temporarily established over the Midlands from time to time. In Yorkshire, three fluvio-deltaic episodes (Fig. 14.14) were each separated by a marine transgression. The typical sequence developed in each cycle is one of abrupt lithological change from marine clastic coarsening-upwards sequences or carbonates into non-marine siltstones and sandstones. The coarsening-upwards successions appear to represent shoreline deposits which are cut erosively by fining-upwards units of fluvial origin. In the Scalby Formation (upper Deltaic Series) an upward change from braided to meandering channel deposits is attributed to eustatic rise during the Callovian transgression (p. 200; Nami & Leeder 1978). The fluvial cyclothems are frequently associated with rootlet beds, plant beds and thin coals. Spectacular meandering palaeochannels are exposed locally (Nami 1976, Nami & Leeder 1978).

Sandstone stringers within the fluvial overbank muds are sometimes traversed by the tracks of dinosaurs, which are particularly well preserved on surfaces with desiccation cracks. Some beds exhibit the abundant moulds of unionid freshwater bivalves that were overwhelmed by rapid influxes of sand, and others display abundant remains of beds rich in the horsetail *Equisetites*. Locally these plants are preserved in their upright position. These types of preservation, involving sudden phases of intense deposition were probably caused by flooding events. Apart from the horsetails, which thrived in the acid soils, an extensive flora of ferns, ginkgoales, conifers and bennettitalean gymnosperms was also established (Hemingway 1974).

In Yorkshire, grainsize trends and palaeocurrents indicate transport from northwest to southeast. The sandy sediments are dominantly mature subarkoses and quartz arenites (Nami 1976) which, along with the restricted suite of heavy minerals, suggest a sedimentary, rather than igneous or metamorphic, provenance. Carboniferous spores sometimes accompany the indigenous middle Jurassic components, indicating that Carboniferous sediments, perhaps on the Mid-North Sea High, provided a considerable proportion of the detritus (Fig. 14.15; Sellwood & Hallam 1974, Nami 1976). Stratigraphic evidence (Hallam & Sellwood 1976*) suggests that the coarser terrigenous-clastic

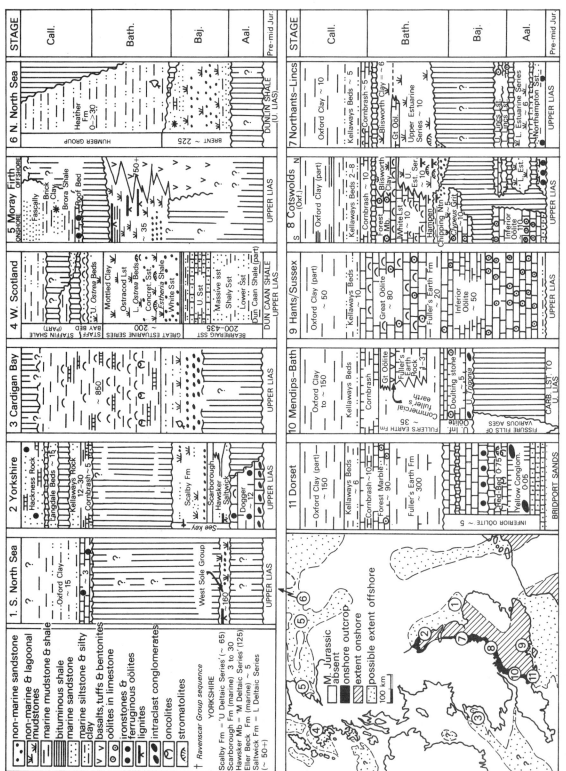

Figure 14.14 Stratigraphic correlation chart for the middle Jurassic. Inverted dishes indicate shelly beds.

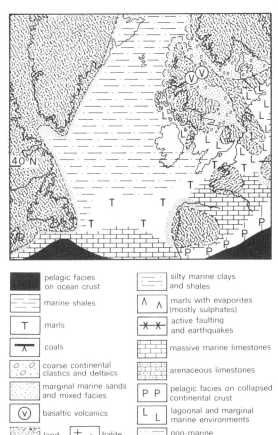

pelagic facies on ocean crust

marine shales

marls

coals

coarse continental clastics and deltaics

marginal marine sands and mixed facies

basaltic volcanics

land   halite

silty marine clays and shales

marls with evaporites (mostly sulphates)

active faulting and earthquakes

massive marine limestones

arenaceous limestones

pelagic facies on collapsed continental crust

lagoonal and marginal marine environments

non-marine silty clays and silts

Figure 14.15 Generalised middle Jurassic (Bajocian) facies and palaeogeography in the British area (after Hallam & Sellwood 1976*).

sequences represent relatively short periods of time by comparison with the intervening marine episodes (Fig. 14.14). Pulses of material may well have been derived during the periodic and sudden phases of uplift that are thought to have accompanied rifting in the North Sea Basin.

In E. Scotland, a fluvio-deltaic sequence developed near Brora, where freshwater sands are overlain by a muddy sequence with fresh- and finally brackish water faunas below a pyritic coal seam. Here, alluviation accompanied transgression and the fluvial abandonment phase led to thick peat accumulation.

NW Scotland received abundant terrigenous-clastic sediments throughout the middle Jurassic. Coastal sand complexes typified the Bajocian, and muddy oyster-rich lagoons developed during the

Bathonian (Hudson & Morton 1969, Hudson 1963). The early middle Jurassic (Aalenian−Bajocian) is represented by a 200−485 m sequence known as the Bearreraig Sandstone. According to Morton (1965), this consists of three coarsening-upwards units in which micaceous marine shales pass up into coarse-grained quartz sandstones. The entire sequence clearly reflects shallowing to high-energy marine conditions. No sedimentological work has yet been published, but the sequence probably represents a series of accretionary shoreface, or barrier island, complexes that were supplied with terrigenous clastics from a Scottish land area (Fig. 14.15). Similar sequences are known offshore in Brent (J. M. Bowen 1975*), and it is possible that north and west of Britain the fluvio-deltaic environments were fringed by higher energy coastlines than those to the south and east. In the North Sea Basin, part of the oil-bearing Brent Sand Formation (Bajocian−Callovian) represents fluvio-deltaic environments both in the Brent field itself (Fig. 14.16) and elsewhere (e.g. the Piper field; Fig. 14.17).

In the English Midlands, modest uplift resulted in the erosion of late lower Jurassic beds and, at the opening of the middle Jurassic, shoreface sediments containing Aalenian ammonites (Northampton Sands) accumulated against the northern fringe of the London−Brabant Massif. In this area, however, sands, silts, variegated clays and rootlet beds succeed the marine sequence. This facies change marks the formation of a lagoonal complex as paralic environments spread southwards from the Yorkshire−North Sea area during the Bajocian. These lagoons gave way southwestwards to a broad shallow shelf upon which a variety of carbonate sediments were deposited. Regressions were usually marked by the progradation of argillaceous and sandy sediments with stronger freshwater influences, whereas transgressions resulted in the eastward spread of shelf carbonates. At other times, carbonates were usually confined to the southern and western areas of Britain (Fig. 14.14).

Marine incursions into the lagoonal belt allowed oyster *(Liostrea)* beds to be established (Hudson & Palmer 1976). At other times, rootlet beds and variegated clays are either unfossiliferous or contain freshwater fossils such as the gastropod *Bathonella* (formerly *Viviparus*), unionid bivalves, the alga

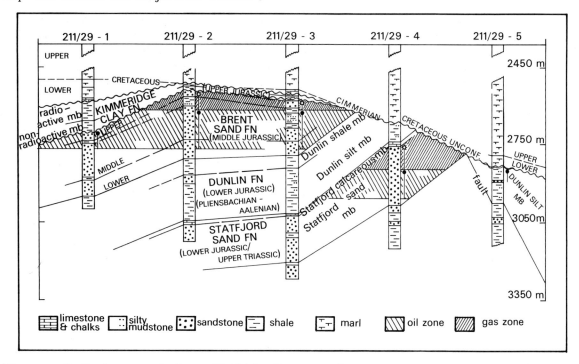

Figure 14.16 Section across the Brent oilfield (after Bowen 1975*).

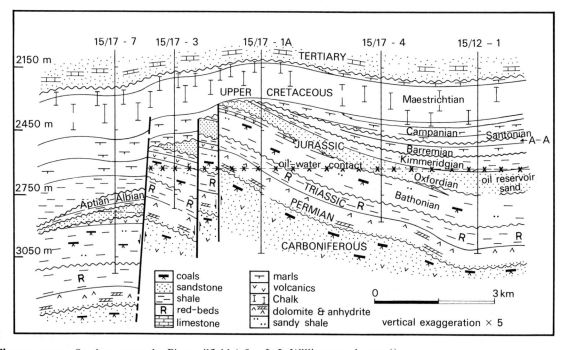

Figure 14.17 Section across the Piper oilfield (after J. J. Williams *et al.* 1975*).

Chara and the arthropod *Euestheria*. Swamp-dwelling dinosaurs are also recorded (e.g. *Cetiosaurus*), and there is complete gradation between facies containing the most restricted euryhaline communities and those with normal diverse marine communities (Hudson 1963, Sellwood 1978; Fig. 14.18). Corals, terebratulids and a few echinoids lived in the least restricted lagoons and were very rarely accompanied by ammonites. In both the English Midlands and NW Scotland, photic lagoonal conditions are marked by the presence of stromatolotic limestones and desiccation features, including 'birdseye' limestones and gypsum pseudomorphs (Palmer & Jenkyns 1975, Hudson 1970). Thus salinity variation and perhaps strong seasonality restricted the fauna in these lagoons.

During the Bathonian too the sea advanced and retreated several times over the Oxfordshire area. The northern margin of the London–Brabant

Figure 14.18 Salinity-controlled fossil assemblages in Bathonian and early Callovian sediments of the Skye area (after Hudson 1963).

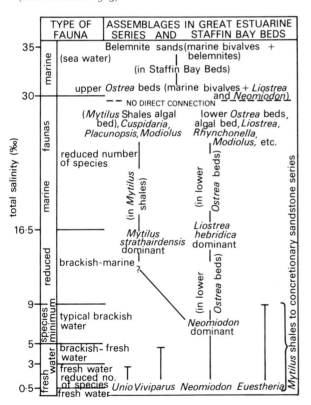

landmass (Fig. 14.15) provided terrigenous clastic material at times of low sea-level stands, and shallow water carbonates formed on offshore shoals. Deepening allowed marine clay deposition but inundated the sourcelands of the sands and temporarily stopped carbonate formation. As the massif itself was transgressed, carbonate production began upon it and continued until the next phase of regression (Sellwood & McKerrow 1974). Ironstones that are mineralogically comparable with those of the early Jurassic formed close to middle Jurassic shorelines during transgressive phases (e.g. Knox 1970, Taylor 1949).

Carbonate facies of the middle Jurassic, as seen in the Inferior and Great Oölite Series of the Cotswolds, range from micrites to grainstones. Oölitic grainstones are commonly cross-bedded on a large scale (sets up to 2 m thick) and represent ripple- and dune-bedded sediments (e.g. J. R. L. Allen & Kaye 1973). Like their modern counterparts, the oöliths probably formed in less than 5 m of water under the influence of strong current activities. Low diversity faunas in the oölites probably reflect the high-mobility high-stress environments of the shifting oölitic substrates (Sellwood 1978). Sheltered shelf-lagoon environments uncontaminated by terrigeneous clastic material were the sites of predominantly lime mud deposition, with calcarenite layers only being introduced during storms.

The montmorillonitic fuller's earth clays of the Bath area, which have been interpreted as bentonites (fine volcanic ash), provide the first evidence of middle Jurassic volcanoes close to Britain (Hallam & Sellwood 1968). Recent drilling in the Forties and Piper oilfields encountered thick sequences (c. 740 m) of undersaturated alkali basalts interbedded with middle Jurassic fluvio-deltaic sediments (Howitt *et al.* 1975). The basalts occur in flows up to 9 m thick and are associated with agglomerates and tuffs. Some of the lava flows were deeply weathered before burial.

The main volcanic centres (Figs 14.6 and 14.15) lay close to the junction of three grabens (the Moray Firth, Central North Sea and Northern North Sea Grabens). It has been suggested that these troughs comprised a triple junction (Whiteman *et al.* 1975, Sellwood & Hallam 1974), with the volcanism and rifting phases of North Sea evolution being directly connected (Howitt *et al.* 1975). This triple junction has even been claimed to mark the site of a mantle plume which caused thermal doming of the crust. The volcanic suite is comparable, in a general way, with other suites of rift site volcanics. Volcanism was certainly associated with uplift, and a major unconformity exists in and around the areas with the greatest development of volcanics (e.g. Figs 14.6 and 14.17). Further to the north, south and west (into the Moray Firth) there are fewer signs of major disturbance, suggesting that uplift was greatest east of Scotland close to the site of the triple junction (cf. Fig. 14.6).

The fuller's earth clays in the Bathonian around Bath (Hallam & Sellwood 1968), and similar montmorillonitic clays in the upper Estuarine Series of the Midlands and in Skye (Bradshaw 1975), were derived from the alteration of airfall volcanic ash. It has been suggested that these bentonites were derived from the proven series of volcanoes in the North Sea. There are, however, serious objections to such an interpretation, particularly the thickness (up to 1 m) of some beds, which would require relatively close volcanic sources. Perhaps more volcanic centres existed south and west of Britain, or even on the London–Brabant Massif.

The youngest middle Jurassic stage, the Callovian, witnessed the commencement of a major transgressive phase, and facies became marine and mainly argillaceous over much of NW Europe. However, Callovian sediments are absent over some structural highs (e.g. Piper) in the North Sea although present in others (e.g. Brent). In Brent, Callovian shales drape over the block-faulted topography and reinforce the idea of phases of tectonic activity within the middle Jurassic.

The initial Callovian transgression spread a relatively condensed coarse limestone known as the Cornbrash over England and condensed glauconitic sandstones into E. and W. Scotland. Subsidence eventually led to a blanketing of the whole region in a series of dark and often bituminous clays, which contain abundant infaunal deposit-feeders (especially protobranch bivalves) and pendant epifaunal suspension-feeders (e.g. *Bositra* and *Meleagrinella*). Environmentally (Duff 1975), these middle Jurassic clays mark a return to restricted muddy marine environments like those of the early Jurassic. In S. Britain the beds are known as the lower Oxford Clay and in W. Scotland as the Dundas Shale (of Sykes

1975). Oxford Clay facies did not develop in N. Yorkshire until late Jurassic times (Fig. 14.14).

14i   Middle Jurassic palaeogeography

The first major indications of local tectonic uplift in the British Jurassic were at the end of the early Jurassic (Table 14.1). Around the margins of the London–Brabant Massif in the E. Midlands, and locally in Yorkshire, sandy and ferruginous Aalenian deposits rest unconformably upon the Lias. In NE England this regression culminated in the development of fluvio-deltaic successions. Marine incursions into this area ceased after the Bajocian until the early Callovian, and thus the Bathonian was probably the time of maximum regression.

South and southwest of the central North Sea area the environments were in turn variable salinity lagoons and, with increasing distance from the clastic sources, dominantly calcareous. To the north and northwest (W. Scotland and the northern North Sea), higher-energy barrier coasts developed.

At the site of maximum uplift in the central North Sea area, basaltic volcanism occurred, possibly synchronously with graben collapse; but following this phase, the area began to subside and transgression began.

14j   Middle Jurassic environments beyond Britain

In the Dutch sector of the Central North Sea Graben there are over 1000 m of non-marine middle Jurassic shales, sandstones and coals, but towards the Netherlands the facies become more marine. Coarse sandstones, kaolinitic clays and coals typify the sequence in S. Scandinavia, where boulder beds provide evidence of contemporaneous uplift. In E. Greenland, early middle Jurassic sediments are dominantly transgressive and may overstep on to Caledonian basement. Sediments generally became muddier, but locally uplift occurred towards the north and, as in Scotland, a continuous supply of coarse clastic detritus was provided (Birkelund *et al.* 1974). On the island of Andøy in N. Norway, deeply kaolinised granitic basement rocks are overlain by fireclays, non-marine bituminous shale, siltstones and cannel coals, giving the only onshore evidence in Norway of the thick offshore succession (Dalland 1975).

West of Britain there is often a major hiatus between the Lias and the early Cretaceous, and so facies reconstructions are speculative. Sandier successions have been encountered in Cardigan Bay (Penn and Evans 1976) approaching the old Manx Horst, but little data are available. Far to the west, middle Jurassic paralic sediments were drilled on Orphan Knoll in the Atlantic, but even here there is a major unconformity between these and the overlying early Cretaceous beds. The significance of this unconformity will be discussed later.

Further west, on the Grand Banks of Newfoundland and the Scotia Shelf, deposition of carbonates and evaporites continued in the south while more terrigenous clastic sequences were deposited towards the north. Facies changes are very rapid. Seismic evidence from this area suggests that pre-Cretaceous grabens and half-grabens were developing and that structures within the individual basins were greatly modified by the movement of early Jurassic salt.

Southwards across Europe, sequences were dominantly calcareous; in the Paris Basin, for example, facies developed comparable to those in S. Britain (e.g. Purser 1975). However, far to the south, many of the carbonate platform areas of the early Jurassic had suffered block faulting and foundering that was locally accompanied by volcanism (Fig. 14.13). Differential subsidence had led to the development of an exaggerated submarine topography of sea mounts and basins which closely controlled facies development (Bernoulli & Jenkyns 1974*).

14k Late Jurassic environments in Britain

Late Jurassic seas reached their maximum extent (Fig. 14.4) during the upper Oxfordian (Table 4.1). This was the culmination of the Jurassic eustatic rise in sea level, and it resulted in a diminution in the availability of coarse-grained terrigenous clastic material. Calcareous deposits spread northwards through Europe to cover a larger area than at any time previously in the Mesozoic. Horst blocks still provided sands and coarser material locally (Figs

14.5, 14.19 and 14.20). The London—Brabant Massif, for example, supplied thick paralic sequences to the Dutch portion of the Central North Sea Graben, and marine sands were produced during the Kimmeridgian transgression in the Boulonnais, on the Piper Structure and on the Mid-North Sea High. More generally, however, the story was one of subsidence accompanied by rapid shale deposition. The extensive development of restricted basinal marine environments allowed the accumulation of laminated bituminous Kimmeridge Shales under stagnant, or near stagnant, marine bottom conditions. Their deposition was probably controlled by a combination of shelf geometry (restricted circulation) and high organic productivity in the surface waters (Gallois 1976).

Mild rejuvenation of existing horst blocks locally accompanied the late Jurassic transgression. A result was the emplacement of a series of transgressive and sometimes glauconitic sands in a facies more typical of regressive situations. In S. Britain, such sands (e.g. the Highworth Grit and Bencliff Grit) are often cross-bedded and were affected by both multidirectional and oscillatory currents. In Oxfordshire and Dorset, flaser bedding is common, and callianassid crustaceans probably produced the pellet-lined *Ophiomorpha* burrows that locally penetrate the sediment. These sedimentary features are common in deposits affected by tidal currents, and modern *Ophiomorpha* structures are well known from modern shoreface environments. U-shaped burrows of *Diplocraterion* are also common, with oysters and trigonid bivalves being the most abundant shells. In Dorset, the Bencliff Grit exhibits a series of well-developed channels which may represent tidal channels (R. C. L. Wilson 1968, Talbot 1973). The sands of the Bencliff Grit are patchily impregnated with oil which forms seepages on the coast.

Under the North Sea, but more richly endowed with oil, is the Piper Sandstone (Fig. 14.17). This glauconitic sand contains a similar suite of structures and faunas to the Oxfordian sands at outcrop. The sand is interbedded with a number of ammonite-bearing shales, and the entire sequence represents a series of regressive shoreface beach-sands (J. J. Williams *et al.* 1975*). Glauconitic and marine sands mark the Oxfordian transgression in Scotland, on the flanks of the Mid-North Sea High and elsewhere around the basin.

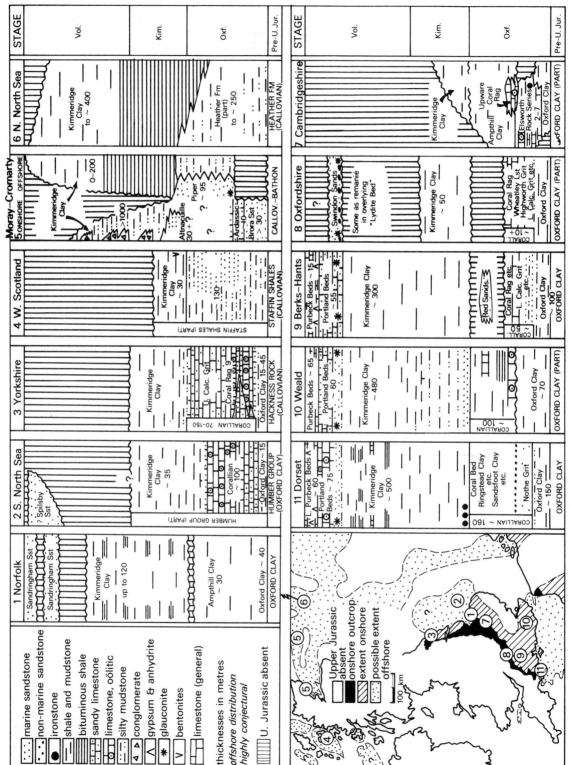

Figure 14.19 Stratigraphic correlation chart for the upper Jurassic.

Early in the late Jurassic, and normally following the deposition of the sandier deposits, coralgal build-ups, oölite shoals and calmer water lime muds formed in a series of shelf carbonate environments. Similar facies were also established towards the end of the Jurassic in the Portland Group of S. England. Environmentally they compared, in a general way, with the middle Jurassic carbonates (Ch. 14h).

The Oxfordian transgression drowned the fringes of the London–Brabant Massif and restricted the availability of quartz sand. Condensed pebbly and transgressive shell-beds pass up into shelly calc-arenites that are sometimes oölitic, and the oölitic beds, in their turn, are overlain by coraliferous limestones from which the Corallian sequence takes its name. The coralgal facies always overlies and interfingers with shelly limestones and oölites,

Figure 14.20 Generalised upper Jurassic (Kimmeridgian) facies and palaeogeography in the British area (after Hallam & Sellwood 1976*).

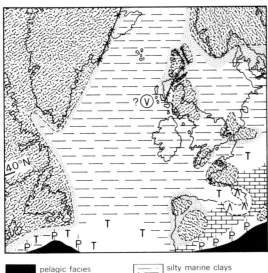

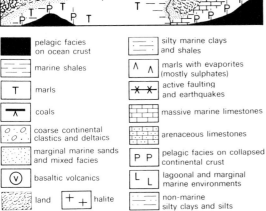

	pelagic facies on ocean crust		silty marine clays and shales
	marine shales	∧ ∧	marls with evaporites (mostly sulphates)
T	marls	✳ ✳	active faulting and earthquakes
∧	coals		massive marine limestones
O . O	coarse continental clastics and deltaics		arenaceous limestones
	marginal marine sands and mixed facies	P P	pelagic facies on collapsed continental crust
Ⓥ	basaltic volcanics	L L	lagoonal and marginal marine environments
	land + + halite		non-marine silty clays and silts

much of this material having been derived from the growing coral patches. Corals in the so-called 'reef' (more correctly 'thicket') association consisted dominantly of *Thecosmilia*, *Thamnasteria* and *Isastraea*, associated with red algae and spongiomorphs. The dead coral heads were often drilled by boring bivalves (e.g. *Lithophaga*), algae, bryzoans and sponges, and pectinid bivalves, brachiopods and a variety of invertebrates inhabited the crevices between corals. Abrasion of corals by physical and biological processes provided debris that filled cavities between coral masses and provided substrates for oyster attachment *(Exogyra)* and invertebrate burrowers. The more sensitive cephalopods are rare in the coral associations, but thick-shelled grazing gastropods provide striking parallels with modern reef associations. Influxes of bentonitic material sometimes partially killed the fauna in both S. England (Ali 1977) and Yorkshire, draping the low relief (a few centimetres) of coral heads in the thickets. Talbot (1973) has interpreted the well-defined Corallian cycles of S. England as the result of rapid eustatic sea-level rise, followed by coastal progradation (Fig. 14.21).

After the main phase of late Jurassic transgression in E. Scotland, the paralic environments of the Moray Firth Basin underwent massive and rapid subsidence, while N. Scotland continued to supply the area with coarse clastic detritus. The line separating the rising block from the subsiding basin was the Helmsdale Fault, an ancient offshoot of the Great Glen Fault (Fig. 14.20).

The initial subsidence phase is marked by massive sands resulting from debris flow. These pass into ammonite-bearing carbonaceous shales containing minor turbidites (Fig. 14.22). Accompanying the turbiditic sands are thick boulder beds, of mass flow origin, containing blocks of Old Red Sandstone, Moine gneiss, Sutherland granite and eroded Jurassic sandstones. Upwards, through the 700 m of Kimmeridgian deposits, there is a gradual reduction in the carbonaceous content of the matrix and an accompanying increase in the calcareous content. Reworked Jurassic debris is absent higher up, and boulders other than Old Red Sandstone are lost. High in the sequence the matrix between boulders consists of fragmental shells including corals, brachiopods and echinoderms. Neves and Selley (1975) have suggested that the sequence can be

(a)

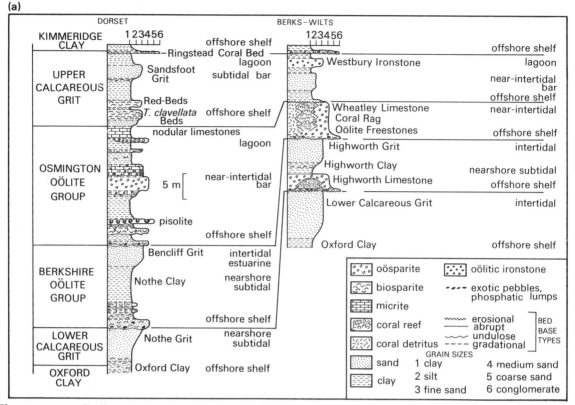

Figure 14.21a Cyclic sequences and interpretation of depositional environments of the Corallian (upper Oxfordian) in S. England (after Talbot 1973).

(b)

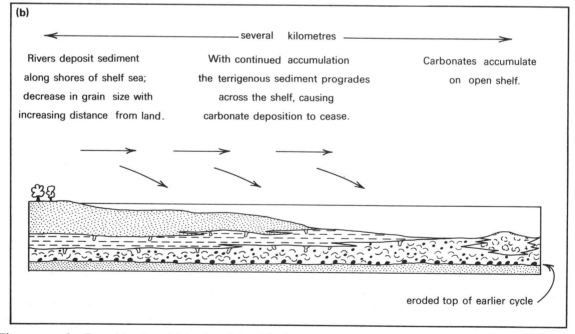

several kilometres

Rivers deposit sediment along shores of shelf sea; decrease in grain size with increasing distance from land.

With continued accumulation the terrigenous sediment progrades across the shelf, causing carbonate deposition to cease.

Carbonates accumulate on open shelf.

eroded top of earlier cycle

Figure 14.21b Depositional model for the origin of the limestone—clay—sand cycle in the Corallian of S. England (after Talbot 1973).

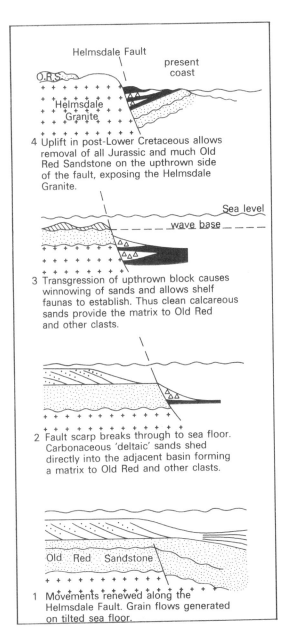

Helmsdale Fault

present coast

O.R.S.

Helmsdale Granite

4 Uplift in post-Lower Cretaceous allows removal of all Jurassic and much Old Red Sandstone on the upthrown side of the fault, exposing the Helmsdale Granite.

Sea level

wave base

3 Transgression of upthrown block causes winnowing of sands and allows shelf faunas to establish. Thus clean calcareous sands provide the matrix to Old Red and other clasts.

2 Fault scarp breaks through to sea floor. Carbonaceous 'deltaic' sands shed directly into the adjacent basin forming a matrix to Old Red and other clasts.

Old Red Sandstone

1 Movements renewed along the Helmsdale Fault. Grain flows generated on tilted sea floor.

Figure 14.22 Model illustrating successive stages in the upper Jurassic evolution of the Moray Firth region adjacent to the Helmsdale Fault (after Neves & Selley 1975).
(1) Movements along the Helmsdale Fault trigger the first mass flows of sediment on the tilted sea floor.
(2) The fault scarp breaks through to the sea floor, allowing carbonaceous sands to be shed directly downslope from a northwestern fluvio-deltaic source. The carbonaceous sand becomes the matrix between clasts of Old Red Sandstone and other rocks, derived from the fault scarp itself.
(3) Sediment source in the northwest is transgressed, allowing shelf faunas to become established and causing winnowed sands with shells to be the new matrix between larger clasts.
(4) Uplift after lower Cretaceous times allows the removal of Jurassic and much Old Red Sandstone sediment north of the fault, exposing the Caledonian Helmsdale Granite.

adjacent to the fault margins clearly reflect the rotational nature of the fault block basins (Fig. 4.23b).

The marine Kimmeridge Shales of the late Jurassic are similar in most respects to those of the earlier Jurassic that precede them. They are often rich in preserved organic material (including kerogen) and contain restricted marine benthos comparable with that of the Oxford Clay. Nektonic animals are represented by ammonites, belemnites and a variety of vertebrates (e.g. ichthyosaurs, plesiosaurs and fish). Rhythmic alternations (on a metre scale) of bituminously laminated shale lacking a benthos and bioturbated shale with a restricted benthos are sometimes developed (as at Kimmeridge itself), and it is clear that bottom conditions oscillated between anaerobic and poorly ventilated from time to time. Fine calcareous laminations within the shales, and locally laminated limestones, are composed of coccolithophorids representing seasonal 'blooms' (Gallois 1976). 'Blooms' may also have been responsible for the mass mortalities of ammonites and fish that plaster some laminae.

A broad, relatively shallow and semi-enclosed shelf (Fig. 14.20) with shallower and deeper zones would have been ideal for the establishment of anaerobic basins. During later burial, organic material within the Kimmeridge Shales matured, due to thermal 'cracking', to provide hydrocarbons which were emplaced in adjacent sands. The Kimmeridge Shales are much thicker per unit time, as measured in ammonite zones, and are much more uniform than other facies developed previously in the

interpreted in terms of a fault-controlled model (Fig. 14.22).

In the Moray Firth area a further horst block existed offshore, the Halibut Horst (Fig. 14.6), and a model for the Moray Firth region at this time could compare with that which developed in E. Greenland during the early Cretaceous (Fig. 14.23a; after Surlyk 1978). In both models active faulting triggered mass movements of sediment, and in the Moray Firth the greater thicknesses of sediment

(a)

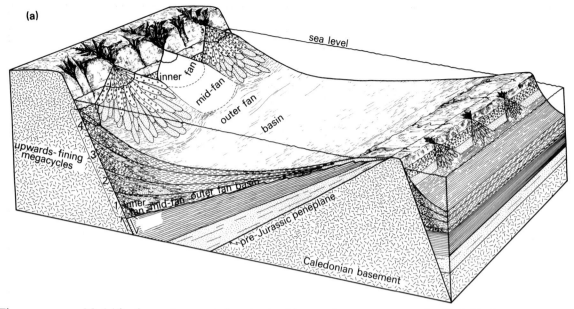

Figure 14.23a   Model for the emplacement of marine boulder conglomerates as a result of fault block movement. The model is based upon the mapped facies relationships observed in the early Cretaceous rocks of E. Greenland (after Surlyk 1978).

Figure 14.23b   Upper Jurassic isopachs in the Moray Firth to northern North Sea area (after Brooks & Chesher 1975).

Jurassic of N. Europe. Much of the area underwent rapid subsidence and, at the same time, received large amounts of argillaceous detritus. It is a paradox that, at a time when so many of the potential sources of sediment were covered by sea (Fig. 14.20), so much mud was still entering the basin. In W. Scotland and in the northern North Sea, pyroclastic material is interbedded with marine mudstones (Knox 1977), suggesting that volcanism was now occurring west of Britain.

South of the London–Brabant Massif the Kimmeridge Clay coarsens upwards into the initially sandy, and subsequently calcareous, Portland Group (Fig. 14.19), reflecting the initiation of Volgian uplift. This restricted further thick Jurassic deposition to the Wessex–Weald area, while thin marine sands were deposited as a northern fringe to the London–Brabant Massif adjacent to the North Sea (especially in S. Lincolnshire and N. Norfolk; see also Ch. 15a). Elsewhere in Britain, uplift produced the late Cimmerian unconformity which separates Jurassic from Cretaceous beds.

The Portland Group contains few corals or brachiopods. Belemnites and crinoids are absent, and the fauna is dominated by bivalves. Ammonites, however, reach monstrous proportions, and

(b)

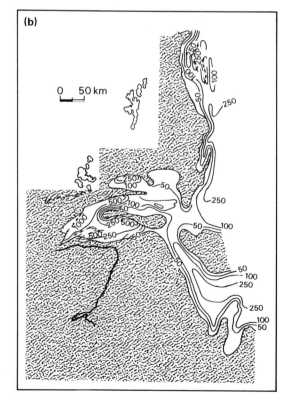

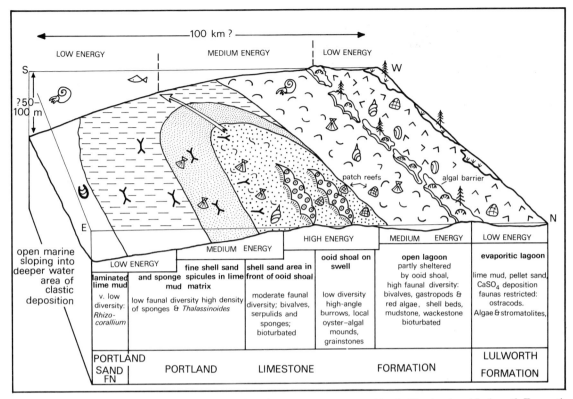

Figure 14.24   Generalised model for depositional environments represented in the Portland and Lulworth Formations of Dorset (after Townson 1975).

Townson (1975) has attributed this restriction of the fauna to hypersaline conditions over the whole basin.

The basal cherty micrites of the Portland Group are rich in sponge spicules and represent deposition in a moderately deep and tranquil marine environment. They pass upwards into higher-energy oölitic grainstones, reflecting a shallowing phase which culminated in the deposition of the lagoonal micrites and evaporites of the Purbeck Group. Townson (1975) has proposed an environmental model (Fig. 14.24) which attempts to show the possible relationships between these late Jurassic facies.

Shallowing probably led to the emergence of most of Britain late in the Jurassic and to a complete restriction of the vestigial Portland sea. Ultimately, a complex of variable salinity lagoons was established over the Wessex–Weald area (Fig. 14.24), and shoreline emergence provided the stromatolitic and pelletal limestones of the basal Purbeck Group. Later, gypsum precipitation began, either on

Figure 14.25   Latest Jurassic palaeogeography of S. Britain and adjacent regions at the Jurassic/ Cretaceous boundary, showing the vestigial marine basin remaining after late Cimmerian uplift (after P. Allen 1972).

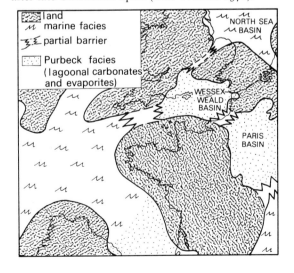

shoreline flats or on the floor of a shallow isolated gulf (Townson 1975, I. M. West 1975). Complete emergence allowed soils to develop, and periodic pluvial phases permitted coniferous forests to grow.

Periodically, marine influences were felt in this basin (Fig. 14.25), but opinions are divided over whether the marine incursions came from the west or south. Casey (1973*) has provided compelling evidence for a northern connection with a vestigial North Sea over the Oxfordshire area (Fig. 14.25) and one particular marine incursion from the north marks the beginning of the Cretaceous period (Rawson et al. 1978*).

14l   Late Jurassic palaeogeography and environments beyond Britain

From drilling sites in the central Atlantic, late Jurassic pelagic facies are known that are directly comparable with deposits of the same age from the Tethyan Alpine−Mediterranean belt (Bernoulli & Jenkyns 1974*). Oceanic crustal basalts at Cat Gap (off the Bahamas) are overlain by a thick sequence of green−grey nannoplanktonic micrites containing radiolarians. These pass up into the slightly nodular, more condensed, red limestones of the Oxfordian−Tithonian (Table 14.1). Aptychi from cephalopods, together with bivalves and replaced radiolarians are common, but ammonite moulds are rare. Interbedded graded and conglomeratic pelagic deposits represent material redeposited by mass flow under deep bathyal conditions.

In the Alpine−Mediterranean belt comparable condensed red nodular limestones are known as 'Knollenkalk' or 'Ammonitico Rosso'. They probably accumulated slowly on submarine highs (Fig. 14.13). Thicker sequences are represented by interbedded grey marls and argillaceous limestones known by a variety of names, including 'Fleckenkalk' and, if cherty, 'Hornsteinkalk'. They are locally interbedded with bituminous sapropelic marls, with associated mass-flow materials indicating the relatively deep basinal origins of these thicker sequences. Although coccolithophorids were present from earliest Jurassic times, they underwent an evolutionary explosion in the late Jurassic and white nannoplanktonic limestones became widespread. Radiolarites developed widely in the Alpine−

Mediterranean belt during the middle to late Jurassic. They generally lack calcareous debris and represent material deposited below the lysocline, the zone which separates well preserved from poorly preserved calcareous assemblages. They too are sometimes associated with turbidites and slumps.

From Texas to Nova Scotia the late Jurassic regression produced a series of shales passing up into high-energy carbonate grainstones comparable with the British Kimmeridge−Portland sequence. In E. Greenland there was a major westward transgression during the Oxfordian, with coarse marine sandstones resting directly upon Caledonian basement (Birkelund et al. 1974). Nearshore sands pass up into dark silty shales which compare with the Kimmeridge Clay Formation of E. Scotland. Late in the upper Jurassic (Volgian) there commenced a phase of tectonic activity that resulted in the rotation of normal fault blocks and the emplacement of mass flow conglomerates derived from the fault margins (Fig. 14.23a).

Towards the end of the Jurassic, worldwide sea level was falling. Simultaneously a new phase of local uplift was under way, but this time west of Britain (Hallam & Sellwood 1976*). This phase was probably related to rifting over the site of the Rockall Trough. In the British area many of the massifs were rejuvenated at this time, and the combination of eustatic lowering and horst uplift produced an end-Jurassic phase of erosion (recognised now as the late Cimmerian unconformity). These same movements promoted that mild uplift of the London−Brabant Massif which caused the periodic isolation of the Wessex−Weald Basin during Portland−Purbeck deposition (Ch. 14k).

14m   Jurassic faunal realms and palaeoclimate

Many of the invertebrate faunas of the Jurassic show increasing provincialism from the late Lias onwards. Arkell (1956*) has recognised three major realms on the basis of global ammonite distributions: Tethyan, Boreal and Pacific. The Boreal realm occupied the northern part of the present northern hemisphere and, except for the Pacific margins, the Tethyan realm occupied the rest of the world. Hallam (1975*) has discarded the Pacific realm because it is not sufficiently distinct from the

Tethyan. The two major realms may themselves be subdivided into a number of provinces.

The concept of realms first arose because it was realised that particular ammonite families and subfamilies were restricted geographically. Ammonites and belemnites are significantly different in the Boreal and Tethyan realms. For most other invertebrate groups the Boreal realm is marked by the omission of Tethyan genera rather than by the development of indigenous faunas (Hallam 1975*).

Since provincialism was first recognised, arguments have raged over its probable causes. The southern limit of the Boreal realm passed through S. Europe, but the boundary was gradational and at various times migrated either northwards or southwards. During the Callovian, Boreal ammonites spread into S. Europe; in the Oxfordian, Tethyan ammonites spread northwards. Tethyan faunas had greater diversity than Boreal ones, and certain groups were largely confined to the Tethyan realm (e.g. rudistid bivalves, radiolarians, tintinnids, dasyclad algae). Of the ammonites, phylloceratids and lytoceratids were very characteristic of the Tethyan realm, although not always confined there. In the past, explanations of provincialism usually fell under four categories: temperature, land barriers, depth of sea and salinity. Earlier hypotheses favoured temperature control, with support for this view coming from the distributions of Jurassic reef corals and carbonates, which both decrease northwards. However, reef corals occur in E. Asia, a region that would have lain in high Jurassic latitudes (Fig. 14.4). Climate was probably very relevant, but other factors such as the overall variability of the environment are likely to have been of greatest importance (cf. Valentine 1973). Where there are strong seasonal differences and high environmental stresses, as in shallow shelf seas, diversities are low by comparison with those in adjacent oceanic areas. In oceans, seasonal variations of all types tend to have less effect. The Tethyan realm, with its high diversities, may reflect the higher environmental stability of areas strongly influenced by oceanic (Tethys and Pacific) waters; the Boreal realm, on the other hand, was mostly confined to the more variable, and thus stressful, environments of the epeiric seas.

As mentioned previously, Jurassic glacial sediments are as yet unknown, and most reconstructions place the poles over oceanic areas (Figs 14.3 and 14.4). Given this configuration, no major permanent ice-cap is likely to have grown, and the vigorous atmospheric circulations of the more recent past are unlikely to have existed. Global climates were probably far more equable than those of the present, and the widespread distribution of reptiles (e.g. dinosaurs) and of plants lacking frost tolerance supports this view.

14n Palaeogeographic evolution and summary

Over much of NW Europe, the Jurassic opened with a marine transgression that was an extension of the late Triassic marine invasion. The sea spread over the heartland of the NW European Craton and formed a vast epeiric (epicontinental) shelf.

Within the epeiric basin, thickness variation in the Jurassic strata can be related to a series of basins and swells resulting from differential subsidence in the Hercynian basement. Where the Jurassic is underlain by thick deposits of Zechstein salt, as in the present North Sea area, the diapiric rise of salt domes and pillows also created a pattern of swells and adjacent basins, with general alignment parallel to the underlying basement grain.

In a simplified way, Jurassic (and even Cretaceous) sedimentation in NW Europe can be related to combination and interplay between two major factors: (a) the elevation and subsequent collapse of rift-related crustal upwarps, superimposed on (b) long-term eustatic changes (Hallam & Sellwood 1976*).

There was a more-or-less progressive eustatic rise in sea level from the earliest Jurassic (Hettangian) to the later Jurassic (upper Oxfordian), followed by a late Jurassic to early Cretaceous fall. Periodic and local upwarps of the lithosphere were accompanied by tension and, occasionally, volcanism. Upwarping occurred first in the early Jurassic over the site of the central Atlantic rift. In the middle Jurassic, doming was centred on the North Sea, with erosion occurring over the main sites of uplift and terrigenous clastic sands being deposited in adjacent areas. During the main phases of uplift, ancient fault blocks in peripheral areas were reactivated by movement along normal faults, which allowed the accumulation of thick basinal marine shales. These shales later matured into ideal oil source-rocks. In the late Jurassic, basinal areas adjacent to active fault-bounded blocks received turbidites and other mass-flow deposits derived from the rising horsts.

Through the Jurassic, major source areas for clastic sediments entering the epeiric seaway lay to the northeast (Scandinavian Shield) and, at times, in the North Sea. At the end of the Jurassic and early in the Cretaceous, upwarping west of Britain, over the site of the present-day Rockall Trough, provided a major western source for later clastics. However, in the British and North Sea areas this was marked by a further rejuvenation and tilting of ancient horst blocks. These processes, combined with an overall lowering of sea level, produced late Jurassic regressive facies in S. England and a major unconformity (the Cimmerian unconformity) between Jurassic and Cretaceous sediments elsewhere in Britain and over the North Sea horsts.

15 The Wealden rivers and chalky seas of Cretaceous Britain

15a Introduction and correlation

The widespread phase of late Jurassic earth movements (late Cimmerian) transformed most of NW Europe into land. In the British area, locally subsiding areas remained, such as those of the Wessex–Weald and East Anglian–North Sea regions. These two basins (Fig. 15.1) were periodically, and at first only temporarily, connected as subsidence over the Oxfordshire–Bedfordshire area allowed marine waters from the North Sea to flood into the Wessex–Weald gulf (P. Allen 1975*). Late Jurassic environments south of the London isthmus consisted mostly of muddy swamps and lagoons depositing limestones and evaporites. At the same time a marine embayment over S. Lincolnshire and N. Norfolk received condensed and glauconitic marine sandstones.

The Cretaceous opened with a short-lived marine transgression which, in S. England, produced the oyster-rich Cinder Bed Member (c. 2 m thick; Fig. 15.2) (Casey 1973*). This oyster bed passes upwards into clays containing the freshwater molluscs *Viviparus* and *Unio*. Around the margins of the Wessex–Weald Basin the Cinder Bed Member passes into marginal sandy facies, but in the north there is faunal evidence of more normal marine salinities approaching the Oxfordshire straits (Casey & Bristow 1964).

Like their Jurassic counterparts, Cretaceous zones are recognised on the basis of ammonites. In S. Britain the lagoonal facies of the end-Jurassic and early Cretaceous contain no ammonites. Thus, correlations with the fully marine beds have been achieved (see Casey & Rawson 1973* and Rawson *et al.* 1978* for reviews) using other organisms (e.g. pollen, spores and ostracods). Casey (1973*) however, has demonstrated that ammonite-bearing marine sequences straddle the Jurassic/Cretaceous boundary in the S. Lincolnshire–Norfolk area (the only area in NW Europe where such a sequence

exists) and that the faunas are of Boreal type, correlating with those of Russia, Greenland and Canada. Ammonite exchange between the North Sea region and the Tethys could not take place directly because of the palaeogeography (Fig. 15.1). Correlations between the Tethyan and Boreal realms are, at the moment, equivocal, being achieved via the Caucasus where Tethyan and Boreal faunas are interbedded.

Figure 15.1 Generalised early Cretaceous (Ryazanian–Barremian) facies and palaeogeography of the British area (after Hallam & Sellwood 1976*).

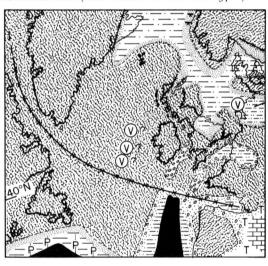

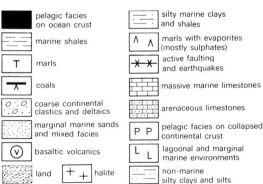

▉ pelagic facies on ocean crust	silty marine clays and shales
marine shales	∧ ∧ marls with evaporites (mostly sulphates)
T marls	✳ ✳ active faulting and earthquakes
⊼ coals	massive marine limestones
o°.°o coarse continental clastics and deltaics	arenaceous limestones
marginal marine sands and mixed facies	P P pelagic facies on collapsed continental crust
Ⓥ basaltic volcanics	L L lagoonal and marginal marine environments
land	+ + halite	non-marine silty clays and silts

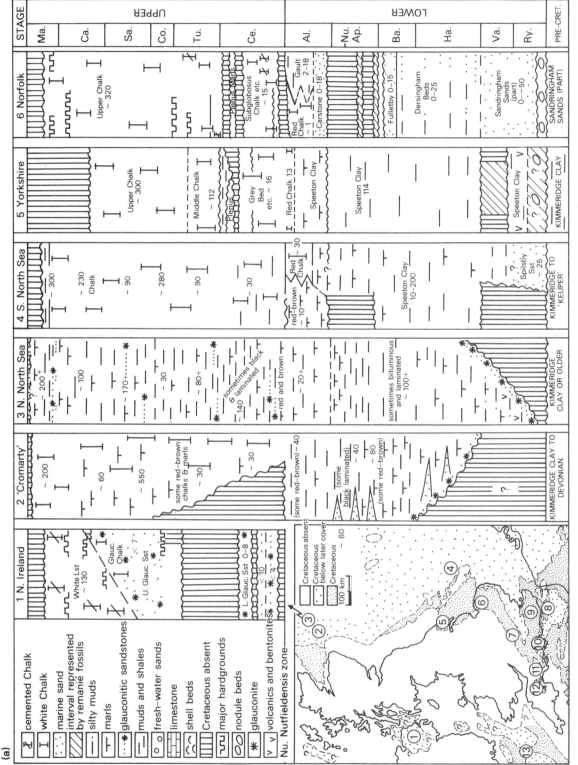

Figure 15.2 Stratigraphic correlation chart for the Cretaceous (after Rawson *et al.* 1978). Numbers (25,300 etc.) refer to average thicknesses (in metres).

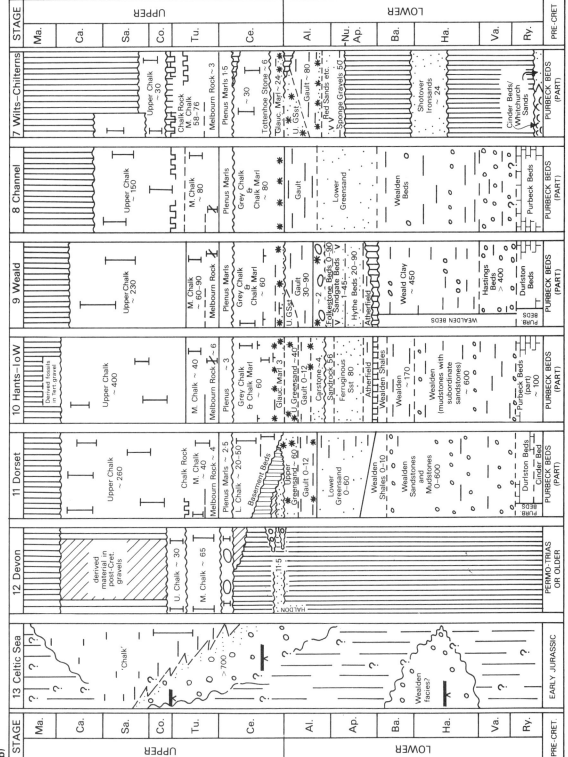

Figure 15.2 (cont.)

System		Stage	Radiometric age (Ma)	Major Geological Events
CRETACEOUS	UPPER		65	Dinosaurs, mosasaurs, ammonites become extinct.
		Maestrichtian	70	End-Cretaceous uplift and inversion begins. Intense compression and flysch emplacement occur in S. Europe.
		Campanian	80	Most of Europe and much of the world are submerged. Chalk facies are widespread. There are signs of tectonic unrest in the Alpine belt as flysch sequences are emplaced.
		Santonian	84	
		Coniacian	87	Regional downwarping largely replaces block faulting as the typical subsidence style in NW Europe.
		Turonian	88	N. European Craton is progressively transgressed from east to west. As the western parts gradually subside, local sands give way to chalk. N. Atlantic opens from Biscay to Labrador.
		Cenomanian	93	
	LOWER	Albian	100	Block faulting enters its final phase. Main transgressive episode starts. Shelf sands in S. and central England suffer strong tidal action. Connection of North Sea and Wessex–Weald Basins is synchronous with volcanism (fuller's earth formation). Graben collapse occurs west of Britain (Rockall Trough). Organic-rich muds are deposited in the North Sea while the world's oceans become oxygen deficient at depth. Central Atlantic reaches one-tenth of its present width.
		Aptian	110	
		Barremian		Weald–Wessex region is a muddy coastal plain fringing a variable salinity lagoon. North Sea area is a marine basin opening northwards into the Arctic. Block faulting continues. Chalky marls are formed.
		Hauterivian		Active block faulting generates sandy coastal plain sands as fan deltas in S. England. An embayment of the North Sea marine basin extends over parts of Lincolnshire, Norfolk and S. Yorkshire. Doming occurs west of Britain (Rockall Trough area).
		Valanginian	120	
				Humidities in the area increase. Evaporite formation ends in the south.
		Ryazanian		Transient marine inundation across the London Platform occurs.
JURASSIC		Volgian	135	Regression and block faulting movements take place in North Sea. London Platform is uplifted, and the North Sea and Wessex–Weald Basins are partially separated.

Table 15.1 Cretaceous stages, time and summary of major geological events. Note: in S. Europe, latest Jurassic to pre-Valanginian rocks are known as Berriasian (data after Casey 1973*, P. Allen 1975, Hancock 1975* and Kauffman 1977).

In S. England, the replacement of limestones and evaporites by freshwater clays, and eventually fluvial sandstones, probably reflects a substantial climatic change towards more humid conditions. P. Allen (1975) has suggested that a causal link may exist between this increased humidity and the Cinder Bed transgression. Run-off, he has suggested, was increased because uplift at the basin margins (particularly over the London isthmus) changed the climate. This occurred synchronously with increased rates of basinal subsidence reflected in the transgression.

Although ammonites form the basis for Cretaceous correlations, the detailed work of W. J. Kennedy (1969, Kennedy & Cobban 1977) has shown that, in the chalk facies of the later Cretaceous, differential preservation has limited ammonites to discrete horizons. Thus zones based upon ammonite assemblages have been erected, but such zones are not directly comparable with Jurassic ammonite zones. Other fossils, including the bivalve *Inoceramus* (Kauffman 1973, 1977) and coccoliths (Perch-Nielsen 1972), may prove to be as good as, or better than, ammonites for biostratigraphy.

Table 15.1 gives the sequence of stages and a summary of the main Cretaceous events in the British area.

15b The Wealden morass

Late Jurassic to early Cretaceous earth movements caused sandy non-marine sequences to develop widely over NW Europe at the opening of the Cretaceous (P. Allen 1967). Such facies are often termed 'Wealden', after the Sussex−Kent area in Britain where the facies has been most exhaustively studied. In this region the Wealden comprises two major groups: the lower sand-dominated Hastings Beds Group and the upper muddier Weald Clay Group (Fig. 15.2).

The Hastings Beds Group reaches a maximum thickness of *c*. 400 m in the Weald. It comprises three major cycles of sedimentation in which claystones and mudstones coarsen upwards into crossbedded sandstones with a capping of frequently bone-rich gravel (Fig. 15.3). Above, there is an eventual return to mudstones after a passage through cross-laminated siltstones rich in the horsetail *Equisetites*. The mudstones often contain freshwater ostracods (e.g. *Cypridea*), algae (charophytes), gastropods (e.g. *Viviparus*) and bivalves (e.g. *Unio*). They sometimes have lenticular shell-beds composed of the brackish water bivalve *Neomiodon*. Until recently (P. Allen 1975*) the Hastings Beds cycles were interpreted as deltaic sequences, with the deltas prograding southwards from an uplifted London−Brabant Massif (London Platform). The muddier facies were interpreted as the prodelta environments and the more sandy sequences as delta shoreface and distributary channel deposits. Eustatically controlled transgressions periodically gave rise to the reworked pebble beds and spread mud facies over the abandoned delta tops.

However, signs of exposure or very shallow water (e.g. soil beds with horsetails, desiccation cracks and reptile footprints) exist in all lithologies throughout the Weald, particularly in those previously interpreted prodelta muds. Other problems

Figure 15.3 Facies sequence and environmental interpretation for the Wealden (Hastings Group) of the Weald (after P. Allen 1975).

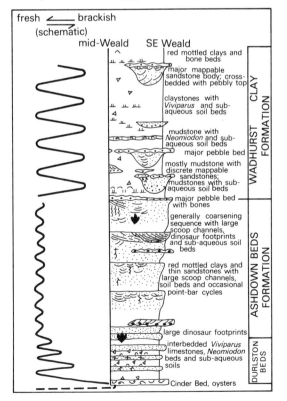

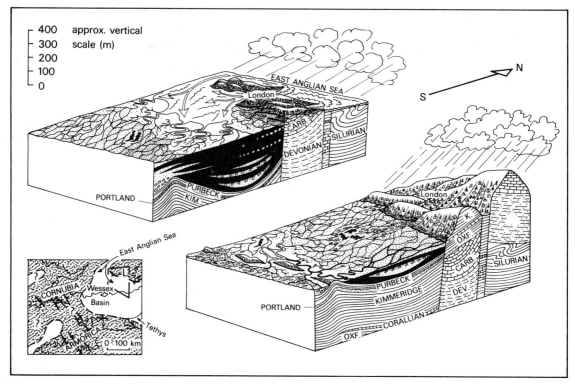

Figure 15.4 Process models for the deposition of the Wealden in SE England: (A) uplift of the London–Brabant Massif generates braided alluvial sandplains; (B) downfaulting and denudation of the massif allows the basin to return to mudplain conditions. Transgression from the north periodically made conditions brackish. In Weald Clay times, Cornubian–Armorican uplift generated braided sandplains in the west (After P. Allen 1975).

concern the transgressions, which appear to have come from the north (the same direction from which terrigenous material was supplied!).

P. Allen's new model (1975*) overcomes these difficulties by reinterpreting all the mud facies as lagoonal to lacustrine mudswamp deposits in which the environmental salinity was controlled by the rate of freshwater run-off. The coarsening sequences are now seen as alluvial braidplain advances into the mudswamp area, with each advance being motivated by fault-controlled uplift of the London Platform. Progressive uplift changed the local climate, increasing run-off and causing braidplain progradation. Subsequent cessation of uplift and continuing subsidence within the Weald Basin resulted in braidplain abandonment and the re-establishment of mudplain conditions. Water depths over the Weald area were always very shallow during Hastings 'times'. Modern horsetails, comparable with those of the Wealden, will not establish themselves in depths of more than about

1·5 m. Desiccation frequently occurred, and stabilised, but waterlogged, sandbars became sparsely colonised by meadows of *Lycopodites* (P. Allen 1976, Harris 1976).

The area of the Weald is seen as a fault-controlled 'trap door' type of basin, its narrow northward connection with the northern sea being via the Oxfordshire–Bedfordshire strait. Supporting evidence for intra-Cretaceous faulting is to be found in the Kent coalfield and under the Thames Estuary (Owen 1971).

The palaeogeology (Fig. 15.4) of the London Platform is revealed by petrographic evidence. Much of the sediment may represent reworked Jurassic material. The pebble fractions contain Portland–Kimmeridge cherts and ammonites, and Carboniferous, Old Red Sandstone and (?)Silurian debris, some of which may have been recycled via the New Red Sandstone.

In the Wessex area rivers were also depositing coarse sediments possessing petrographic affinities with sources in the west including Cornubia, Armorica and NW Iberia (P. Allen 1972). This was the first time since the Triassic that a major supply of coarse terrigenous clastic sediments was derived from western sources, and it hints at a major phase of uplift in this area (Hallam & Sellwood 1976*). Material from Wessex failed to get further east than Winchester, which supports the notion that two partially isolated sub-basins existed south of the London—Wales area at this time. Partition is believed to have been achieved along a linear swell (the South Downs Swell of P. Allen 1975*). This may mark the position of the elevated margin of a tilted fault block in the Hercynian basement (Fig. 15.4).

The transition to the Weald Clay Group is marked by a passage through sequences containing well-developed upward-fining fluvial cycles. These represent the deposits of established meandering channels rather than the sheet-and-stream-flooding deposits of empheral streams. This change probably reflects the gradual geomorphic decay of the London Platform. The basal beds of the Weald Clay are often intensely bioturbated and include crustacean burrows (Ophiomorpha), which suggest considerable marine influence. Many of the non-marine ostracods die out, and the brackish water Neomiodon is replaced by Filosina. The group has a maximum thickness of 450 m, although in places some 150 m have been removed at the top. The facies indicates a reversion to mudswamp environments, with occasional, short-lived and local, arenaceous interruptions (P. Allen 1975*).

Most sands subsequently entered the basin from the west, via Wessex, which may reflect an increase in the competence of the western rivers, coupled with the over-running of the South Downs Swell. Salinities were very low in the east and brackish in the west. Thus once again marine influences seem to have been greatest in the direction of strongest fluviatile influence. This suggests that a delicate balance existed, with normal marine salinities being prevented over most of the basin by the dominance of fluvial run-off. Towards the top of the Weald Clay there is a reversion to marine deposition. The return of oysters and echinoids heralds the beginning of the great transgressive phase of the Cretaceous.

West of Britain there is often an unconformity between the lower Jurassic and the Wealden beds,

which suggests a major phase of pre-Cretaceous uplift. This occurs in the Celtic Sea, Western Approaches and upon the E. Shetlands Shelf. Uplift in the west (Fig. 15.1) may have permitted the dinosaur *Diplodocus*, recently discovered on the Isle of Wight, to have migrated into Europe from N. America. This phase of uplift was succeeded by an inferred graben collapse in the offshore areas mentioned above, and also over the site of the Rockall Trough (Fig. 15.1) where, later in the Cretaceous, oceanic crust was emplaced.

It is tempting to apply the Jurassic model (Fig. 14.5) to this new site of rifting and to view the 'Wealden facies' as detritus resulting from the uplift phase associated with graben formation.

15c The early Cretaceous North Sea

In Norfolk, Lincolnshire and their immediately offshore areas in the North Sea, the Kimmeridge Clay Formation is overlain by the thin (c. 10 m) sequence of glauconitic and phosphatic sandstone known as the lower Spilsby Sandstone. This is equivalent to the Portland and Purbeck sequences south of the London Platform (Casey 1973*). Variable sequences of sand extend as far south as Oxford but are poorly exposed. The Kimmeridge Clay is truncated, and erosion surfaces exist within the lower Spilsby Sandstone itself. These marine sands, with their indigenous and reworked faunas, represent deposition on the southern fringes of a basin that extended over the North Sea and opened into the Arctic Ocean. Transgression from this northern sea produced the Cinder Bed Member in the south (Fig. 15.2).

North of the London Platform the transgression is marked by a nodular bed of cemented sandstone containing abundant reworked Jurassic ammonites and indigenous ammonites of the earliest Cretaceous (Ryazanian) stage. These glauconitic nearshore sands are very condensed, with the whole Ryazanian (Fig. 15.2) being represented by a mere 15 m of sediment. Indeed, in the Norfolk—Lincolnshire area the total thickness of strata equivalent to the Wealden of Sussex (c. 850 m) is seldom more than 70 m. In S. Yorkshire, although the very basal Cretaceous is absent, the earliest Cretaceous is represented at outcrop by about 60 m of the marine

Speeton Clay, which locally thickens to 175 m off-shore (Rhys 1974).

It is clear that deposits thin on to the northern margin of the London Platform. Transgression periodically left deposits upon the platform, but these were systematically reworked by later trans-gressive−regressive episodes. The northward pass-age into muddy beds and the thinness of sands in the area shows that the London Platform was not providing the same volume of terrigenous clastic material here as it was in the south. Perhaps palaeoslopes were directed southwards in this area, and/or only indurated Lower Palaeozoic (or older) rocks were exposed on this part of the massif with more easily eroded sourcerocks (e.g. Carboniferous Coal Measures) being more limited in distribution (cf. Wills 1973).

Eastwards, into the Dutch sector of the North Sea, thicker early Cretaceous sequences (up to 800 m) developed after the late Cimmerian uplift as the sea transgressed the area from the south (Heybroek 1975). In the central and northern North Sea a different picture emerges. Earliest Cretaceous deposits are only known from a few localities, and in the northern North Sea there was possibly an early Cretaceous igneous event. The oilfields are mostly sited upon fault-controlled 'highs' where the earliest Cretaceous (Ryazanian) is mostly unrepresented. They were, however, progressively transgressed, and Cretaceous deposition frequently begins with transgressive marine sands of Valanginian or Hauterivian age (Fig. 15.2). Marine deposition was possibly continuous within the basinal areas that have not yet been drilled.

In general, the early Cretaceous deposits of the North Sea consist of mudstones, marls and some limestones. An appreciable contribution to the car-bonates comes from coccoliths, which had extended their range northwards since the Kimmeridgian. New arrivals include planktonic foraminiferans (*Hedbergella*), and other organisms typical of these beds were siliceous sponges and radiolarians. In view of the biota, a relatively deep-marine basinal environment is suspected.

The Moray Firth area illustrates a number of interesting features and so will be considered in more detail. The present shoreline is virtually the same as that for the early Cretaceous, but offshore lay a Palaeozoic massif (Halibut Horst of Fig. 14.6)

that was comparable, in some respects, to the London Platform. It subdivided the Moray Basin into two separate sub-basins.

Intense tectonic activity occurred at the Jurassic/Cretaceous boundary, accentuating horsts and accelerating the subsidence of adjacent basins. The Halibut Horst has a core of Old Red Sandstone (and older rocks) overlain by Carboniferous deposits (Calciferous Sandstone Group). It had a southward tilt, and phases of uplift resulted in the rapid generation of sandy detritus derived mostly from the poorly cemented Carboniferous rocks. Most of the sands were shed southwards into the S. Moray Firth sub-basin. To the north, deposits were mostly clays and marls. But at the fringes of the Scottish−E. Shetland Platform (and around minor horsts within the sub-basin) slide conglomerates and mass flow deposits were generated as sub-marine fans against active fault scarps. These fans were comparable with those of the Kimmeridgian (Ch. 14k). Thus, within the Moray Firth, uplifted horsts with a capping of friable Carboniferous sedi-ment generated sandy turbiditic material. The Scotto-Shetland Massif with its indurated Old Red Sandstone and crystalline rocks seems to have gen-erated little sandy material but supplied coarse local conglomerates.

Source area evolution for individual submarine fans is reflected in the changing nature of clasts within the conglomerates (Fig. 15.5). In certain areas the hinterlands began by providing virtually unwinnowed carbonaceous and muddy fluvial detritus, but later transgression of the source area caused winnowing and cleaning of the supplied material. Clasts of shelly benthic faunas and glauconitic material record this change. Eventually, clasts of coccolithic and sponge-rich pelagic chalks were supplied, indicating that the source area had become a submerged sea mount in relatively deep water (Fig. 15.6).

The basinal environments mostly received muds which, in addition to pelagic microfossils, also con-tain ammonites and belemnites. Anaerobic condi-tions within these basins periodically caused the formation of bituminous shales, which frequently retain undisturbed coccolithic laminae.

Basinal environments within the northern North Sea are represented by thick sequences of shales and clays. Over horst areas these give way to thinner

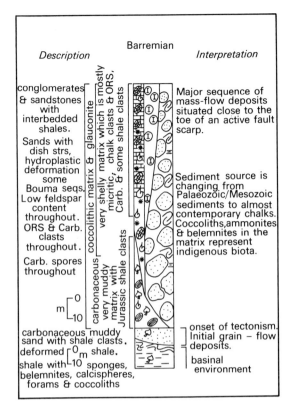

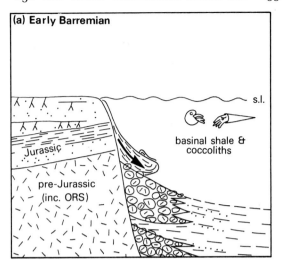

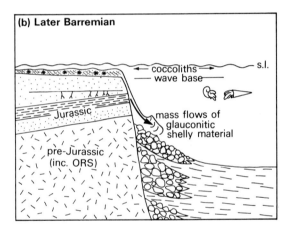

Figure 15.5 Clast and matrix evolution in an early Cretaceous marine conglomeratic sequence from the Moray Firth region. Vertical lines within the section do *not* imply erosion surfaces but merely indicate compositional trends.

coeval sequences of marls which, in the crestal areas, become even thinner chalks and red nodular chalks. For individual stages, thickness may change from 300 m in shales to 10 m or less in the chalks. Such facies persisted in the North Sea and, as shown below, developed more extensively in the later Cretaceous.

Figure 15.6 Models illustrating facies evolution in the Moray Firth region during the early Cretaceous. (a) Early Barremian: carbonaceous fluvio-deltaic sands are introduced directly into fault-controlled basins. (b) Later Barremian: transgression of the source area causes 'cleaning' and winnowing of the sediment, and glauconitic sand now forms the matrix to slumped conglomerates. (c) Even later Barremian: contemporary chalk clasts are now included within the conglomerates, suggesting that the source area has become a submarine high or sea mount.

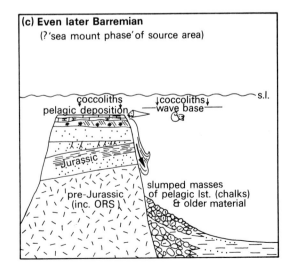

15d   Transgression begins: green sands, grey clays and red chalks

The northern fringes of the London Platform suffered transgression from the North Sea in earliest Aptian times (Table 15.1). Although these deposits are not preserved, reworked faunas laid down at this time were incorporated into slightly later beds (Casey 1961*). In S. Britain the mudplain environments of the Weald Clay Group began to be transgressed a little later in the Aptian (Fig. 15.2). This transgression, however, is thought to have come from the south, via the Paris Basin and English Channel, simultaneously flooding areas from the Isle of Wight to W. Kent. E. Kent was flooded a little later.

The northern and southern basins remained divided by the London Platform until upper Aptian times (*nutfieldensis* Zone of Casey 1961*), when faunal and mineralogical interchange first occurred. A westward extension of marine conditions into Dorset took place at this time, over terrain that had probably suffered faulting in the earlier Cretaceous.

From Wiltshire to Buckinghamshire, deposits of *nutfieldensis* Zone age overstep on to Jurassic rocks ranging down to the Corallian. Volcanogenic fuller's earth sediments occur locally (Ch. 15e) in rocks of this zone (Fig. 15.2).

Onshore in Britain, the Aptian and Albian Stages (Fig. 15.2) are represented by a complex series of muddy and sandy sediments (the Lower Greensand Group) laid down in shallow marine and shoreline environments. Rich marine faunas are often present, especially oysters and other bivalves. Gastropods, brachiopods and echinoids may also occur. The inclusion of cephalopods (including ammonites) attests to the marine conditions and allows precise correlations to be made (Casey 1961*).

In the Isle of Wight the Lower Greensand is 250 m thick. After the deposition of a transgressive shell-bed, the Perna Bed, most of the remaining Aptian (Fig. 15.2) is composed of a succession coarsening from offshore marine mudstones, the Atherfield Clay Formation, into the series of glauconitic muddy sandstones of the Ferruginous Sands Formation. The sequence may represent a prograding delta wedge that advanced from the west into the southern basin. The youngest Aptian and oldest Albian Sandrock Formation comprises coarsening sequences a few metres thick. These commence in lenticular bedded clays which pass up into flaser-bedded, and finally herringbone cross-bedded, sands (Dike 1972). The sands are frequently burrowed by *Ophiomorpha*, and the sequences may represent the bar or barrier deposits that formed, under tidal influences, as delta abandonment occurred (during the *nutfieldensis* transgression?). The overlying Carstone Formation is a fossiliferous coarse-grained marine sandstone. Its base is slightly erosive, containing phosphatised nodules which record a further transgressive advance of the Cretaceous sea, the sediment itself being an offshore shelf sand.

To the east (Fig. 15.2) the Lower Greensand varies greatly in thickness. It again commences in a basinal marine mudstone (the Atherfield Clay) which rests erosively upon the Weald Clay. Above there is a progressive coarsening into the glauconitic Hythe Beds Formation, representing sands that probably formed under tide-dominated shelf conditions (Narayan 1963). The overlying Sandgate Beds Formation is strongly transgressive (*nutfieldensis* Zone). Sediments are very variable, consisting locally of muddy glauconitic sands; elsewhere they comprise cross-stratified sands. They overstep the Hythe Beds and in the Kent coalfield lie directly upon the Atherfield Clay. The Sandgate Beds and its equivalents both north and south of the London Platform, contains the main developments of fuller's earths. Upwards, the Sandgate Beds are overlain by the Folkestone Beds Formation, which often consists of white or yellow quartz sands, cross-bedded on a large scale (set thickness >5 m).

The Folkestone Beds are thought to have been deposited as megaripple or sandwave complexes preserving unidirectional currents flowing southeastwards (Narayan 1971; Fig. 15.7). Currents acted in a pulsatory manner, with slack water episodes sometimes allowing drapings of clay to form over the megaripple structures and upon individual foresets (Allen & Narayan 1964). Herringbone sets are frequently encountered, and most of the structures in the Folkestone Beds can be closely compared with those from tide-dominated situations such as in the modern North Sea.

North of the London Platform, a condensed sequence of lower Aptian marine sands occurs in Bedfordshire and Cambridgeshire (Fig. 15.2). It con-

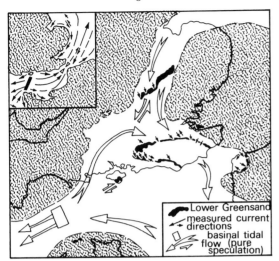

Figure 15.7 Speculative comparison between current trends on the modern tide-dominated shelf around Britain (inset) and those on the late Aptian to early Albian shelf of Britain (current directions after Narayan 1971).

tains Palaeozoic pebbles probably derived from the platform to the south. Following the upper Aptian (*nutfieldensis* Zone) transgression, the Woburn Sands Formation developed as a major subtidal sandwave complex (de Raaf & Boersma 1971). It retains structures, mineralogical composition and current directions similar to its equivalent (the Folkestone Beds) south of London. The surface morphologies of the individual sandwaves can be mapped locally, their crests being marked by ironpans produced during a phase of early Albian exposure.

In NW Europe the activity of tidal currents is recorded locally from earlier Mesozoic sediments. However, their startling expression in the British Lower Greensand records a radical change in shelf energetics. Presumably the break-down of the London Platform (*nutfieldensis* Zone) would have allowed major exchange of currents between the northern and southern basins (via the Oxfordshire–Bedfordshire straits). But these straits had existed earlier without strong tidal results (e.g. Cinder Bed Member). The southern basin actually began to undergo strong tidal influence while the two basins were still separated (e.g. Hythe Beds).

Palaeomagnetic evidence suggests that by the end of Wealden times the central Atlantic was about one-tenth of its present width. It is possible that graben collapse in the Western Approaches at about

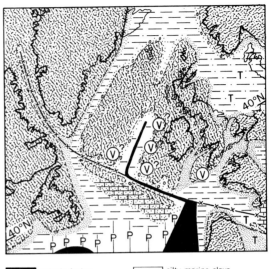

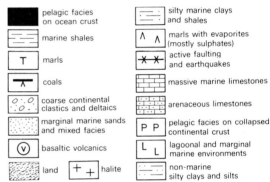

Figure 15.8 Generalised lower Cretaceous (upper Aptian) facies and palaeogeography of the British area (after Hallam & Sellwood 1976*).

this time (Fig. 15.8) provided a corridor for the penetration of tides from this new ocean. Enhancement of tidal processes occurred later when the northern and southern basins were connected (giving a Dover Straits-type situation as in Fig. 15.7). If this hypothesis is correct, we may expect nearshore high-energy deposits to have formed nearby (Fig. 15.7). These deposits may be represented by gravelly facies along the western outcrop of the Lower Greensand (e.g. the Faringdon Sponge Gravels; cf. Krantz 1972).

From S. Yorkshire into the North Sea the equivalents of the Lower Greensand are mostly represented by relatively deep water shales and claystones. However, in the Moray Firth area tectonic rejuvenation of certain horsts was marked by the

emplacement of turbiditic sandstones. The early
Aptian shales are sometimes bituminous, like their
time equivalents in the Atlantic Ocean, but in the
upper Aptian better circulation was established,
leading to the deposition of more clacareous sediments with abundant planktonic foraminifera (e.g.
Hedbergella). Increased circulation in this area may
reflect the break-down of the London Platform in
nutfieldensis times. Grey clays in the basins give way
to grey and red nodular chalks over submarine
highs and foreshadow facies developments onshore
in the later Albian.

Over much of S. and SE England, Lower Greensand deposition was terminated by transgression
and deepening during the early Albian (Table
15.1). Sedimentation reverted to that of basinal
marine mudstones (the Gault Clay Formation).
Reduced rates of terrigenous clastic supply during
the initial transgressive advance were marked by
the development of condensed and phosphatic
nodular horizons. The Gault Clay oversteps on to
progressively older formations and is the earliest
Mesozoic unit known to extend across the
Palaeozoic London Platform (Owen 1971). Westwards, however, the clay facies passes laterally into
the glauconitic sands of the Upper Greensand Formation (Fig. 15.9), which also oversteps earlier
deposits. This overstep in nearshore arenaceous
facies shows that clastic supply continued from the
west even though transgression was progressing. As
will be shown, sand supply continued until these
western source areas were finally transgressed by
the encroaching sea in the upper Cretaceous (Hancock 1969).

Later, in Albian times, glauconitic shelf sands of
Upper Greensand facies prograded eastwards as far

Figure 15.9 The westward overstep of Albian–
Cenomanian sediments in S. England and their associated facies changes.

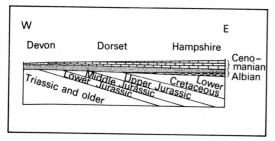

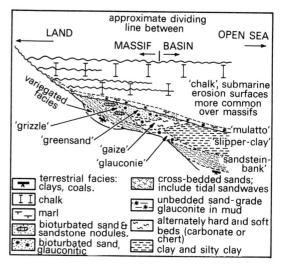

Figure 15.10 Generalised relationships between the
main facies represented in the Albian–Maestrichtian sediments of N. Europe. Names in inverted commas are commonly used by Cretaceous stratigraphers (after Hancock
1975[*]).

as Sevenoaks (Owen 1975) and may represent a
minor regressive phase. Little sedimentological
work has been published on the Upper Greensand
facies. It is a variable, and often bioturbated,
deposit which may contain wave ripple laminations
and abundant glauconite. Rich marine faunas are
present locally, being dominated by bivalves (especially the oyster *Exogyra* and clusters of the serpulid
Glomerula). Concentrations of opaline sponge
spicules in some beds provided silica which formed
chert bands during diagenesis. In Britain this
Greensand facies seems to represent a shelf sand
that accumulated at or about the limits of wave
effectiveness (Fig. 15.10). By comparison with modern glauconite occurrences, the abundance of this
mineral not only reflects relatively slow rates of
deposition but also indicates water depths of more
than 50 m (cf. Porrenga 1966, Hancock 1975[*]).

On the northern margin of the London Platform
the clastic facies of the Albian are locally condensed,
as in the Cambridgeshire Greensand, and further
north terrigenous facies pass laterally into the condensed carbonates of the Red Chalk Formation
(Fig. 15.2). This sediment is directly comparable
with the red chalks that formed over submarine
highs in the North Sea throughout the early Cretaceous. At outcrop the Red Chalk consists of very
condensed red marls and micritic coccolithic lime-

stones containing red chalky nodules. The upper surface of the formation may be bored or laminated (C. V. Jeans 1973). The base of the formation is usually conformable but may rest unconformably upon Jurassic rocks or disconformably upon earlier Cretaceous sediments (e.g. the Carstone and Speeton Clay). The main area of the Red Chalk distribution coincides with the location of the Market Weighton Horst, its structure being demarcated, as late as the Albian, by active faults (C. V. Jeans 1973). The scenario was geologically comparable with that for North Sea and E. Greenland red chalk and marine red-bed development (Fig. 15.11). The red colouration (due to limonite) probably reflects the low rate of sedimentation upon essentially pelagic highs. Under these conditions, little organic material survives into the burial phase, and oxidising conditions continue into the sea bed. The nodular appearance, comparable to that of later condensed chalks and earlier deep-water limestones (e.g. Ammonitico Rosso of the Tethys),

resulted from early diagenetic carbonate dissolution and redistribution upon the sea floor. The carbonate balance of the Cretaceous seas, which would have controlled this sea floor diagenesis, awaits further study.

Continuing transgression in the upper Cretaceous led to most of N. Europe being blanketed in the coccolithic ooze that now forms the Chalk.

15e Tectonism, volcanism and Atlantic rifting

As indicated above, upper Jurassic regression was soon followed by an influx of coarse clastic sediments (the Wealden), some of which were derived from new sources in the west. In SW England a phase of intra-Cretaceous folding has long been recognised (Fig. 15.9), and, more recently, intra-Cretaceous block faulting has also been demonstrated on the London Platform (Owen 1971). In the North Sea, particularly in the Moray Firth region, the tilted fault blocks that developed during the earlier Mesozoic persisted. They were locally accentuated across the Jurassic/Cretaceous boundary, remaining as recognisable entities until the end of the lower Cretaceous. An exactly comparable structural evolution occurred simultaneously in E. Greenland (Surlyk 1975). The later Cretaceous history of the region saw the elimination of the fault block topography and the blanketing of the region in pelagic chalk. Indeed, through the early Tertiary too, it seems that relatively gentle downwarping and upwarping of broad regions occurred, rather than the exaggerated differential movement of local blocks that had been the keynote of the earlier Mesozoic (Kent 1975).

Volcanism, probably west of Britain, had already begun in the late Jurassic (Knox 1977). Local sites in the Northern North Sea Basin were sporadically active through the late Jurassic (P. A. Ziegler 1975*) and into the early Cretaceous. At Zuidwal (the Netherlands), a volcanic plug of very earliest Cretaceous age (Cottençon *et al.* 1975) may account for the impure fuller's earth known from the Hastings Beds Group at Maidstone (Fig. 15.12). The main phase of lower Cretaceous igneous activity was, however, of Aptian age (*nutfieldensis* Zone),

Figure 15.11a Cartoon illustrating the possible scenario for the development of red chalk facies in the North Sea.

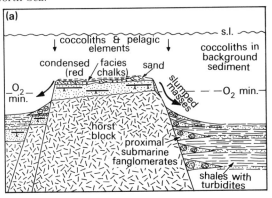

Figure 15.11b A North Sea horst block reconstructed for the upper Cretaceous at a time when a red chalk facies accumulated upon its crest.

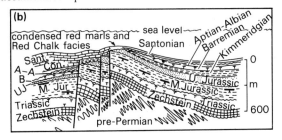

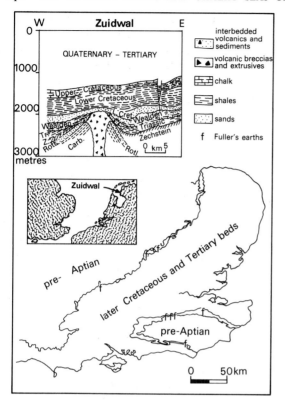

Figure 15.12 Distribution of fuller's earths in the lower Cretaceous of S. England, with insets showing the location of the Zuidwal contemporary volcanic vent and a section through it (after Jeans *et al.* 1977 and Cottençon *et al.* 1975).

recorded by the extensive development of (montmorillonitic) fuller's earths (Fig. 15.12). Hallam and Sellwood (1968) revived earlier speculations concerning a volcanic origin for these deposits, and this view has later been confirmed by Jeans *et al.* (1977) with the recognition of glass shards and well-preserved pyroclastic material of possible trachytic composition. The actual volcanic sites, however, remain a mystery. The Wolf Rock Phonolite of Cornwall (Fig. 15.8) represents the foundered wreck of an early Cretaceous volcano, but it may be too old (*c.* 131 Ma; J. G. Mitchell *et al.* 1975) to have supplied the material. However, other and as yet undiscovered sources might have lain west of Britain, and still others may await discovery on the London Platform itself.

The whole picture deduced above points to an intensification of tectonic, volcanic and transgressive processes during the early Cretaceous. During

the Aptian the Rockall Trough, which separates the Rockall Bank from the rest of continental Europe, began to form (Fig. 15.8). This major event marked the initiation of sea floor spreading in the N. Atlantic and the eastward rotation of Iberia (D. G. Roberts 1975, Laughton 1975). Thus a causal link seems to have existed between the more-or-less synchronous events in areas to the east. Spreading in the Rockall Trough was a temporary affair and was succeeded by collapse both of the graben and its margins. One repercussion of the uplift and succeeding collapse phase was the unconformable overstep (Fig. 15.9) of progressively younger Cretaceous strata over the western regions (Hallam & Sellwood 1976*).

15f The Chalk sea

As the late Cretaceous sea encroached on to the craton, the sources for terrigenous clastic material were progressively eliminated. Pelagic deposits became dominant and gradually advanced westwards (Table 15.1). The bulk of this pelagic material is the debris from coccolithophorid planktonic algae, occurring mostly in separate micron-sized plates. However, some plates are still in their original rings, called coccoliths (Hancock 1975*).

Most of the Chalk was probably deposited as magnesium-low calcite, which is relatively stable. The rarity of early lithification may indicate that little magnesium-high calcite or aragonite was originally present (Hancock 1975*). In areas of low heat flow, and where deep burial has not occurred (i.e. most onshore outcrops), initial porosities have been preserved. However, under the thick Tertiary overburden of the North Sea, and beneath Tertiary basalts in N. Ireland, the Chalk has been cemented and porosities have greatly diminished.

Like the Gault Clay and Upper Greensand, the Chalk is a diachronous facies. Its base is normally marked by a condensed marly deposit with a variable quartz and glauconite content (frequently termed the 'basement bed'). It contains reworked phosphatised fossils in addition to an indigenous fauna. Above, the Chalk normally commences in rhythmically alternating chalky limestones and interbedded marls arranged in sequences similar to those described from the Lias (Fig. 14.9; Ch. 14d).

Their rich burrow assemblage (W. J. Kennedy 1967, 1970) proves the depositional origin of the rhythms. There is a progressive loss of detrital clays upwards, and later chalks are normally about 98 per cent $CaCO_3$ (Hancock 1975*).

Normal salinities in the Chalk sea are indicated by the presence of groups such as echinoderms, brachiopods and cephalopods. However, substrate conditions must have been rather soft and therefore inhospitable to many benthic taxa. Certain bivalves (e.g. species of *Inoceramus*, *Pycnodonte* and *Spondylus*) became specially adapted to the soft bottom conditions. Certain *Inoceramus* species had greatly inflated left valves which allowed them to 'float' on the substrate; species of other genera developed spines or buoyant cavernous shells. Encrusting epifaunal organisms that needed firm substrates sometimes took advantage of these 'pioneer' species, and encrustations a few centimetres high can be found upon large *Inoceramus* shells. In view of the intense bioturbation (W. J. Kennedy 1967, 1970; Kennedy & Garrison 1975), it is possible that the actual sea bed was in a thixotropic state, being permanently firm only at depths of a few tens of centimetres.

Over the submerged massifs (Fig. 15.13), chalk facies thin and more stratigraphic gaps are present in sequences. These gaps are frequently associated with signs of contemporary hardening by cementation of the chalky sea floor (hardground development). Encrustation, phosphatisation and glauconitisation are all features seen in these chalk hardground horizons (e.g. Chalk Rock). R. G. Bromley (1967) has shown how cementation a few centimetres below the sea floor caused crustaceans to modify their burrowing activities. Hardground chalks have higher magnesium values than normal white-chalk facies. This probably reflects early cementation by magnesium-high calcite remobilised from the shells of the richer benthos associated with the more stable substrates. There is, however, no evidence for exposure, and the early cementation seems to have taken place under wholly submerged marine conditions.

The European Chalk is characterised by the presence of irregularly shaped flints which mostly follow, but sometimes cut across, bedding planes. Flint consists of micron-sized, randomly arranged, quartz crystals with interstitial water-filled microcavities. Flint nodules frequently envelop fossils and burrows; and, although they did not form penecontemporaneously, since flint pebbles do not occur within the Chalk, they clearly formed before the earliest Tertiary. The silica was probably derived wholly from biogenic sources (e.g. sponge spicules), although some small contribution from distant volcanic sources is possible. Flints probably grew in several stages, with their formation being aided by the carbonate-rich diagenetic environment. The first phase was probably the mobilisation of opaline silica and the subsequent concentration of amorphous opal around sites of organic concentrations. The opal later transformed into low temperature cristobalite during solution and reprecipitation. In its turn cristabolite transformed into chalcedonic quartz (flint), although the precise requirements for this are still a matter of dispute (See Keene 1976 and Wise & Weaver 1974 for recent discussions on silica diagenesis).

In the clear water of the modern tropics, coccolithophorids live at depths of 50–200 m. Their tiny skeletons accumulate from the limits of wave effectiveness to depths in excess of 3 km (calcium-carbonate compensation depth). There is usually little evidence to suggest strong scouring and current action on the floor of the Chalk sea, and accumulation in deepish water is suggested. Fossils provide conflicting evidence for the depth of the Chalk sea. Certain foraminiferans appear to indicate shallow water (a few metres), but the presence of hexactinellid sponges seems to indicate depths of 80–100 m (Reid 1968). The maximum depth is difficult to ascertain, but the well-preserved coccoliths in outcropping Chalk suggest that little etching has occurred. Etching occurs today for coccoliths accumulating below about 1 km. Provided that the carbonate balance of the Cretaceous seas was not vastly different from that of the present, deposition between 100 m and 600 m seems likely (Hancock 1975*). The ancient massif areas were probably more shallowly submerged (Fig. 15.13).

Estimates of Chalk deposition rates vary because of the difficulty in assessing the amounts of both compaction and time. However, Table 15.2 provides a guide to sediment thicknesses for each stage in a number of places. Average figures for the white-chalk facies fall within the range of 20–40 m Ma^{-1}.

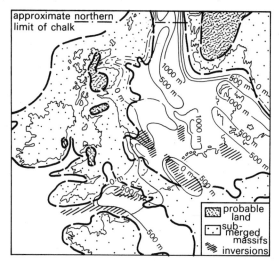

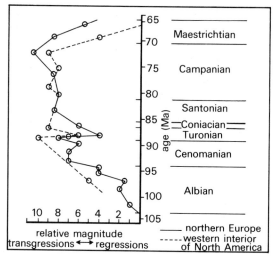

Figure 15.13 Cretaceous isopachs and the distribution of islands and submerged shoal areas in the Chalk sea in NW Europe. The Chalk facies passes into marine shales in the north. Positions of late Cretaceous inversion axes are also indicated (isopachs after Selley 1977, palaeogeography after Hancock 1975*).

Figure 15.14 Age and relative magnitudes of transgressions and regressions in the late Cretaceous (after Hancock 1975*).

Figure 15.15 Generalised upper Cretaceous (Campanian) facies and palaeogeography for the British area.

The depth of the Chalk sea must also have been affected by eustatic sea-level changes. As in the Jurassic, eustatic fluctuations are interpreted from the evidence of synchronous transgressive and regressive events that correlate over vast areas (Fig. 15.14). There is no evidence for polar ice caps in the Cretaceous, and eustatic fluctuations were probably controlled by the periodic elevation and collapse of ridge systems during successively active and quiescent phases of central Atlantic sea-floor spreading which, on available evidence (Pitman & Talwani 1972), was already well under way in the late Cretaceous (Fig. 15.15).

In Europe, climatic conditions influenced both the rate of organic productivity within the sea and the nature of terrigenous sediment supplied during progressive inundation of the craton. Although sands were generated locally from submerging massifs, clay availability became severely restricted. Terrigenous plant debris is not found onshore in NW Europe after the Cenomanian, and phosphatic nodules in the Chalk are comparable with those from modern marine areas influenced by more arid climates (Hancock 1975*). Northwards, however, into the Viking Graben of the North Sea (Fig. 15.13), the Chalk facies is replaced by silty shales.

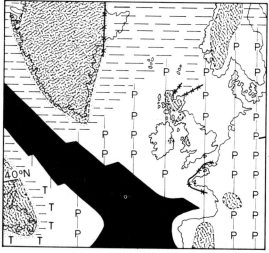

Table 15.2 Thickness in metres (and thickness per million years) of the Chalk by stages in Britain (after Hancock 1975).

Stage and Duration		Isle of Wight	Portsdown	N. Norfolk	Yorkshire Coast	Southern North Sea
Maestrichtian 6 Ma	upper	—	—	—	—	325·6 (54·3)
	lower	—	—	26·4 (c. 14)	—	
Campanian 11·5 Ma	upper	144·8 (c. 50)	—	99·1 (17·2)	—	119·0 (20·7)
	lower	104·5 (18·2)	—	79·2 (13·8)	99·7 (c. 29)	62·3 (10·8)
Santonian 4 Ma		116·4 (29·1)	100·6 (25·2)	94·5 (23·6)	143·3 (35·8)	154·7 (38·7)
Coniacian 1 Ma		16·2 (16·2)	18·3 (18·3)	22·9 (22·9)	36·9 (36·9)	54·9 (54·9)
Turonian 2·5 Ma		63·7 (25·5)	82·3 (32·4)	56·4 (22·6)	105·5 (42·2)	135·4 (54·2)
Cenomanian 4·5 Ma		55·7 (12·4)	103·6 (23·0)	16·2 (3·6)	29·4 (6·3)	34·6 (7·7)
	Total	501·3	304·8	394·7	414·8	886·5

This may reflect a change to a more humid climate, with the bulk of the clastics being derived from Greenland (Hancock & Scholle 1975).

15g 'Dies irae'? – requiem for the dinosaur

The lower Cretaceous fault block topographies in the North Sea and over the ancient Hercynian massifs were eliminated as general downwarping and draping took place during the upper Cretaceous. Differential subsidence continued, so that by the end of the period basinal regions had received up to 500 m of upper Cretaceous sediment in contrast to the 100–200 m that had accumulated over the submerged massifs. Some troughs within the North Sea (Fig. 15.13) received exceptional thicknesses (c. 1·5 km), which may relate to redistribution of Zechstein salts.

Late in the Cretaceous, some of the basins began to undergo uplift (inversion) in areas peripheral to the main North Sea Basin (e.g. the Weald, English Channel, etc.; Fig. 15.13). Secondary basins formed on the flanks of these newly inverted basins. The major phase of inversion took place from the Maestrichtian (Table 15.1) to the Palaeocene (early Tertiary) and correlates with the last phase of downfaulting in the central and northern parts of the Central North Sea Graben (P. A. Ziegler 1975*). These movements were accompanied by the emplacement of submarine mass flows containing semi-consolidated chalks (Selley 1976).

In S. (Alpine) Europe, the tensional regime that had typified most of the Mesozoic was replaced by a phase of compressive stresses attending the closure of the Tethys. Intense differential movements in the upper Cretaceous were accompanied by the emplacement of Alpine flysch sequences. This change of regime probably provided the driving forces behind the formation of inversion 'axes' on the cratonic foreland. These movements presaged the more general uplift that was to affect the area at the close of the Cretaceous, but these and the more intense Tertiary events will be considered later (Ch. 16).

In the seas, some ammonite families (e.g. the *Hoplitidae*) underwent diversification, and certain genera (e.g. *Turrilites*) adopted a nekto-benthic mode of life, developing asymmetrically coiled shells (Kennedy 1978). The real molluscan success story of the Cretaceous, however, is that of the rudistid bivalves, which constructed colossal build-ups with a morphology similar to that of modern coral reefs. They spread throughout the Tethyan belt, from the Pacific and Gulf of Mexico, through S. Europe, to the Middle East and Asia.

Modern bony fishes evolved during the Cretaceous. However, the large marine reptiles of the Jurassic (e.g. ichthyosaurs and pliosaurs) began to decline, and in the Cenomanian they were replaced by other carnivores such as the mosasaurs (a group of lizards) and long-necked plesiosaurs (Halstead 1975).

On the land, great changes occurred in the composition of the late Cretaceous flora, with angiosperms (modern flowering plants) achieving dominance over the other groups (particularly gymnosperms) by the end of the Cenomanian (Table 15.1). Placental mammals appeared in the early Cretaceous, although dinosaur reptiles dominated the land until the end of the period.

In the air, pterosaurs had greatly advanced and by the late Cretaceous were complex, lightly built and highly efficient flying machines. Some may even have been hair covered. These animals were accompanied by the birds, dinosaur derivatives which first appeared in the upper Jurassic *(Archaeopteryx)*. Their fossilisation potential is low, so that they have left a thin record. However, it is clear that during the Cretaceous many familiar modern forms, such as owls, cormorants and certain waders, had already evolved (Halstead 1975).

At the end of the Cretaceous a major phase of extinction affected many disparate groups. Of the molluscs, ammonites, inoceramids and rudistids died out. The ammonites commenced a steady decline in the Campanian, and there are few records of upper Maestrichtian forms. Belemnites might have just survived into the Tertiary but then soon expired. Other invertebrate groups suffered losses of genera and some of whole families (e.g. planktonic foraminiferans, bryozoans and echinoderms). Major extinctions also affected marine phytoplankton populations. The land flora had already suffered its greatest change earlier in the Cretaceous.

Of course, the greatest popular appeal surrounds the extinction of the dinosaurs, but their demise can only be viewed in the context of the whole wave of end-Cretaceous extinctions. Catastrophic explanations for these extinctions are just as unfounded as they were for the extinctions at the end of the Permian (Ch. 13d). Instead, causes must be sought in the total of Earth disruptions that took place at the end of the Mesozoic era (65 Ma ago). The long phase of global environmental stability was rudely interrupted by a plexus of change. This produced repercussions throughout the world's food chains. The continents continued a phase of rapid migration and dispersal (Smith & Briden 1977*) which must have totally altered the global climatic regimes. Regression on a worldwide scale at the end of the Mesozoic severely restricted the extent of shelf seas and increased the competition for ecological niches on the narrow continent-dominated shelves that remained. The net result was the extinction of the least adaptable groups at virtually every trophic level. Early Tertiary niches were more rigorous and became filled by the descendants of genera 'delivered' by their own adaptability from end-Cretaceous oblivion.

16 Cainozoic basins of Northern Europe

16a Introduction

A general phase of uplift affected the British area in the late Cretaceous. The Chalk was gently folded, and most of the British Isles – Irish Sea area became land at the end of the Mesozoic. In the North Sea this episode of latest Cretaceous to early Tertiary uplift was marked by the inversion of some of the former sedimentary basins. Subsequently the North Sea began to subside rapidly and remained a marine basin throughout most of the Tertiary and Quaternary. In the west, the central Atlantic Ocean and Labrador Sea continued to open during the early Tertiary, but no further oceanic crust was emplaced in the Rockall Trough after the early Cretaceous. The earliest oceanic crust separating Greenland and Rockall is of Palaeocene age (*c*. 60 Ma) and, although the Labrador Sea stopped widening in the Eocene, active spreading continued in the newly formed N. Atlantic from Palaeocene times to the present day.

Biostratigraphic correlation of Tertiary marine beds is best achieved using microfossils, such as coccolithophorids and planktonic foraminiferans (e.g. *Globigerina*). These organisms provide a fairly precise stratigraphic control, and beds containing them can be correlated with marine shelf deposits that contain shelly fauna such as the benthic foraminiferan *Nummulites*. It is even possible to correlate these marine sequences with terrestrial successions zoned on mammalian remains. Recent work has allowed the various biostratigraphic schemes to be integrated with the absolute time scale based upon both radiometric ages and the magnetic polarity reversals of the Earth (Berggren 1972, Le Brecque *et al.* 1977; Table 16.1).

16b The Palaeogene North Sea Basin

During Tertiary and Quaternary times the North Sea area underwent profound and rapid subsidence (Fig. 16.1). Tertiary sediments are unknown in E. Scotland and W. Norway which remained the main areas of sediment supply. S. England and northern continental Europe were periodically transgressed from the North Sea Basin during the Palaeogene (Table 16.1). Although the Western Approaches and Celtic Sea Basins (Fig. 16.2) persisted as marine embayments throughout the Palaeogene, marine incursions into the Irish Sea and Bristol Channel areas did not occur until very late Neogene or Quaternary times.

Along the axial zone of the North Sea the beginning of the Tertiary was marked by a phase of flexuring. Fault lines were reactivated, and locally the Chalk was eroded from scarps to provide mélange deposits (J. R. Parker 1975*). Over most of the area, however, this period of activity rejuvenated ancient faults, and the cover rocks responded by forming drape folds (Blair 1975). These Tertiary events have been interpreted as a renewed phase of rifting (Selley 1976).

During the Palaeocene (Table 16.1) the axial zone of the North Sea began a major phase of subsidence, and simultaneously the Scotland–Shetland Platform was uplifted along its western margin. Volcanism accompanied this tilting, and a large volume of sediment was supplied to the axial trough. Palaeocene sands generated by these processes provide the reservoirs in both the Forties and Montrose oilfields. In the Forties field (Fig. 16.3) a sandy Palaeocene sequence with debris flow and turbidite deposits is overlain by monotonous marine shales and mudstones representing much of the remaining Cainozoic (J. R. Parker 1975*).

The Forties Sandstones contain lithic clasts compatible with a Scottish origin (Thomas *et al.* 1974), and the interbedded and overlying shales contain microfossils indicating a pelagic environment (e.g. globigerinids, diatoms and radiolarians). Regionally, these sandstones form part of a major

Table 16.1 Cainozoic epochs, time and summary of major geological events

Era	Period	Epoch	Radiometric age (Ma)	Major Geological Events
CAINOZOIC	QUAT-ERNARY	Holocene	10,000 a	Low altitude glaciations take place in NW Europe, Greenland and N. America. Snowline in equatorial regions is lowered.
		Pleistocene	1·8	
	TERTIARY / NEOGENE	Pliocene	5	Hominids begin to make tools in Africa. Main uplift of the Jura in Europe and deposition of the Crags in East Anglia take place.
		Miocene	22·5	Messinian evaporites form in the Mediterranean. Main folding in S. Alps and widespread regression in NW Europe occur. Reactivation of basement faults causes drape-folding of later cover rocks in S. Britain. Glaciation commences in Antarctica.
	TERTIARY / PALAEOGENE	Oligocene	37·5	Thick molasse sequences are emplaced in the central Alps.
		Eocene	53·5	Main metamorphism and deformation occur in the E. Alps. Deep water sedimentation takes place in the North Sea where rapid subsidence occurs. Periodic transgressions from the North Sea inundate the adjacent subtropical lowlands. Volcanism ceases in Britain and becomes centred on the Atlantic Oceanic ridge.
		Palaeocene	65	Thermal doming associated with rifting between Greenland and Rockall is reflected by regional regression and emplacement of turbiditic sands in the North Sea. Intense igneous activity occurs in Greenland, Faroes and NW Britain. Rifting between Greenland and Rockall culminates with the emplacement of new oceanic crust (c. 60 Ma) and the N. Atlantic Ocean is born. Mammals rapidly diversify after the extinction of the dinosaurs.

submarine-fan complex constructed from the northwest (Figs 16.3 and 16.4). Seismic sections show the 'bottomset' turbidites passing northwestwards into 'foreset' shales, which themselves grade into 'topset' sands. The 'topset' beds include shelly and glauconitic sediments, and this association is interpreted as a coastal plain complex. J. R. Parker (1975*) has suggested that the axial basin was already 900 m deep in the early Palaeocene and this, if it is true, can be taken as a reflection of accelerated rates of Tertiary subsidence. The North Sea Basin

was fringed by coastal plain environments throughout the remaining Cainozoic, but outside the Palaeocene and Eocene few sandstones were emplaced in their 'bottomset' beds. Thus reservoirs comparable with those of Forties were not formed.

Tuffs and volcanic glass shards are widely distributed in the early Tertiary sediments of the North Sea (Jacqué & Thouvenin 1975). They were probably derived from the Thulean igneous province (Ch. 16c) to the west and northwest and first appear interbedded with lower Palaeocene sediments.

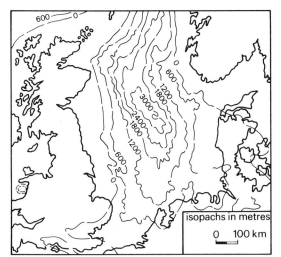

Figure 16.1 Tertiary isopachs for the North Sea Basin (after Selley 1977).

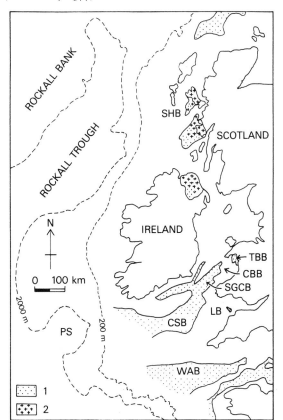

Figure 16.2 Major Tertiary basins in W. Britain. CBB = Cardigan Bay Basin; CSB = Celtic Sea Basin; LB = Lundy Basin; PS = Porcupine Seabight; SGCB = St George's Channel Basin; SHB = Sea of the Hebrides Basin; TBB = Tremadoc Bay Basin; WAB = Western Approaches Basin.

1 = sediments; 2 = volcanics.

Tuffs are most abundant in the upper Palaeocene but are absent after the lower Eocene (Table 16.1); they thus provide a precise chronology for the igneous activity. In the North Sea, phases of maximum tuff production coincide with diatom blooms, which may have been triggered by sudden influxes of this volcanically derived siliceous material.

Initial shallowing of the North Sea at the opening of the Tertiary was succeeded by a long phase of rapid subsidence. Over much of the basin, sands were of little importance, and muds with pelagic fossils were the more typical deposits until Miocene–Pliocene times (Table 16.1). Neogene regression produced lignitic sequences around the North Sea margin that are comparable with those that formed coevally in N. Germany, and the emplacement of Neogene sands probably resulted from the uplift of adjacent land areas, which may have been an Alpine effect (Ch. 16g).

16c The N. Atlantic igneous province and Atlantic sea-floor spreading

Igneous activity began in the W. Greenland portion of the N. Atlantic igneous province (Fig. 16.4) during the upper Cretaceous, but large scale activity did not commence elsewhere until the Tertiary (Table 16.1). From both the isotopic ages of lavas and the dating of tuffs in the North Sea the climax of this activity was in the Palaeocene (Table 16.1), and in the British area most activity had ceased by the early Eocene. This age range is small considering the size of the province (Bell 1976*). As the Atlantic opened, later activity became centred upon the new spreading ridge and particularly at the location that was eventually to become Iceland. A great variety of igneous rocks are represented in the province, ranging in type from nephelinites (E. Greenland only) through alkaline lavas and tholeiites to rhyolites.

In Britain, igneous activity was localised at a number of major centres (Fig. 16.5). Each centre was constructed by a number of overlapping volcanic piles in which the focus of activity periodically shifted. Within a single complex (e.g. Skye) phases of igneous activity were separated by periods of quiescence, and within the active phases themselves the magma chemistry evolved from initially basic to more acidic. Intrusion and extrusion occurred, and

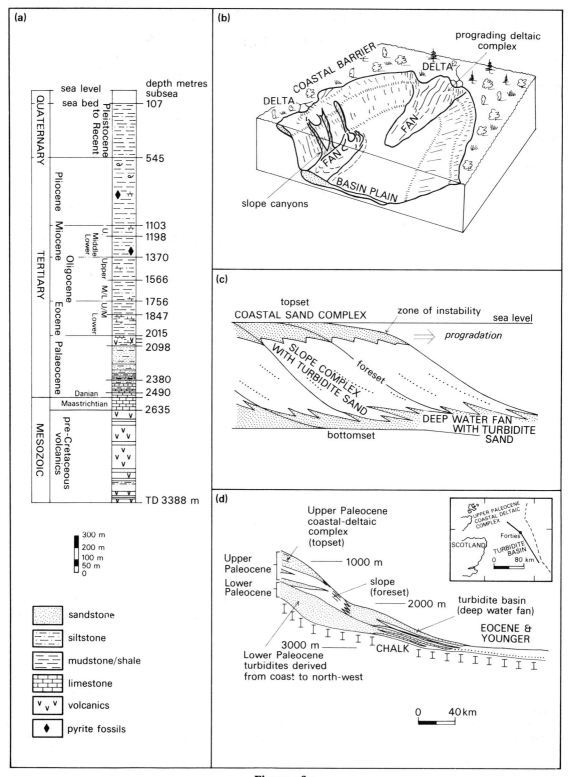

Figure 16.3

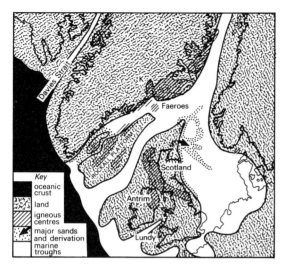

Figure 16.4 Palaeocene palaeogeography of the British area, showing part of the N. Atlantic (Thulean) igneous province. Thule of classical times was probably the Shetlands (compiled from Bell 1976* and Selley 1976 and references cited in text).
Tongues of dots in the North Sea indicate the distribution of basinal Palaeocene sands (see Figure 16.3).
K = Kangerdlugssuaq.

complex structures developed. In the Cullin Complex of Skye, for example, there were at least five episodes of dyke injection and three episodes of cone sheet emplacement. However, the style of igneous activity generally seemed to follow broad trends, with: (1) the eruption of plateau basalts along with some palagonites (altered glassy basaltic lavas) and tuffs, (2) the intrusion of plutonic complexes, and (3) the formation of dyke swarms. Of course, there are many individual exceptions to these trends.

Prodigious volumes of igneous material were produced. In the Hebridean Basin a 1·8 km thickness of basalts has been estimated (Binns *et al.* 1975), and in E. Greenland at least 6 km of basalts were erupted locally (Deer 1976). The preserved lava piles merely represent the erosional remnants of originally thicker sequences. Geophysical models for centres such as Skye suggest a predominantly gabbroic and ultrabasic mass at depth, with a volume of 1500−3500 km³ extending to a depth of between 5·5 and 14 km. It is certainly true that most

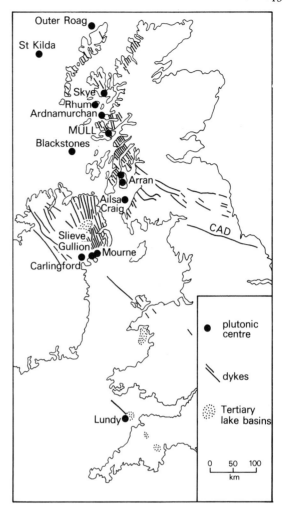

Figure 16.5 Distribution of major Tertiary igneous centres in Britain, and the trend of the main dyke swarms (after M. Brooks 1973 and earlier works).
CAD = Cleveland−Armthwaite dyke.

of the exposed rocks are of basic composition (olivine basalts, tholeiites and gabbros); acid rocks, particularly granites, comprise around 10 per cent of the total volume (Bell 1976*). A dispute still rages over the origin of the granites. Such vast volumes of basaltic magma could have melted large amounts of country rock, so that the granites, which include peraluminous and peralkaline types, could have

Figure 16.3   (a) Generalised stratigraphic sequence of the Forties oilfield (after Thomas *et al.* 1974). (b) Possible mode of emplacement of the Palaeocene sandstone in the Forties area (after J. R. Parker 1975*). (c) Idealised cross-section through the prograding deltaic pile that deposited the Forties sandstones (after J. R. Parker 1975*). (d) Generalised cross-section through the Palaeocene central North Sea (after J. R. Parker 1975*).

resulted from the partial melting of Lewisian and Torridonian basement rocks. During crystallisation such volumes of basic magma could also have produced some acidic material by fractionation, but isotopic evidence suggests that an origin from basement partial melting is more likely (Bell 1976*).

In the N. Ireland–Hebridean region, plateau lavas lie in shallowly faulted basins, and individual igneous centres are associated with major faults that were reactivated ancient structures (e.g. Skye on the Camasunary Fault and Mull on the Great Glen Fault). Perhaps phases of igneous activity coincided with periods of faulting. During quiescent periods, individual lava flows (some up to 15 m thick) were weathered and lateritic soils developed (Ch. 16e). Some volcanoes formed subsidence calderas which, when inactive, contained lakes with rich floras that included lotus, a species that today inhabits warm humid areas in Asia.

The igneous centres of W. Scotland and N. Ireland occur within a wide belt of dykes with a NW–SE trend (Fig. 16.5). Magnetic anomalies in the Irish Sea indicate a northward continuation of the Irish dykes (M. Brooks 1973) which strike in the direction of Anglesey and N. Wales where similar basic dykes occur (Fig. 16.5). These and the Cleveland–Armthwaite dyke (Fig. 16.5) have provided convincing Tertiary radiometric ages (Evans et al. 1973), and their emplacement clearly implies a major tensional phase with stresses operating in a NE–SW direction (i.e. oblique to the Atlantic continental margin).

In the Bristol Channel, the Lundy Complex was a southern outpost of igneous activity (Figs 16.5 and 16.6). This centre lies on the seaward extension of the NW–SE-trending Sticklepath–Lustleigh wrench fault, which (p. 254) was a reactivated Hercynian structure. From geophysical evidence (M. Brooks 1973) it is clear that the Lundy granite is merely the highest portion of an intrusion complex that becomes more basic at depth. Thus it is comparable with the igneous centres of the Hebrides. Many mineral lodes in SW England give Tertiary dates (Edmonds et al. 1975), and further Tertiary igneous centres may await discovery offshore.

Although the volumes of igneous material are much greater, activity in E. Greenland was generally comparable with that in Britain. However,

additional igneous suites occur, including syenitic and nephelinitic varieties, and activity was associated with the major monoclinal flexure of the Kangerdlugssuaq Dome (Fig. 16.4), which was initiated in the late Cretaceous (C. K. Brooks 1973). Uplift of this structure might have occurred at the same time as the rifting between the present E. Greenland coast and the Rockall Plateau. C. K. Brooks (1973) has interpreted the dome, with its associated alkaline igneous rocks, as the site of an early Tertiary hotspot which eventually became the proto-Icelandic plume.

If this interpretation is correct, Palaeocene igneous activity in the area probably marks the culminating phases of the main rifting events. Igneous activity had occurred sporadically throughout the Mesozoic, notably in the North Sea (middle and upper Jurassic); in SW Greenland (upper Jurassic); and in SW England, the Rockall Plateau and the Netherlands (lower Cretaceous). However, the age of the earliest emplacement of oceanic crust between Greenland and Rockall (magnetic anomaly 24) took place in the Palaeocene and is dated at c. 60 Ma. This event was accompanied by vigorous igneous activity in the adjacent areas. Eventually, activity

Figure 16.6 SW England showing the relationship between the Lundy igneous complex, Tertiary basins and Sticklepath–Lustleigh Fault belt. Other major wrench faults of probable Tertiary age are indicated. Inset shows the possible mechanism for sag pond basin development along the line of a wrench fault.
BB = Bovey Tracey Basin; PB = Petrockstow Basin.

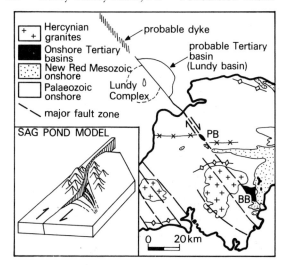

became focused on the newly created Reykjanes–Iceland–Jan Mayen ridge as the N. Atlantic evolved from its rifting into its spreading phase (Bell 1976*).

Events deduced for the Tertiary rifting phase compare in a general way with those of earlier rifting phases (Fig. 14.5). First there was an initial phase of doming, succeeded by a rifting phase accompanied by volcanism. Rifting ended in the early Palaeocene with the emplacement of oceanic crust. Finally, ocean floor spreading was accompanied by a protracted phase of subsidence which affected the new continental margins and continues to the present time.

In E. Greenland, Kangerdlugssuaq Fjord has been interpreted as the northwestern arm of a Y-shaped rift system in which the other rift elements run parallel to the present coast. A major dyke swarm parallels the coast and may represent the onshore remnants of oceanic crustal dykes emplaced during the rifting and early spreading phase. In Britain, however, there is no monoclinal flexure, and the regional dyke swarms run NW–SE. The newly formed continental margin runs roughly at right angles to the NE–SW-trending ridge-and-basin structure in the Hebridean area (Hallam 1972). One explanation for this geometrical discrepancy is that when separation was finally achieved between Greenland and Eurasia, opening was swifter in the south than in the north (Bell 1976*). The result was a skewing of the two plates, which produced a system of tensional fissures that were roughly perpendicular to the rift structure, into which the dykes were injected. However, the NW–SE trend of dyke swarms in Britain is roughly parallel to the axis of the Labrador Sea and may reflect the Tertiary invasion of tensional lines inherited from the Cretaceous opening of this oceanic area. Clearly, further work is needed to integrate the structural and petrological aspects of this story.

16d Tertiary environments in the British area

Much of the British Isles became land during the earliest Tertiary, and the almost ubiquitous blanket of Chalk began to be eroded. Remnant outliers exist in Ireland and W. Britain. At the opening of the Palaeocene (Table 16.1) S. Britain lay at about

40°N and with the rest of cratonic Europe was steadily drifting northwards. In the late Miocene (10 Ma ago) the area lay at nearly 50°N (Smith & Briden 1977*).

Britain probably began to assume its present tectonic style early in the Tertiary, with the uplift of areas in the north and west and the subsidence of areas in the southeast. The western areas provided terrigenous clastic sediments to both the subsiding eastern regions and to fault-controlled basins that developed in the west (e.g. Bovey Tracey, Petrockstow, Celtic Sea, Bristol Channel, Cardigan Bay, Lough Neagh, Sea of the Hebrides). Deposition on the southeastern fringes of early Tertiary Britain records the interplay between marine transgressions from the North Sea and the subsequent progradation of terrigenous coastal-plain sequences into the newly formed marine area. Repeated transgression and regression resulted in the cyclic Tertiary sequences of the London–Hampshire (and Paris) Basins (Fig. 16.7).

In Britain the earliest transgressive deposits are of Palaeocene age and are confined to the area of E. Kent. However, later (Eocene) beds overstep earlier ones, and at its maximum the sea seems to have penetrated beyond W. Berkshire (Fig. 16.8). It is tempting to relate these transgressive–regressive episodes to phases of British tilting, but the possibility of Tertiary eustatic changes cannot be discounted (Vail 1977).

In each cycle the transgressive beds tend to be bioturbated glauconitic sands, containing flint pebbles derived from the Chalk and diverse marine faunas dominated by bivalves and gastropods. More euryhaline environments were inhabited by oysters and, where marine beds directly overlie the Chalk, the contact is sometimes infested with borings. Early Eocene transgressive deposits are mostly thin (Fig. 16.7), but they give way upwards to thicker regressive sands and red-mottled kaolinitic clays. The red-beds probably represent reworked lateritic clays (Montford 1970) deposited in lagoons and marshes. In the west, sandy sequences such as the Reading Beds, with strongly unidirectional cross-bedding, represent alluvial deposits derived from western source areas.

The most widespread marine advance during the Eocene extended marine conditions from the North Sea over the whole of E. Britain from East Anglia to

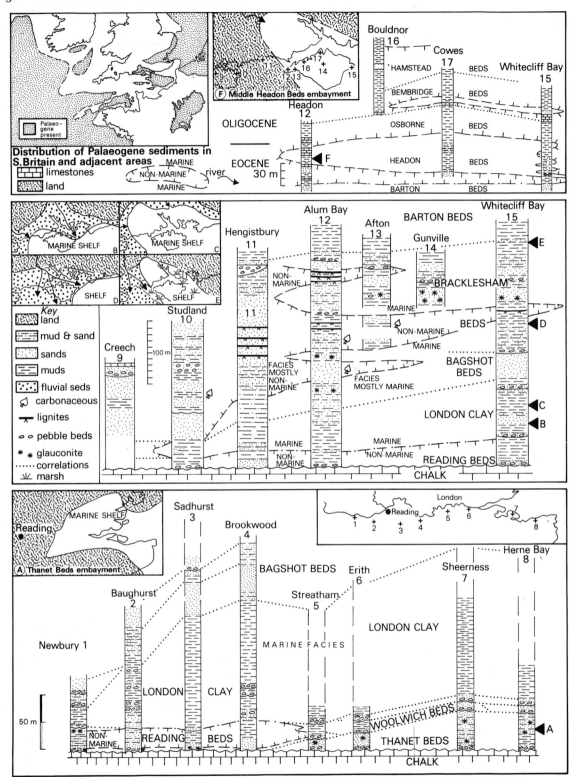

Figure 16.7 Stratigraphic correlation, facies and sedimentary sequences in the Palaeogene of S. England, with insets to show the palaeogeographic situation at selected times (compiled from Curry 1965 and 1966 and Murray & Wright 1974).

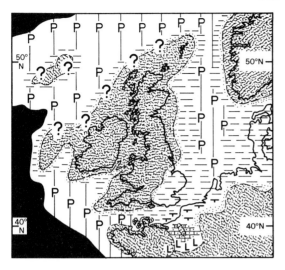

Figure 16.8 Generalised palaeogeography of the British area during the Eocene. For key see Figure 15.15.

Dorset and southwards over the Low Countries, N. France and NW Germany (Fig. 16.8). The western English Channel was also transgressed at this time, but from the direction of the Atlantic. The effect of the transgression was to spread a typical North Sea-type shale facies, the London Clay, over E. Britain. The London Clay is a fairly monotonous and sometimes intensely bioturbated mudstone containing an abundant and diverse fauna of molluscs which, although dominated by bivalves and gastropods, also includes nautiloids. Crustaceans, fish and reptiles (crocodiles and turtles) represent the more exotic macrofauna, the microfauna includes essentially pelagic forms, such as diatoms, radiolaria and globigerine foraminiferans.

Although many of the early Tertiary deposits contain plant debris, the London Clay has a wealth of terrigenous vegetable material (>500 species). This includes pollen, spores, logs and fruits, all of which provide evidence for the early Tertiary climate (Ch. 16e).

The London Clay grades upwards into sandier beds (Fig. 16.7), with flaser bedding and cross-bedding indicating currents of variable direction and magnitude. These transitional sandy beds frequently exhibit complex crustacean burrows (*Ophiomorpha*) and are comparable with sediments from some modern tide-dominated shoreface envi-

ronments. As regression continued, shoreface sands gave way, diachronously, to non-marine sands showing strongly unidirectional cross-bedding structures that indicate derivation from the west. On the Isle of Wight, marine beds in the east are seen to pass into non-marine facies in the west. Further west, in Dorset, a series of sands and flinty gravels (the Barton Beds of Fig. 16.7) represents the deposits of braided streams which drained the Cornubian and western uplands. In Devon, gravels capping the Haldon Hills (near Exeter) comprise the western outpost of these gravels. Kaolinitic clays were derived from intense weathering and were deposited within the alluvial sequences from which they are now extracted as ball-clays.

In the later Eocene, marine transgression allowed the British debut of the nummulites. These large benthic foraminiferans were more abundant in S. Europe where they contributed to substantial carbonate build-ups (Heckel 1975). In the Bracklesham Beds of the Isle of Wight they occur as monospecific associations within thin current-winnowed laminations. They were probably near the northern limits of their range, and a more normal Tertiary fauna of marine molluscs occurs in adjacent beds.

The competitive interplay between marine and non-marine environments continued until the end of Eocene times. Latest Eocene and early Oligocene sediments are only preserved onshore in Britain in the Isle of Wight–Hampshire area where, at the end of the Eocene, marine beds were replaced by units containing the freshwater molluscs *Viviparus* and *Unio*. However, in the succeeding Oligocene the freshwater ponds inhabited by these molluscs were invaded by the returning eastern sea. Thus the Oligocene sequence of the Isle of Wight illustrates the continuation of transgressive–regressive cycles. Some beds are packed with freshwater snails (*Galba* and *Planorbis*); others contain terrestrial faunas that include mammals; still others contain barnacles, serpulids and oysters. But throughout there is a persisting theme of decreasing marine influence westwards.

Devon, Cornwall and the western lands had clearly supplied a large amount of sediment to the Hampshire area during the early Tertiary. Denudation of these western areas must have been accompanied by phases of uplift, otherwise the supply of sediment would have progressively diminished. Clearly, some tectonic control would have promoted a periodic rejuvenation of the sediment source areas. Evidence of contemporary tectonism in Devon is provided by the existence of great thicknesses of Oligocene fluvial and lacustrine sediments in the fault-bounded basins of Petrockstow (c. 660 m) and Bovey Tracey (c. 1200 m). Both of these basins, and further basins in the Bristol Channel, lie along the Sticklepath–Lustleigh Fault zone (Fig. 16.6). This NW–SE-trending dextral wrench fault was initiated, along with a number of others (Fig. 16.6), during Hercynian movements, but it became reactivated in the Oligocene when a cumulative dextral movement of about 30 km took place in SW England. The Bovey Tracey and Petrockstow Basins developed as sagponds that are comparable, in a small way, with those occurring along the line of the modern San Andreas Fault (Fig. 16.6).

In the Bovey Tracey Basin the Tertiary deposits rest upon a preserved outlier of glauconitic Cretaceous sands. Marginal and basal deposits consist of cross-bedded fluvial gravels rich in reworked Cretaceous cherts. Later beds comprise sandy clays, lignites and high-quality kaolinitic ball-clays. The kaolinite results from the deep weathering of the surrounding terrain, which included exposed granite on Dartmoor, Carboniferous Culm, Devonian shales and limestones, and Cretaceous Greensand. The lignites are largely composed of fragments of *Sequoia* accompanied by *Nyssa* (a 'gum') and *Cinnamomum* (cinnamon); some probably represent foundered masses of originally floating vegetation. Animal remains are very scarce, being limited to freshwater molluscs; the absence of a rich fauna is somewhat enigmatic.

In continental Europe, particularly in Belgium and the Paris Basin, early Tertiary beds are represented by terrigenous clastic deposits including sands, clays and thick lignites. Sequences are comparable with those of S. England, and periodic transgressions and regressions occurred. However, a major difference in the Paris Basin is the greater frequency of limestones and of evaporites in the Oligocene (e.g. *Marnes supragypseuses* Formation). These evaporitic beds may reflect both climatic change and the more enclosed nature of the embayment towards the south (Fig. 16.8).

16e The Palaeogene climate of the British area

Early Tertiary deposits of the British area, such as the London Clay, contain a wealth of fossil plants whose descendants survive at the present time. Snags arise, however, when simple floral comparisons are made between the Tertiary and the Recent. About 47 per cent of the flora is typical of modern lowland tropical regions; however, the remainder includes taxa capable of living in both lowland and extratropical conditions, and of these about 11 per cent represent purely extratropical genera (Daley 1972*).

Paralic sequences and nearshore muds of the early Tertiary are reputed to contain the fruits and pollen of mangroves (Montford 1970), and the fruits of the palm *Nipa* have long been recorded. Such plants clearly compare with those of modern tropical mudflats and at least reflect a frost-free climate. In the past it has been popular to interpret the 'extratropical' genera as the remains of an upland flora that had been transported into the offshore basins. These remains include *Metasequoia*, *Magnolia*, *Engelhardtia* (a sort of walnut), *Betula* (birch) and *Cedrus* (cedars). Daley (1972*) has suggested that this floral anomaly can be explained if the climate was variable but mostly warm, being seasonal (but frostless) and very humid in the low-lying regions. The extratropical genera would have lived away from major waterbodies under less humid conditions, but not necessarily at high altitudes.

Support for a warm climate is derived from the presence of the abundant kaolinitic (ball-clay) sediments of SW England, as kaolinite forms most readily under humid tropical conditions. Even in the northern volcanic districts red lateritic soils developed between extrusive episodes, and, as noted above, lotus grew in the caldera lakes of Skye.

Later, in the Oligocene, humidities decreased, gypsiferous evaporites began to form in the Paris Basin to the south, while on the Isle of Wight calcrete soils formed, recording a decrease in rainfall. In the succeeding Miocene (Table 16.1), climates in S. Europe became very arid, and during this period (Messinian Stage) the last remnants of the Tethys (the Mediterranean) underwent repeated phases of desiccation (p. 256).

16f Alpine effects

During the late Mesozoic, Europe was affected by compressional forces from the south. The Tethys began to narrow as its oceanic crust was subducted, and this ocean, which had separated Europe from Africa through much of Mesozoic time, is now represented by mere remnants of oceanic crust in the E. Mediterranean. The Alps and Carpathian chains are believed to represent the remains of the peripheral mobile belt that formed on the southern margin of the European Craton. Chains such as the Dinarides of Yugoslavia and the Hellenides of Greece may be the equivalent belts of the African Craton.

In Alpine Europe, Mesozoic and early Tertiary pelagic deposits grade upwards into major sequences of terrigenous turbidites called flysch. Throughout the Alpine belt the emplacement of flysch marks the onset of compression and uplift. Flysch sequences were themselves involved in Alpine deformation and thus represent synorogenic deposits. In some parts of the belt (e.g. the S. Alps) compression culminated in the Miocene. However, deformation and metamorphism occurred earlier (upper Eocene to lower Oligocene) in the E. Alps at a time when flysch was still accumulating further west. In the Jura fold belt the main movements took place in the Pliocene.

In the most intensely deformed zones, thick piles of thrust sheets and recumbent folds (nappes) developed. Uplift might well have initiated gravity sliding as the main mode of nappe emplacement. In the 'internal zones', deformation was so intense that crystalline basement rocks were reworked by both thrusting and regional metamorphism (especially of the low-temperature high-pressure type). Within the orogenic belt, less intense deformation was suffered by the rocks of the 'external zones', where the soles of thrust sheets were lubricated by the presence of Triassic evaporites. Trümpy (1973) has noted that the most intense orogenic movements took place over c. 5 Ma, during the late Eocene and early Oligocene. A crustal shortening of at least 300 km has been estimated for this brief period, providing an average rate of N−S movement of 5−6 cm a^{-1}.

Late orogenic uplift was marked by a Europe-wide restriction of Neogene and post-Tertiary deposits. Uplift provided vast amounts of alluvial sediments of molasse type, which accumulated in newly formed troughs adjacent to the rising mountain tracts (cf. Ch. 9g). Repeated phases of uplift and outbreaks of volcanism occurred, which have, in places, continued until the present time.

While these events were taking place in S. Europe, the Atlantic underwent a vigorous phase of sea floor spreading which deprived NW Europe of its cratonic 'support' in the northwest. Upon the craton itself, presumed Alpine effects were mostly limited to the rejuvenation of pre-existing basement structures (e.g. the Great Glen and Sticklepath −Lustleigh Fault belts) and to broad warpings of the crust (e.g. the North Sea). Movement along basement faults sometimes caused folding of the cover sediments. In N. (cratonic) Europe it is the timing of such movements that provides the main evidence of their 'Alpine' pedigree. However, it is not entirely certain that crustal movements as far north as Britain resulted predominantly from Alpine compression. They could reflect crustal readjustments following the Palaeocene departure of Greenland due to sea floor spreading. In any case, 'Alpine' effects in Britain were probably magnified as a result of early Tertiary Atlantic opening.

In Britain, Miocene sediments are unknown onshore, which is believed to indicate that the Miocene was a time of a major phase of uplift. Support for this view comes from the Tertiary basins offshore (Figs 16.1 to 16.3), which contain sequences that become more sandy in the Miocene and Pliocene (p. 257). Onshore in S. England, beds up to middle Oligocene in age have been affected locally by folding (e.g. Isle of Wight−Hampshire Basin; Fig. 16.9). In other structures the precise age of folding is unknown (Fig. 16.9), but because they have the same general E−W trend they are believed to have formed at about the same time. Some are cut by presumed Pliocene planation surfaces, and thus a Miocene deformation age has been inferred for all. Geophysical work at Reading University, and drilling on individual structures (e.g. the Kingsclere Anticline; Fig. 16.9) have revealed that Tertiary movements were merely the latest in a long history of disturbances. The Kingsclere−Pewsey structures (Fig. 16.9), for example, lie close to the line of the postulated Hercynian thrust front. Other structures in the Mesozoic cover lie on the eastward projec-

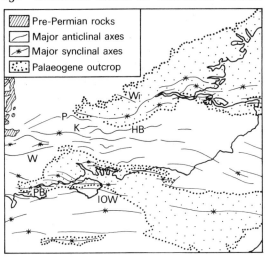

Figure 16.9 Major Tertiary anticlinal and synclinal structures and structural trends in S. England.
IOW = Isle of Wight; P. = Pewsey; W. = Wardour; K. = Kingsclere; HB = Hog's Back; PB = Purbeck structures; Wi. = Windsor.

tions of major structures in the Hercynian basement (e.g. the Vale of Wardour Anticline; Fig. 16.9).

More tangible evidence for Tertiary faulting in Hercynian basement rocks is provided by the dextral wrench-fault system of SW England (p. 250, Fig. 16.6). The Sticklepath–Lustleigh line can be related to the Lundy igneous complex (M. Brooks 1973). In Scotland, the Skye igneous centre lies upon the Camasunary Fault, and Mull lies upon the Great Glen line. These structures, which cut the Caledonian and older basement rocks, were themselves active during the early Tertiary. Cardigan Bay (Fig. 16.2) contains about 650 m of Oligocene lacustrine sediments and appears to have been bounded by the Mochras Fault. These are a few examples of proven large-scale Tertiary fault activity, there are many other examples whose effects are not so dramatic. It is not yet entirely clear whether other major structures, such as the Bala and Church Stretton Faults, were active, but it seems probable that they were.

Cloos (1939) suggested that the overall structure of the British Isles was a horst-and-graben system in which the axial depression was represented by the Hebridean and Irish Seas. Hallam (1972) has reaffirmed this interpretation and related the Palaeogene timing of the main movements to N. Atlantic opening. Viewed in such a way, the present

gross structure of Britain, with its 'highland' regions in the west and northwest and its 'lowland' and subsident regions in the southeast, can be seen as a legacy of N. Atlantic-rifting events. A more easily recognisable rift now contains the River Rhine between the uplifted massifs of the Vosges, Black Forest and Ardennes. Graben collapse took place here in the early Tertiary after a broad phase of uplift which occurred from the late Mesozoic. This rifting was also associated with igneous activity in the form of alkali basalts, trachytes, phonolites and tuffs, which were erupted in areas as far afield as the Massif Central, Black Forest, Auvergne and Swabia. The formation of the Rhine Graben can be fitted into a 'thermal-doming model', but was it related to 'Alpine' or 'Atlantic' events? Whatever the origins of these and similar movements in Britain, their effects still linger on. Modest earthquakes still occur along old structural lines, such as the Great Glen, and perhaps it is no coincidence that parts of Britain's western coastline lie close to, and parallel with, mapped faults in rocks of all ages.

The last major event of the Alpine Orogeny was a period of widespread uplift which culminated in early Miocene times. This produced an early Miocene regression in many parts of Europe, as well as the fault-controlled folding in SE England and N. France (Ch. 16f), and resulted in a palaeogeography not unlike that of today. Whereas in Oligocene times shallow seas flooded across Europe, linking the North Sea Basin with the Tethys, the Alpine uplift had, by Miocene times, pushed the sea back towards the present limits of the North Sea (Fig. 16.10) and, with the formation of the Alps, produced a vast new source of detritus which ever since has fed some of the major N. European rivers. As a result of Alpine tectonic movements, the Tethys also shrank back towards the boundaries of the present Mediterranean Sea and became progressively isolated from both the Atlantic and Indian Oceans during the Miocene. By the Messinian, the youngest stage of the Miocene, the Mediterranean was completely isolated from the adjacent oceans and had become a giant evaporating basin. Sea level fell, and extensive sabkha-type evaporites were deposited on the basin floor several kilometres below the sea level of the adjacent open oceans. The Mediterranean Basin was reflooded at the beginning of the Pliocene and has not changed significantly since.

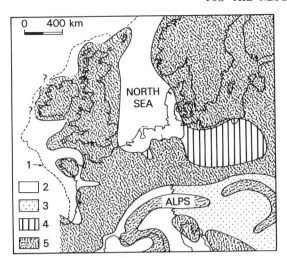

Figure 16.10   Miocene palaeogeography of NW Europe (after Pomerol 1973).
1 = present-day 200 m isobath; 2 = sea; 3 = para-Tethys (a shallow sea partly isolated from the Tethys); 4 = basins of continental sedimentation; 5 = land.

In the British Isles the Miocene uplift produced a major gap in the stratigraphic record, stretching from *c.* 30 Ma (the youngest Oligocene in the Hampshire Basin) to *c.* 3 Ma (the East Anglian Pliocene). This period is represented on land by only very local terrestrial deposits (e.g. cave fills in Derbyshire and N. Wales). In the adjacent shelf areas and in N. Europe, the gap is either far shorter or non-existent, and thick Neogene (Miocene and Pliocene) sediments may be found. Around the British Isles, Neogene to early Pleistocene sediments (the Pleistocene and the Holocene together make up the Quaternary) were deposited in two main areas: (a) the North Sea Basin and the adjacent European mainland from Denmark to France, with an embayment stretching through Germany into E. Europe; and (b) the Tremadoc Bay and Western Approaches Basins and the deeper waters beyond the edge of the shelf (Fig. 16.2). These areas will now be discussed separately.

16g   The Neogene to early Pleistocene North Sea Basin

The North Sea area continued as an actively subsiding sedimentary basin throughout the Neogene and Quaternary. The thickness of Neogene sediments is over 1 km at the basin centre and up to 600 m in the Netherlands. Neogene isopachs show a similar pattern to those for the whole Tertiary (Fig. 16.1). A plentiful sediment supply for the basin was provided by the centripetal drainage pattern resulting from the Alpine uplift. The predecessors of such rivers as the Rhine and Meuse must have contributed large quantities of sediment to the North Sea since their inception in the Miocene. In Britain, the Alpine uplift is thought to have tilted the land eastwards, producing a predominantly E-flowing consequent drainage pattern of which such rivers as the Thames, Humber, Tyne and Forth are descendants. Under these palaeogeographic conditions we would expect the North Sea Neogene to consist largely of marine muds fringed with a belt of sandy nearshore sediments, but as no oil reservoirs have been found in the North Sea Neogene, few details of these sediments are available. However, muds seem to predominate, and there are occasional carbonate horizons and sands. The latter, indicating shallower water, become commoner upwards into the Pliocene. Similar sediments are found on the European mainland, and here the sandy littoral deposits pass southeastwards into fluvial sediments. During Miocene times a basin of largely continental sedimentation, in which extensive peat deposits accumulated, extended southeastwards from the North Sea (Fig. 16.10). These now form lignite horizons which are extracted from huge open pits in Germany and Poland.

There is no apparent break between the Miocene and the very similar underlying Oligocene sediments in the centre of the North Sea Basin, shelf conditions prevailing throughout. Thus, the early Miocene regression recognised at the basin margins did not significantly reduce water depths at the basin centre. However, the shallowing towards the end of the Miocene and into the Pliocene, suggested by the coarsening of the sediments, is recognised throughout the basin. There is no lithological break across the Pliocene/Pleistocene boundary in the central North Sea, although the boundary can be defined by the incoming of serveral cold-water species of foraminiferans. Here the early Pleistocene deposits, termed the Basal Beds by Holmes (1977), are up to 200 m thick and are truncated by a major erosion surface of late Pleistocene age.

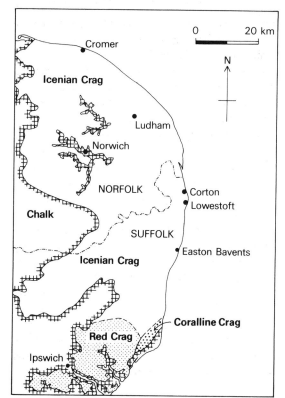

Figure 16.11 Distribution of the Crag (Pliocene and lower Pleistocene) under the later Quaternary deposits in East Anglia (after R. G. West 1968).

The situation on land in the British Isles is somewhat anomalous. Although there is an absence of marine Miocene, Pliocene and lower Pleistocene marine sediments are found in East Anglia where they form a succession less than 70 m thick collectively known as the Crag (Fig. 16.11). The Pliocene comprises 30 m of cross-stratified sands containing a temperate to warm fauna of bryozoans, bivalves, gastropods and brachiopods, somewhat misleadingly known as the Coralline Crag. These sands were deposited by strong tidal currents which formed large sandwaves similar to those in the present North Sea. The tidal characteristics, and therefore the basin geometry, of the Pliocene North Sea were probably similar to those of today. However, the connection between the southern North Sea and the English Channel was not necessarily made via the present Strait of Dover, and a connection through the London and Hampshire Basins is more likely.

The Coralline Crag is of uppermost Pliocene age and, as in the case of the North Sea Pliocene, is overlain by lithologically similar sediments, here known as the Red Crag. These are assigned to the lower Pleistocene on the basis of characteristic pollen assemblages and a marked increase in the proportion of sub-Arctic molluscs, which indicate a significant cooling of the climate. The Red Crag is much more extensive than the underlying Pliocene (Fig. 16.11). It overlaps on to the Chalk, where its base is marked by NNE–SSW-trending channel-like depressions up to 6 km wide and 45 m deep. Although these troughs may be of tectonic origin, it is also possible that they were eroded by tidal currents flowing parallel to the shoreline. The Red Crag contains a marine fauna and shows cross-stratified sets of shelly sand and gravel, several metres thick, that are typical of tidal shelf deposits, overlain by finer flaser-bedded intertidal sands and silts.

There is probably a major stratigraphic gap between the Red Crag and the overlying sediments, which have been described from a borehole at Ludham in Norfolk. These are a series of alternating sandy and clayey units of shallow marine and estuarine origin deposited during the Ludhamian to Baventian Stages (Fig. 17.2). The sands are characterised by a temperate flora and fauna, the clays by a cold climate assemblage. By the time these sediments were deposited, uplift to the south had probably severed the connection between the North Sea and the English Channel.

Unconformably overlying the Red and Coralline Crags in S. Suffolk and the younger Baventian sediments on the Norfolk coast are sands and clays assigned to the Pastonian Stage; these are collectively known as the Icenian Crag. A study of their molluscan fauna shows that they were deposited in a range of open shallow-marine to estuarine or tidal flat environments (West & Norton 1974). Beach gravels, known as the Westleton Beds, prograded southeastwards over these marine sediments in the north. The Icenian Crag was deposited following an important transgression which cut sediments of a variety of ages from Tertiary to Baventian. Correlation of the Baventian and Pastonian Stages with the more complete Dutch succession suggests that the break in sedimentation terminated by the Pastonian transgression may be of the order of a million years long (Fig. 17.2). During this long period of time a considerable part of the lower Pleistocene sequence could have been eroded, leaving a major gap in the British Pleistocene succession.

More details of the early Pleistocene are available from the Netherlands, where the location of the coastline throughout the Pleistocene was controlled by the balance between the fluvial sediment input, from such major rivers as the Rhine, and the rate of tectonic subsidence, modified by eustatic changes in sea level. Here, marine conditions predominated in the northwest of the country during the upper Pliocene and persisted for some time into the Pleistocene. However, by the time represented by the Baventian–Pastonian hiatus in Britain, fluvial sediments were prograding out into the area of the present North Sea, forming a large delta.

16h The Western shelves and ocean basins during the Neogene to early Pleistocene

The differences between the Palaeogene of the North Sea Basin and the Tertiary basins on the western side of Britain persisted into the Neogene. The Neogene sequence of the latter areas is much thinner and more calcareous than that of the former, because of lower subsidence rates and lesser amounts of fluvial sediment input.

As a result of early Miocene uplift the western shelves were probably exposed and subaerially eroded, the sea retreating to near the present shelf edge. The succeeding transgression advanced into the Western Approaches of the English Channel, from which Miocene silts containing *Globigerina* have been dredged. It is not known if the Miocene sea advanced far into St George's Channel; but in the Mochras Farm borehole at the eastern margin of the Tremadoc Bay Basin (Fig. 16.2), 520 m of Tertiary non-marine lignitic sands and muds have been found. These are probably, at least in part, of Neogene age. The Tremadoc Bay Basin is terminated at the Welsh coast by a series of major faults which where probably active from Mesozoic to Neogene times. The preservation of the Tertiary section here, as elsewhere between Ireland and Wales, is due to a combination of syndepositional faulting and regional downwarping.

On the western shelves, as in the North Sea, the Pliocene/Pleistocene boundary is not marked by any change in lithology and cannot be recognised on seismic profiles. It is possible that early Pleistocene marine sediments are included in what has been interpreted as Neogene on the basis of seismic profiles and that, as in the North Sea, the Neogene palaeogeography persisted into the Pleistocene. Although Neogene sediments have not yet been proved on the western shelves outside the areas of the Tertiary basins referred to above, they may be present on the outer shelf west of Ireland and Scotland, as implied by the palaeogeography sketched in Figure 16.10.

Throughout the Tertiary the subsidence rates on the western shelves were slower, and the resulting sediments thinner, than in the North Sea Basin. However, beyond the shelf edge, sediment thicknesses were controlled not by subsidence but by sediment availability. The thick Tertiary sequence in the Rockall Trough includes over 600 m of Neogene sediments. On the eastern side of the trough these comprise two large submarine fans built from terrigenous sediment, which may have been fluvially transported to near the shelf edge when sea level was low. To the west, Rockall Bank contributed little Neogene detritus to the trough, having been below sea level since Miocene times, and the thick sequence here was deposited by oceanic currents flowing southwards from the Norwegian Sea.

The first appearance of relatively cool water molluscs and foraminiferans has been used to define the local base of the Pleistocene in Europe. Palaeomagnetic work dates this at 2·5 Ma in NW Europe. However, the internationally agreed type section in Italy for the Pliocene/Pleistocene boundary has been dated at 1·8 Ma. If confirmed, this would mean that a large part of the lower Pleistocene sequence described above should be included with the Pliocene. Such boundary problems are common in geology, where an event used for the definition of a stratigraphic boundary, because it is assumed to be contemporaneous over large parts of the world, is subsequently shown to be diachronous. For example, a cool water fauna appeared in NW Europe quite a long time before there were any significant faunal changes in the Mediterranean.

Alternating cool and temperate climatic intervals

are recognised throughout the Pleistocene in Europe and used to define local climatic stages (Ch. 17c; Fig. 17.2). The expansion and contraction of glaciers at higher latitudes during this time would have produced eustatic changes in sea level, which might have resulted in minor regressions and transgressions. Indeed, changes in the volume of glacier ice might also have produced significant sea-level changes throughout the Tertiary (Ch. 17b). However, climate does not seem to have deteriorated enough for glaciers to have reached the southern North Sea before quite late in the Pleistocene, the first evidence for the presence of ice in East Anglia being found in the sediments of the Beestonian Stage (Fig. 17.2). The evolution of Britain during the glacial later part of the Pleistocene will be discussed in Chapter 17.

16i Summary

In the British area, large scale inversion (uplift) of some sedimentary basins began during the Senonian Stage of the late Cretaceous, and in the southern part of the Central North Sea Graben this process continued into upper Palaeocene times (Table 16.1). Elsewhere former basins inverted much later (e.g. the Miocene inversion of the Weald). The intensity of inversion tectonics decreased northwards, and for this reason compressive Alpine stresses have been suggested as the main motive force.

Before the deposition of the earliest Tertiary beds the Chalk had already suffered gentle folding. An associated phase of uplift transformed most of Britain into a low-lying chalky downland with a warm and humid climate. This land was rapidly colonised by deciduous forests which, although including genera more typical of the modern tropics, also included plane, oak, hazel and magnolia. The terrestrial fauna was dominated by mammals and birds. Uplift also affected the North Sea, and late Cretaceous fossils were reworked, probably by submarine processes, during a phase of shallowing.

During the Cainozoic the main depositional basins in the British area were in the North Sea, Western Approaches, Celtic Sea and Rockall Trough. In the North Sea up to 3·5 km of sediment accumulated, and it was probably during the Tertiary subsidence, and consequent heating, that hydrocarbon formation and migration occurred. In the other offshore basins very thick sequences accumulated (often >1 km). By contrast, the onshore basins of England, France, Flanders and N. Germany normally have sequences less than 600 m thick, and only in the Rhine Graben is 1 km exceeded.

Basaltic volcanism began in W. Greenland before the end of the Cretaceous (c. 77 Ma ago), but in E. Greenland, N. Ireland and W. Scotland igneous activity began in the Palaeocene (c. 65−60 Ma ago). Palaeocene volcanism and uplift were broadly synchronous and probably related to the process of N. Atlantic rifting.

Throughout the Cainozoic the North Sea area remained a marine basin, and transgressions periodically flooded the peripheral areas of S. England and NW Europe. Each transgressive pulse was succeeded by regression as the coastlines prograded. Although fairly complete sequences of Tertiary deposits are found in the offshore basins, the British onshore sedimentary record is broken in the late Oligocene, and Miocene deposits are absent. This hiatus was due to the uplift of Britain and is conventionally interpreted as an Alpine effect. Throughout the Tertiary, the parts of S. Europe now comprising the Alpine chains were undergoing deformation of varying intensities. Tertiary folding and faulting in Britain have, in the past, been viewed as 'the outward pulses of the major Alpine Storm' (Owen 1976). This may be a simplification.

At the opening of the Tertiary, S. England lay at about 40°N, but by late Miocene times (c. 10 Ma ago) the European Craton had drifted northwards by almost 10° (Smith & Briden 1977*). This northward drift transported the area into a cooler climatic belt. Glaciation began in Antarctica at about this time, and as the Miocene drew on this glaciation intensified. There was a worldwide cooling, which eventually culminated in the Pleistocene ice ages.

17 Climate and sedimentation: the Quaternary glacial interlude

17a Introduction

The major structural and palaeogeographic elements of the NW European continental margin have changed little since Miocene times. The British Isles have remained a positive area, separating the rapidly subsiding North Sea Basin from the smaller and usually less actively subsiding basins to the west. From Miocene to lower Pleistocene times the only changes consisted of minor and undramatic transgressions and regressions, together with minor faulting.

Against this fairly simple structural background the effects of the Quaternary ice age can be viewed as an interlude during which a whole new set of climatically controlled processes came to complicate the geological evolution of the area. Glaciations can be thought of as periods during which the usual equilibrium between tectonics, erosion and sedimentation is thrown temporarily out of gear. Rapid changes in sea level, isostatic movements of the crust, erosion in normally depositional areas and deposition in normally erosional areas are all examples of the geologically unusual aspects of glacial periods which, together with the more obvious effects of ice action, produce atypical conditions at the Earth's surface. The cessation of glacial activity is, however, followed by a return to normality, during which many of the effects of glaciation are reversed or removed by active river and coastal erosion and mass movement on steep slopes, leaving only certain aspects to be permanently incorporated into the geological record.

The Quaternary glaciation is not unique; it is only the latest in a long series of glacial episodes that have occurred periodically through time. The Permo-Carboniferous glaciation of Gondwanaland, the Ordovician glaciation in N. Africa and the almost global late Precambrian glaciation (recorded in the Dalradian; Ch. 4c) are well documented. There were probably also several earlier Pre-

cambrian glacial periods (Fig. 1.2). Short glacial periods seem to have recurred at intervals of $c.$ 200—300 Ma, between which glacier ice largely disappeared from the Earth's surface.

What causes glaciations? Theories abound. Anything that can affect the total amount or the distribution of heat reaching the Earth's surface is a possible contributor to climatic change. However, before we can even attempt to understand the causes, it is necessary to investigate the way in which climate has changed.

17b Isotopes and organisms: evidence for climatic changes

The extremes of climatic change can be identified relatively easily in the geological record. The presence of tillites or extensive evaporite deposits indicates the existence of cold glacial conditions or of a hot arid environment respectively. Milder fluctuations require more subtle techniques for their detection, those most used at present being palaeontological and isotopic.

Most organisms have a restricted ecological range, one of the most important restrictions being temperature. Therefore it is possible to use the presence of fossil organisms that have known present-day temperature ranges to infer ancient temperatures. This is fairly straightforward in the recent past, as for most organisms it can be assumed that their optimum conditions for life have not changed significantly since, say, the early Pleistocene. However, going back further in time the number of species that are still alive today decreases, and the assumption that their ecological range has not changed becomes increasingly tenuous. The most useful organisms for palaeoclimatic studies are those which have narrow temperature ranges, and are therefore sensitive to change, and which are small, common and abundant, so that they can be collected in large numbers. The most used are marine micro-organisms (especially foraminiferans), the Coleoptera (beetles) and the pollen of terrestrial plants (i.e. trees, bushes and grasses).

It should be noted that this use of fossils is quite different from their use in stratigraphic zonation and correlation. However, the two uses are not mutually exclusive. Within a given deposit, some fossil species may be used for stratigraphic zonation and others can be used as palaeoclimatic indicators. Changes in the fossil assemblages through a vertical succession due to evolution and extinction are relatively slow, allowing biozones of the order of a million years long to be recognised. Changes resulting from climatically induced migration in the Quaternary are much faster, allowing climatic stages of the order of thousands or even hundreds of years long to be identified. Most of the palaeontological changes seen in the Quaternary are therefore due to changes in climate, and these changes cannot be used for palaeontological correlation in the sense they are used throughout the earlier parts of the geological record. However, these changes can be used for local correlations under certain conditions, as discussed in Chapter 17c.

The oxygen isotope ratio ($^{18}O/^{16}O$) in the calcite shells of marine organisms can be used to estimate past climates. This ratio depends on both the temperature and the isotopic composition of the sea water in which the shells were formed. As water containing the light isotope (^{16}O) is preferentially evaporated from the oceans to be stored as glacier ice, an increase in the volume of ice will lead to a rise in the $^{18}O/^{16}O$ ratio of ocean water. Also, as temperature falls the proportion of heavy oxygen incorporated in the shells increases. The two effects work in the same direction but as the former dominates, $^{18}O/^{16}O$ ratios can be used to give an estimate of the volume of glacier ice (Dansgaard & Tauber 1969).

Both the palaeontological and isotopic techniques were first applied to Mesozoic and Cainozoic deposits on land. The distribution of land plants on several continents and isotopic studies on marine molluscs has suggested that, following the fairly warm global temperature of the Cretaceous, the Tertiary was a period of climatic deterioration. A slow cooling from Oligocene to middle Miocene times was followed by rather rapid cooling from the late Miocene until the Pleistocene (Fig. 17.1a).

Because of the discontinuous nature of the geological record on land, such analyses provided only scattered data points. However, more recent work on deep sea cores that penetrated the slowly, but almost continuously, deposited Tertiary to Quaternary pelagic sediments has enabled the detailed timing of events and climatic history of the ocean basins to be worked out. In these cores, chronology can be ascertained both by radiometric dating and by comparing the palaeomagnetism of the cores with the palaeomagnetic time scale deduced from radiometrically dated terrestrial lava-flows (Cox 1969). Estimates of climate can be obtained from oxygen isotope analyses of calcite shells and the relative abundances of planktonic foraminifera of known temperature range (Imbrie & Kipp 1971). Also, the presence of glacially-fractured quartz sand grains within pelagic muds has been used to infer the presence of icebergs, and hence glaciers. All these data have been used to construct continuous graphs of temperature change for many oceans (e.g. Fig. 17.1c). These all show a general cooling over the last 10–15 Ma, since late Miocene times. Climatic oscillations during the Palaeogene were severe enough to establish temporary valley glaciers in Antarctica. However, the major climatic deterioration seems to have started in late Miocene times, when a large continental ice sheet became established in Antarctica. This has persisted until the present day. From Miocene until Pleistocene times the polar oceans continued to cool, but glaciers remained restricted to high latitudes. Icebergs first appeared in the N. Atlantic about 3 Ma ago. The first major glaciation in N. America may be as old as 1·5 Ma. In NW Europe, palaeontological evidence for major cooling is not evident until the local base of the Pleistocene (c. 2·5 Ma BP, i.e. before present). Here Quaternary climate alternated between temperate and cold phases, and there were extensive glaciations during only the coldest of these, towards the end of the period. Thus in NW Europe the Quaternary can be divided into an earlier preglacial (c. 2·5 to c. 0·6 Ma BP) and a later glacial (<0·6 Ma BP) part.

An interpretation of pollen data from the Netherlands in terms of climatic change is given in Figure 17.1b. This shows a complex pattern of cold and temperate phases, with the cold phases becoming more severe towards the present and with the period

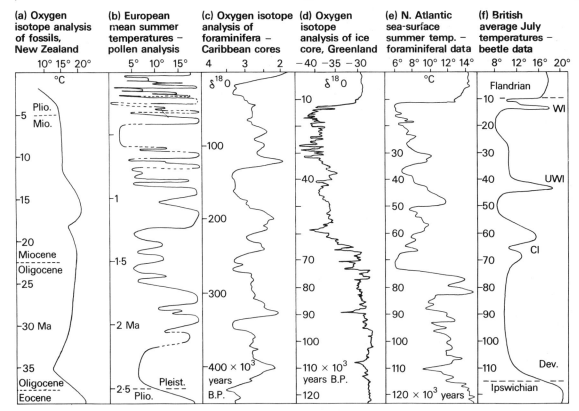

Figure 17.1   Climatic curves for parts of the Tertiary and Quaternary, calculated from various types of data. (a) Fossil data from New Zealand (after Devereux 1967). (b) Pollen data from the Netherlands and East Anglia (after Zagwijn 1974). (c) Foraminiferal data from Caribbean deep-sea cores (after Hays *et al.* 1976). (d) Ice core data from Camp Century, Greenland (after Dansgaard *et al.* 1971). (e) Foraminiferal data from a deep sea core off Ireland (after Sancetta *et al.* 1973). (f) Beetle data from Britain (after Coope 1975*). Curves showing $\delta^{18}o$ values (a measure of the oxygen isotope ratio $^{18}o/\,^{16}o$) are plotted so that deflections to the right correspond to a warming climate. All dates are BP. Mio. = Miocene; Plio. = Pliocene; Pleist. = Pleistocene; Dev. = Devensian.
CI = Chelford interstadial; UWI = Upton Warren interstadial; WI = Windermere interstadial.

of temperature fluctuations becoming shorter. Cold periods, when the land was at least partially covered by ice, are called glacials. The intervening temperate episodes are known either as interglacials, if the mean annual temperature reached that of the Flandrian climatic optimum, or as interstadials, if they were either too short or too cool to allow the development of the interglacial vegetation characteristic of the area. The Flandrian climatic optimum was that part of the Flandrian stage (c. 7000–5000 years BP) during which mean annual temperatures were about 2°C above those of the present.

17c   The preservation and correlation of the Quaternary record in the British Isles

The preglacial and glacial parts of the Quaternary are preserved in very different ways. The distribution of the preglacial sediments broadly follows that of the Neogene. Quaternary subsidence continued within the areas of the Tertiary basins, preserving mostly marine sediments; fluvial deposits accumulated only where major rivers drained into these basins. The present land areas were then, as now, mostly areas or erosion. Eustatic changes in sea level led to minor transgressions during warm phases and regressions during cold phases, although these did not significantly alter the position of the coastline of the British Isles (Ch. 16g).

The preservation of sediments deposited during glacial and interglacial cycles is more complex. During each ice advance, tills were deposited in a range of environments from the highlands to what is now the continental shelf. Fluvial and coastal erosion during the intervening periods, together with glacial erosion during the following ice advances, tended to remove these deposits from the present land areas and the shallower parts of the shelf. Clearly, although the most continuous sequence should be found in the deeper water areas beyond the shelf edge, the most complete record on the continental shelf will be found in the areas of maximum subsidence, i.e. in the central parts of the sedimentary basins. Here, tills deposited from grounded or floating ice will tend to be interbedded with marine sediments, as in the case of the Precambrian Port Askaig Tillite (Ch. 4c), and terrestrial deposits will be less common.

Within the sedimentary basins a fairly complete stratigraphic sequence should be preserved from which the alternation of cold and temperate stages, the climatic record, can be read. The uppermost part of this record can be readily calibrated, as radiocarbon dating is practical back to about 50000 years BP. However, for the older deposits the usual techniques are inapplicable. K–Ar radiometric dating is restricted to volcanic rocks, and palaeontological zoning is far too crude. The most promising method, the one that has proved successful in dating deep sea cores, is palaeomagnetic dating. It should be noted that this can only be applied to continuous vertical sections or borehole cores, not to isolated deposits.

The climatic curve in Figure 17.1b, constructed from pollen analytical data and calibrated by radiocarbon and palaeomagnetic dating, is for the Quaternary of the Netherlands, an area within the North Sea sedimentary basin. If the Dutch Quaternary stratigraphic column could be used as a standard sequence for NW Europe, it should be possible to correlate the isolated deposits around the North Sea Basin by comparing their climatically controlled fossil assemblages with the standard basinal sequence, provided (a) that the standard sequence is a complete sedimentary record (i.e. it includes all the cold and temperate stages), (b) that it is possible to distinguish palaeontologically between each warm stage in the standard sequence, and (c) that the standard sequence and deposits of unknown age are geographically sufficiently close and similar in their depositional environment to have experienced a similar faunal and floral history. However, as the palaeontological differences between climatic stages are fairly subtle, this requires a very detailed palaeontological knowledge of both the standard and unknown sequences, and problems remain in applying such correlations even during the younger interglacials within NW Europe.

Outside the sedimentary basins, only the youngest deposits, i.e. those amenable to radiocarbon dating, can be dated absolutely. In the absence of a standard basinal sequence it is usual to erect a composite local sequence, with stages based on different type localities. This composite sequence is unlikely to be complete; and as it is based on an interpretation of the stratigraphy, which is itself based on assumptions about the climatic history, it may be of doubtful validity.

The climatic stages of the preglacial part of the British Quaternary scheme (Fig. 17.2) are defined from a borehole (Ludhamian to Antian) and coastal sections (Baventian to Cromerian) in East Anglia. There can be little doubt that they are in the right stratigraphic order but, as they are based on deposits from the edge of the North Sea sedimentary basin, there are likely to be major gaps in the record (Ch. 16g). The glacial Quaternary stages are based on type localities in East Anglia and the Midlands. The validity of these stages cannot yet be checked against a standard basinal sequence, and there is controversy regarding the correlation of events with these stages in the British Isles (Ch. 17d). However, there is little doubt that the last glacial stage, the Devensian, correlates with the last glaciation in many other parts of the world, such as the Weichselian in N. Europe, the Würm in the Alps and the Wisconsinian in N. America. The last interglacial, the Ipswichian, can also be correlated with the Eemian of the European scheme.

17d   The pre-Devensian record

On land in the British Isles there is evidence for four

	Dates in years B.P.	Stage	Climate	Examples of typical deposits
QUATERNARY: Recent or Holocene		Flandrian	temperate	silts & peats in Lincolnshire, East Anglia & Lancashire; estuarine deposits – Thames etc.
QUATERNARY: PLEISTOCENE	10 000	Devensian	cold, glacial at end of stage	head in S & SW England; Skipsea Till, Irish Sea Till; Four Ashes Gravel; Dimlington Silts.
	115 000	Ipswichian	temperate	organic deposits near Ipswich; beach deposits at Sewerby & in Devon
	128 000	Wolstonian	cold, in part glacial	Welton Till Fremington Till
		Hoxnian	temperate	organic deposits at Hoxne, Suffolk; Marks Tey, Essex
		Anglian	periglacial and glacial	Lowestoft Till, North Sea Drift, Corton Sands Barham Sands and Gravels; Barham Arctic Structure Soil
		Cromerian	temperate	Valley Farm Rubifield Rubified Sol Lessivé; peats & freshwater sediments – Norfolk
		Beestonian	periglacial and glacial	Kesgrave Sands & Gravels
	~ 600 000	Pastonian	temperate	Westleton Beds Icenian Crag including { Weybourne Crag, Chillesford Clay, Norwich Crag }
	hiatus			
	~ 1 600 000	Baventian	cold	marine silts & clay – Ludham & Easton Bavents
		Antian	temperate	shelly sand } marine sediments in Ludham borehole and Suffolk
		Thurnian	cold	silt } marine sediments in Ludham borehole and Suffolk
	~ 2 050 000	Ludhamian	temperate	shelly sand } marine sediments in Ludham borehole and Suffolk
	hiatus			
	~ 2 450 000	'pre-Ludhamian'	temperate	Red Crag
	~ 2 500 000			
PLIOCENE			warm temperate	Coralline Crag

Figure 17.2 Quaternary stages, dates, climatic conditions and some typical deposits (see G. F. Mitchell *et al.* 1973 for more details).

glaciations (Fig. 17.2). The ice limits of the last glaciation, which took place towards the end of the Devensian cold stage (between 26000 and 10000 years BP), and of the glaciation of maximum extent are shown in Figure 17.3 (D. Q. Bowen 1977*). Within the Devensian ice limit, Devensian till blankets most of the lowlands and large parts of the highlands too. This till usually lies directly on bedrock, and older Quaternary deposits are only locally preserved beneath the Devensian till, as at Dimlington Cliff on the Yorkshire coast (Fig. 17.5) and Shortalstown, Co Wexford. In the S. Irish Sea pre-Devensian till is preserved beneath Devensian and Ipswichian deposits, where ice partly scoured out areas of soft Tertiary rocks. This is unusual, as within the Devensian ice limit on the continental shelf Devensian deposits unsually directly overlie the Tertiary or lowermost Pleistocene. Clearly, Devensian glacial erosion has been important in removing the record of earlier glaciations in these areas.

Outside the Devensian glacial limit, pre-Devensian glacial and interglacial deposits are common, although they have suffered more erosion than the Devensian deposits in the lowland areas to the north and are most extensive where they have escaped erosion on the interfluve areas of East Anglia and the Midlands.

The earliest evidence for glaciation in the Quaternary sequence in East Anglia is found in the sediments of the Beestonian Stage, which overlie the Pastonian marine, estuarine and beach deposits (Ch. 16g). The Beestonian includes a degraded till

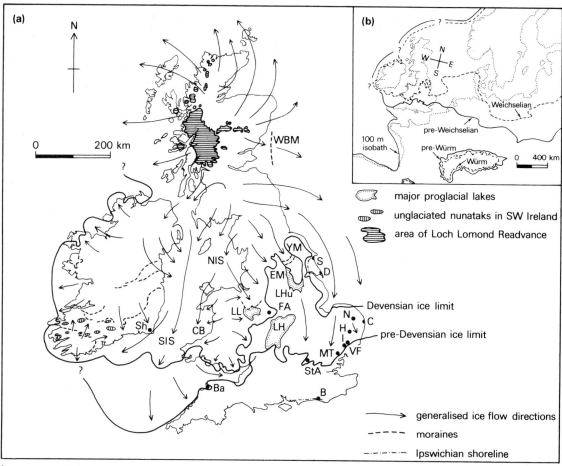

Figure 17.3 (a) Ice limits and ice flow directions in the British Isles (after D. Q. Bowen 1977*). (b) Ice limits for Europe (after Flint 1971).
Localities: B. = Brighton; Ba. = Barnstaple; C. = Corton; CB = Cardigan Bay; D. = Dimlington; FA = Four Ashes; H. = Hoxne; I. = Ipswich; LH = Lake Harrison; LHu = Lake Humber; LL = Lake Lapworth; MT = Marks Tey; N. = Norwich; NIS = N. Irish Sea; S. = Sewerby; Sh. = Shortalstown; SIS = S. Irish Sea; StA = St Albans; VF = Valley Farm.
Moraines: EM = Escrick; WBM = Wee Bankie; YM = York.
For sections N–S and E–W in (b), see Figure 17.6.

remnant in the SW Midlands and extensive fluvial sands and gravels (Kesgrave Sands and Gravels) in East Anglia and Essex (Fig. 17.4b). Abundant ice-wedge casts and involutions indicate that the latter were deposited under periglacial conditions, and palaeocurrent measurements show that, in Essex and S. Suffolk, they were deposited from a major NE-flowing river north of the course of the present River Thames (Rose & Allen 1977*).

During the following Cromerian temperate stage, estuarine silts and freshwater muds and peats accumulated. These are now exposed on the Norfolk coast. Further south, the Cromerian is represented by the Valley Farm Rubified Sol Lessivé, a palaeosol formed under humid warm temperate conditions. Evidence for cold conditions follows with, in Norfolk, the deposition of freshwater and estuarine sediments, cut by ice wedge casts and containing an arctic fauna and flora, and, in Suffolk and Essex, the formation of an arctic palaeosol (the Barham Arctic Structure Soil), loess and blown sand deposits. These features are attributed to a periglacial episode that preceeded the advance of the glaciers that deposited the overlying Lowestoft Till, and they are included with this till as deposits of the Anglian cold stage.

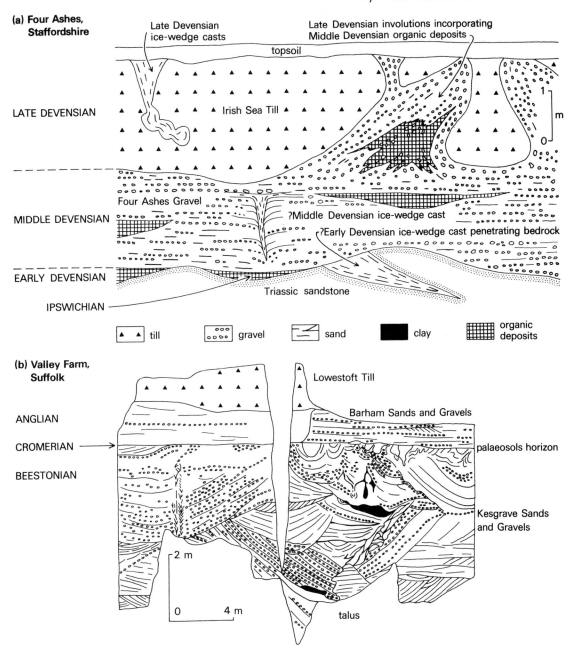

Figure 17.4 Examples of Quaternary stratigraphic sequences. (a) Four Ashes pit, Staffordshire, the type locality for the Devensian: schematic diagram showing the alternation of interglacial and interstadial organic deposits with glacial and periglacial sediments and structures (after Morgan 1973). (b) Valley Farm pit, Suffolk: sketch of north face; the palaesols horizon includes the Anglian Barham Arctic Structure Soil and the Cromerian Valley Farm Rubified Sol Lessivé which it deforms (after Rose & Allen 1977*).

The Lowestoft Till occupies an extensive area in East Anglia, the E. Midlands and S. Lincolnshire. The bulk of the clasts are chalk, the matrix is dominated by clay, and Perrin *et al.* (1973) have suggested that the North Sea is a likely source area. In NE Norfolk there are up to three tills containing Scandinavian erratics interbedded with sands, which are collectively known as the North Sea Drift. Fabric analysis suggests that these tills were deposited from ice that flowed from the northwest, north and northeast; and although the Lowestoft Till overlies the North Sea Drift south of Corton, all these deposits are thought to be of broadly the same age, i.e. Anglian, but to have been deposited from different ice lobes between which fluvial, lacustrine and estuarine sediments accumulated. Along the N. Norfolk cliffs the North Sea Drift exhibits spectacular deformation structures, including great thrust planes, domes, basins, overfolds and chalk rafts up to 10 m thick. With the exception of the chalk rafts, these are probably load and diapiric structures resulting from the deposition of the overlying outwash sands and gravels (Banham 1975).

The major river that flowed northeastwards through Essex and Suffolk during the Beestonian started to migrate south and downcut during this stage. By the Anglian it occupied the position of the present Thames drainage system. Indeed, the major features of the present drainage system of SE England date from the Anglian. Anglian ice produced only a minor southward diversion of the Thames in the Vale of St Albans area. Several subsequent episodes of aggradation and incision have left behind a series of river terraces in the Thames Basin.

All the pre-Devensian tills in East Anglia can be assigned to the Anglian stage. They are overlain by Hoxnian biogenic interglacial deposits. In the Midlands, Hoxnian deposits underlie a pre-Devensian till defined as being of Wolstonian age (Fig. 17.2), although Bristow and Cox (1973) have suggested that it is equivalent to the Lowestoft Till. Thus the deposits of the penultimate glaciation in East Anglia and the Midlands belong to different stages, the Anglian and Wolstonian respectively. It seems either that Wolstonian ice never reached East Anglia, or that the correlation of the Hoxnian deposits in East Anglia with those of the Midlands is invalid, the Anglian and Wolstonian tills both being deposited during the same glacial stage. In view of the fundamental difference between these interpretations it is difficult to produce a synthesis of Anglian—Wolstonian history in Britain, as the interpretation of events in other areas depends on our understanding of the type areas of East Anglia and the Midlands.

In the Midlands, Wolstonian tills with erratics derived from Wales, N. England and S. Scotland overlie Hoxnian interglacial lake deposits and are in turn overlain by Ipswichian fluvial terrace gravels. Tills thought to predate the Wolstonian (i.e. of Anglian age) are of restricted occurrence. As the Wolstonian ice advanced it blocked the Severn Valley, forming a lake (named Lake Harrison) which stretched from Stratford-upon-Avon to Leicester. As the ice advanced further it over-rode the lake, but small lakes might have formed later in the same area when the ice ablated. Outwash from the Wolstonian ice poured southwards into the Thames and other drainage basins, depositing northern erratics. During the following interglacial the valley fills were partly eroded, leaving behind river terraces. Outwash sands and gravels, subsequently terraced, were also deposited in valleys to the north as the ice ablated.

Further west, Wolstonian ice just impinged on the northern coast of Devon and Cornwall, depositing a shelly till, the Fremington Clay, and glaciofluvial deposits around Barnstable. Giant erratics of granite and gneiss, weighing up to 50 tonnes, are also found overlying wave-cut rock platforms along this coast. These may be the remains of a Wolstonian or older till long since washed away, although, as similar giant erratics are also found along the English Channel and S. Irish coasts, it has been suggested that they were deposited from icebergs.

In NW Europe three glaciations are recognised: the Weichselian, Saalian and Elsterian, in order of increasing age. Saalian ice extended further south that that of the other two glaciations in the Netherlands, although Elsterian ice was the most far reaching in Germany and Poland (Fig. 17.3). The Quaternary succession of the Netherlands can be traced westwards into the southern North Sea. Here, during the Elsterian glaciation, tills were deposited in the north, bordered to the south by fluvial and lacustrine sediments. The damming of the North Sea during each glaciation impounded a

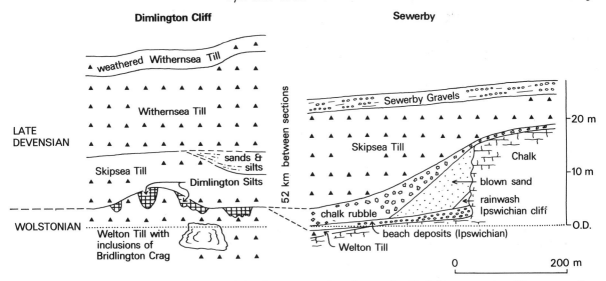

Figure 17.5 Stratigraphic relationships of the Quaternary deposits on the Yorkshire coast (see Figure 17.3 for localities). Heights are in metres above OD (Ordnance Datum). The late Devensian, but preglacial, Dimlington Silts are dated at *c*. 18 400 years BP; peat in kettleholes overlying the Devensian tills is dated at *c*. 13 000 years BP. The rainwash, blown-sand and chalk-rubble solifluction deposits date from the cold early Devensian stage, following the withdrawal of the Ipswichian sea. The Bridlington Crag is a Hoxnian or Wolstonian shelly marine clay eroded from an offshore area by the advancing Wolstonian ice (after Catt & Penny 1966, with terminology after Madgett & Catt, in press).

lake, fed by both outwash streams and glacially diverted preglacial rivers, which eventually drained southwestwards along the bed of the English Channel. During the succeeding Holsteinian interglacial the sea transgressed southwards into the North Sea, and marine sediments were deposited as far south as a coastline which ran E–W at about the latitude of Norfolk. The glaciers of the following maximum glaciation (the penultimate or Saalian of the European scheme) scoured the floor of the southern North Sea, deposited tills, and deformed the underlying interglacial deposits into a series of ice-pushed ridges which are also found in Holland. The marine transgression of the last interglacial was more far reaching than its predecessor. It crossed the present coastline in Yorkshire and Lincolnshire, where a buried Ipswichian cliff line cut in the Chalk along the eastern side of the Wolds is preserved beneath Devensian till (Fig. 17.5), and around the southern coast of England, where wave-cut Ipswichian platforms are found, for example, at Brighton. The continent was clearly cut off from Britain at this time.

Pre-Devensian tills are widespread in Ireland, where all but a few nunataks in the southwest were ice covered during the glacial maximum. The ice was mostly of local derivation, except along the eastern and southern coasts where Scottish ice locally pushed ashore. The paucity of interglacial sites in Ireland makes it difficult to date pre-Devensian events; and although most of the penultimate glacial tills are often assumed to be Wolstonian, there is little firm evidence for this at present.

South of the limit of the maximum glaciation, cold periods were marked by the formation of extensive solifluction deposits known as 'head' or, on the Chalk, as 'coombe rock'. Such periglacial features as involutions, stone stripes and polygons are also found.

17e The story of the last glaciation: the Devensian

In contrast to the confusion surrounding pre-Devensian events, there is a fair consensus of opinion about the detailed history of the last glaciation. This is because Devensian deposits are fresh and extensive, have suffered little erosion and are largely within the compass of radiocarbon dating. A detailed understanding of Devensian events may therefore help in elucidating the more obscure pre-Devensian record.

Climatic curves based on N. Atlantic deep-sea core data show a cooling c. 110000 years BP (about the Ipswichian/Devensian boundary) and a further marked cooling 73000 years BP. After this, the climate oscillated, reaching its most severe deterioration during a short cold phase between 30000 and 11000 years BP (Fig. 17.1e). Peat deposits in kettle-holes on top of Devensian tills in Scotland give radio-carbon dates ranging up to 13000 years BP, while organic remains underneath such tills give dates as young as c. 27000 years BP. Thus it appears that the advance of the Devensian ice was a short event right at the end of the Devensian, lasting c. 10000 years, culminating at c. 17000 years BP, and preceded by a long, usually cold but non-glacial phase.

During the Ipswichian interglacial, which preceded the Devensian, pollen and beetle data imply that summer temperatures were several degrees warmer and winter temperatures were milder than those of today, giving rise to a woodland vegetation. Sea level would have been slightly higher than that of today. The Ipswichian cliffline preserved in Yorkshire lies at just about the present sea level (Fig. 17.5), although the higher-level raised wave-cut platforms along the southern coast of England must have suffered subsequent tectonic uplift. As temperatures fell at the beginning of the Devensian, sea level would have fallen slightly as glaciers expanded at higher latitudes (Fig. 17.7), and the woodland would have been replaced by a sparse tundra vegetation. During the early Devensian the British Isles appear to have been very cold but, probably because of a relative lack of precipitation, the lowlands remained ice free. Strong winds would have cut across this inhospitable cold desert landscape. The abrasive action of blown sand formed ventifacts and undercut rocks. The climate was also conducive to the widespread formation of periglacial features, such as cryoturbations, ice-wedge polygons and solifluction deposits.

The study of fossiliferous horizons within fluvial terrace deposits in the Midlands and S. England (Fig. 17.4a) shows that the climate ameliorated briefly twice, resulting in the Chelford interstadial c. 60000 years BP and the Upton Warren interstadial c. 43000 years BP (Fig. 17.1f). During the former, temperatures warmed to within a few degrees of those of today and a pine and spruce forest vegetation returned. However, the latter warm interval was so short that, although summer temperatures briefly exceeded those of today, the climate deteriorated again before forest had spread northwards into Britain (Coope 1975*). Sea level might have continued to fall throughout this period, with the major rivers flowing along roughly their present courses and cutting across the shallow parts of the present continental shelf. However, the falling sea level did not expose the northern North Sea, where up to 300 m of marine clays with a shallow cold-water fauna, the Aberdeen Ground Beds, were deposited (Fig. 17.6). These overlie an irregular surface, possibly cut by Wolstonian glacial action, which truncates the lower Pleistocene Basal Beds. Dropstones in the Aberdeen Ground Beds indicate the presence of icebergs, which may have calved from the Scandinavian glaciers known to have existed at this time. Lignitised wood samples within the beds may date from the Upton Warren interstadial (Eden et al. 1977).

Some time after 30000, and possibly as late as 25000 years BP, the British climate became less continental and more maritime, allowing a sufficient increase in precipitation for glaciers to move out of the mountains and spread southwards. Flow directions deduced from the distribution of erratics and the orientation of glacial landforms are shown in Figure 17.3. Deep-sea core studies show that the glaciation reached its peak globally c. 17000 years BP, and it is likely that glaciers extended to the Devensian ice limit by this time. Ice thicknesses probably exceeded 2 km in parts of Scotland during the glacial maximum. As with earlier glaciations, the advancing ice impounded proglacial lakes where it cut off the lower reaches of valley systems. For example, the mouth of the Humber was blocked, forming a large lake (Lake Humber) in the Trent lowlands. Gaunt (1976) has suggested that the leading edge of a glacier extending down the Vale of York may have partly floated in this lake, the maximum extent of the ice being indicated by a belt of glaciolacustrine sands and gravels.

The Devensian ice limit is only rarely indicated by good terminal moraines. Usually the Devensian till has a feather edge and the limit is not easy to find. Within the limit the lithology of the tills varies from gravelly clays to sandy clays, clays and even

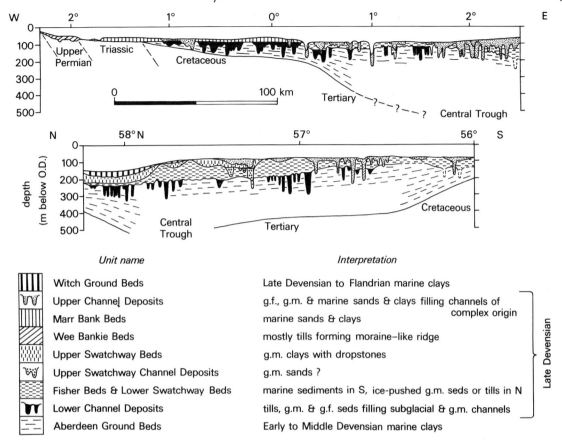

	Unit name	Interpretation	
	Witch Ground Beds	Late Devensian to Flandrian marine clays	
	Upper Channel Deposits	g.f., g.m. & marine sands & clays filling channels of complex origin	Late Devensian
	Marr Bank Beds	marine sands & clays	
	Wee Bankie Beds	mostly tills forming moraine–like ridge	
	Upper Swatchway Beds	g.m. clays with dropstones	
	Upper Swatchway Channel Deposits	g.m. sands ?	
	Fisher Beds & Lower Swatchway Beds	marine sediments in S, ice-pushed g.m. seds or tills in N	
	Lower Channel Deposits	tills, g.m. & g.f. seds filling subglacial & g.m. channels	
	Aberdeen Ground Beds	Early to Middle Devensian marine clays	

Figure 17.6 Stratigraphic relationships of the Quaternary deposits in the central North Sea, shown along the W–E and N–S sections of Figure 17.3b (after Holmes 1977).
gm = glaciomarine; gf = glaciofluvial; seds = sediments; OD = Ordnance Datum.

sands, depending on the local source material. Several lithologically different tills, all dating from the Devensian, may occur in a vertical section. These may represent deposition from different ice lobes with different source areas, although the topmost till is often just a weathered decalcified variant of the underlying one (Fig. 17.5). Glaciofluvial and glaciolacustrine sands, gravels and clays are often interbedded with, or incorporated into, the tills. These are now accepted as being broadly contemporaneous with the tills, the fluvial and lacustrine environments existing under, within and on top of the ice. However, in the past these deposits led to confusion, as it was thought that sands interbedded with tills indicated interglacial conditions separating two glaciations.

Devensian tills have been widely recognised in offshore areas. In the S. Irish Sea, overlying pre-sumed Ipswichian interglacial deposits, Garrard (1977) has found up to 70 m of till, containing shells and local Palaeozoic and Mesozoic erratics together with far-travelled Scottish erratics. In Cardigan Bay, Irish Sea ice over-rode local Welsh ice flowing westwards out of the mountains. Lateral moraines deposited between the Welsh ice streams in Cardigan Bay are now preserved as offshore ridges known as sarns. Similar tills are found in the N. Irish Sea and on the shelf west of Scotland, except that here the youngest underlying sediments are of Neogene to (?)lower Pleistocene age.

In the central North Sea the situation is more complicated (Fig. 17.6). Here a channelled erosional surface, probably cut by glacial scour and subglacial streamflow modified by marine erosion, truncates the early Devensian Aberdeen Ground Beds in central areas and Mesozoic rocks near the coast. This is overlain by a complex pattern of tills, glaciomarine and glaciofluvial clays, sands and gravels cut by further channels, all attributed to the late Devensian glacial and bracketed by radiocarbon dates at c. 32 700 and c. 17 700 years BP. Stratigraphic and facies relationships are complex within these deposits, and deposition under grounded ice, floating ice, open marine and terrestrial conditions is inferred, with oscillation of the margins of both the Scottish and Scandinavian ice sheets (Holmes 1977). Possible ice-pushed ridges have also been recognised.

Foraminiferal evidence shows that the N. Atlantic at the latitude of the British Isles started to warm c. 13 500 years BP, and oxygen isotope work shows that global ice volumes were decreasing after 17 000 years BP. In the British Isles, ice had probably started to ablate from the lowlands by 14 500 years BP, leaving in places deposits of hummocky till, sand and gravel released from the melting ice. The ice front probably did not retreat in an orderly fashion as it does in mountain valleys. Rather, the ice ablated away in place, leaving dead ice masses impounding small lakes in which laminated clays were deposited. Recessional or readvance moraines of this date are rare. The Escrick and York moraines (Fig. 17.3) record some retreat up the Vale of York. However, numerous other features recorded as Devensian readvance moraines have been largely discredited and are now thought to be areas of topographically controlled hummocky ablation till.

As the ice ablated, the sea advanced across glaciofluvial outwash in the south and till spreads in the north. In the S. Irish Sea the transgression at first turned a major meltwater channel running parallel with the coast into an estuary in which bedded clays and silts were deposited. Further north the till is overlain by laminated muds and fine sands containing dropstones. Clearly the transgression had flooded considerable parts of the shelf while extensive glacier ice was still present. This is confirmed in the Forth Approaches area of the North Sea (Fig. 17.6), where the moraine-like Wee Bankie Beds may mark a recessional or readvance ice front that existed while shallow marine sands with an arctic fauna, the Marr Bank Beds, were being deposited to the east (Thompson & Eden 1977). The latter appear to overlie a plane of marine erosion cut by the transgressing sea. The Marr Bank and Wee Bankie Beds are cut by a series of channels, possibly eroded by the subglacial outflow of water from the Wee Bankie ice front together with tidal scour. Late-glacial marine clays with dropstones, the St Abbs Beds, which pass conformably upwards into Postglacial muds and sands, the Forth Beds, were deposited in these channels and against the moraine.

Although it does not seem to have left a record in the offshore sediments, a short cold period followed the warming and ablation of the main Devensian ice. Between 11 000 and 10 000 years BP ice advanced out of the Scottish mountains, but this time only as far as the Midland Valley, depositing till, outwash and leaving behind a good terminal moraine in many places (Fig. 17.3). This episode is known as the Loch Lomond readvance. Although there is debate about whether ice completely disappeared from the British Isles or just retreated to the high mountain areas during the preceding warm interval, the Late-glacial or Windmere interstadial, palaeontological evidence suggests that the interstadial was marked by temperatures as warm as, or warmer than, those of today (Coope 1975*). The Loch Lomond readvance is equated with a period of periglacial activity, during which Devensian tills and Late-glacial sediments outside the readvance limit suffered cryoturbation and blown sand deposits were formed. The intensity of this periglacial activity is demonstrated in SW Scotland by the main Late-glacial wave-cut rock platform which, although up to tens of metres wide, was cut in a relatively short time c. 11 000 years BP.

17f The Flandrian transgression

The melting of the Devensian and equivalent ice sheets throughout the world produced a rise in sea level of about 120 m between 17 000 years and 7500 years BP. The resulting marine transgression was the last major natural geological event to affect the British Isles, although its effects were felt around the world.

If it is assumed that the oxygen isotope curve measures changes in ice volume, a sea level curve can be calculated. This shows an unsteady slow fall during the Devensian, followed by a very rapid rise

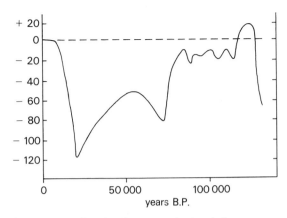

Figure 17.7 Sea level curve calculated from oxygen isotope data. Depths are in metres relative to present sea level (after Shackleton & Opdyke 1973).

(Fig. 17.7) starting *c.* 17 000 years BP. The repeated sawtooth pattern of the oxygen isotope curve (Fig. 17.1c) implies that similar sea-level changes occurred during each glacial–interglacial cycle. The evidence from offshore areas shows that sea level had already risen considerably before the ice disappeared from the continental shelf and that this rise continued, without any significant reversal during the Loch Lomond readvance, into the post-glacial Flandrian Stage. A more detailed curve for Flandrian sea-level rise can be constructed from, for example, the depth of dated peat and intertidal shell samples. This shows the rapid rise continuing until *c.* 7000 years BP, with only a slow subsequent rise of a few metres to the present day.

Although this transgression was much faster than the type of transgression that is familiar from the geological past, some of its geological effects were similar. As first the beach environment and then a strongly tidal shelf environment extended landwards, the top of the till spreads were reworked. The fines were washed out, leaving behind a thin layer of lag gravel. The fluvial sediments of river and meltwater channels were overlain by estuarine and then fully marine sediments, while, upstream, rivers aggraded.

In N. Britain the effect of the eustatic rise in sea level is complicated by the isostatic uplift of the crust following the removal of the ice. This uplift took the form of a doming, with the rate of rebound decreasing radially away from the centre of maximum ice loading in SW Scotland. Around the Forth

Estuary a whole series of raised shorelines indicate a progressive emergence of the area during the ablation of the main Devensian ice, culminating in the widely recognised late Devensian shoreline that immediately predates the Loch Lomond readvance. The eustatic rise in sea level eventually gained the upper hand, leading to a transgression which terminated *c.* 6700 years BP with the formation of the main Flandrian shoreline. Subsequent isostatic uplift has led to a regression in N. Britain, where the Flandrian shoreline now stands as high as 15 m above sea level.

17g Theories for glaciation

Having discussed the nature of climatic change and its geological effects during the Quaternary, we may now briefly consider its causes.

Both terrestrial and extraterrestrial causes have been suggested for glaciations. The constancy of the sun has often been assumed, but the suggestion that its radiation output may vary with time is now taken seriously. The reception of energy from the sun may also vary as the solar system moves through parts of the galaxy with different concentrations of dust. The well-known periodic variations in the Earth's orbit around the sun may lead to a varying energy input to the Earth. Among the terrestrial factors are the distribution of continents and oceans. The transfer of heat from the equator to the poles is effected partly by ocean currents. Oceans oriented N–S are therefore more effective at warming the poles than those oriented E–W, and the motion of continental plates could lead to climatic change. The formation of mountain chains affects wind circulation patterns and provides areas for the growth of glaciers. The injection of volcanic dust into the atmosphere could affect the amount of absorbed radiation, as could the composition of the atmosphere itself with changes in carbon dioxide resulting from biological changes on Earth.

All these factors could influence climatic change, acting singly or in any combination. Does the nature of Quaternary record give any clues as to which of the many hypotheses for climatic change are the most likely? A striking feature of climatic curves, especially oxygen isotope curves, is their almost mathematical regularity (Fig. 17.1c). Each glacial

—interglacial cycle seems to follow a similar pattern. Are any of these hypotheses likely to produce such cyclicity? Are any amenable to analysis? The factor that does produce a calculable cyclicity is the astronomical variation in the Earth's orbit. Milankovitch calculated the variation in radiation received at a latitude of 60°N that was due to changes in the Earth's orbit and suggested that it was of the right amplitude and periodicity to produce glaciations. His calculations have been successively refined and have now been compared with oxygen isotope results (Hays *et al.* 1976). The similarity of the curves is evident, and it even stands up to close statistical scrutiny. Hays *et al.* (1976) therefore believe that this shows that the Milankovitch mechanism, as it is known, is the main cause of short-term climatic change, even though the actual links between a change in radiation received and the expansion of glaciers are not known.

Does this mechanism help with the interpretation of older glaciations, such as the late Precambrian or Permo-Carboniferous? Certainly an oscillating climate during such glaciations is suggested by the geological evidence (e.g. Ch. 4c), although it is not yet known whether the long term cyclicity indicated by the recurrence of major glaciations every few hundred million years could result from astronomical or geological causes.

17h Summary

The understanding of Quaternary history requires the integration of geomorphological, sedimentological, palaeontological and isotopic results from the land, the shelf seas and the ocean basins. The Quaternary record on land is very fragmentary, and the chronology of events will only be fully understood by reference to the more complete history preserved in the ocean basins and the more rapidly subsiding shelf sedimentary basins. The most complete Quaternary succession around the British Isles is found in the southern North Sea and the Netherlands, i.e. in the central part of the North Sea Basin. Here at least seven major cold—warm climatic oscillations are recorded, only the last three or four being cold enough to bring glaciers into lowland Europe. Outside this area, both on land and on the continental shelf, it is only Devensian (last-glacial) deposits that are common, and they usually lie erosively on preglacial sediments or rocks. Clearly, the effects of each glacial advance tend to be removed from the land and the shallow shelf seas by erosion during the subsequent interglacials and interstadials and by the glacial erosion of the next ice advance itself. It is only in the centre of rapidly subsiding sedimentary basins that glacial deposits have a good chance of being preserved.

18 The present . . . and the future

18a Geological processes in the British Isles today

With the exception of those changes resulting from man's activities, such as the felling of forests and ploughing of land, the last major geological event to affect the British Isles was the Flandrian transgression. This does not mean that the British Isles are no longer undergoing geological change. Although one may not consider NW Europe to be geologically active in comparison with such areas as Japan or Iceland, many changes are still taking place in this area at rates comparable to those inferred from studies of the past.

Thermal processes in the British Isles are now restricted to the activity of a few hot springs, such as the one at Bath. Tectonic movements, although not as dramatic as those seen at plate margins, are still having important effects. The major sedimentary basins, such as the North Sea, are probably subsiding at rates at least as fast as during the Tertiary and Pleistocene. Allowing for the effect of eustatic changes in sea level (Ch. 18b), it appears that — with the exception of most of the Scottish mainland and Ulster, which are still isostatically rising at about 0·5 mm a^{-1} in response to the removal of the Quaternary ice load — the whole of the British Isles is undergoing a relatively rapid tectonic subsidence of about 1 mm a^{-1} (= 1 km Ma^{-1}) (Rossiter 1972). The effect of this subsidence has already become critical in the London area, where a high subsidence rate, accelerated by man's activities to 2 mm a^{-1}, has necessitated the construction of a barrier across the Thames to prevent flooding.

The effects of erosion and deposition are very obvious. Inland the soft Quaternary sediments and soils on harder rocks are being eroded by streams, mass wasting and, especially on arable land, rainwash. The effects of coastal erosion are seen in the steep unstable cliffs of S. and E. England, cut in soft Mesozoic–Quaternary sediments. These are receding at a rate of over 1 m a^{-1} in many places. Most of the sediment produced by this erosion is transported along the coast and accumulates in such sediment traps as estuaries. Most of the fluvial input is similarly trapped in estuaries or along the coast in beaches, barriers and dunes. Very little sediment is at present reaching the continental shelf. With certain exceptions, the present thin veneer of shelf sediments around the British Isles is largely composed of reworked glacial and glaciofluvial deposits. These exceptions include parts of the northern North Sea and Irish Sea (where over 10 m of Flandrian mud have accumulated), the English Channel (where a bare rock floor is common) and off the northwestern coasts of Scotland and Ireland (where recent temperate-water carbonate sediments are common). Deep glacially-scoured marine and freshwater lochs are such efficient traps for fluvially derived sediments that little clastic sediment is now being added to this northwestern shelf, although in some places glacial deposits are still being reworked. Debris composed of molluscs and barnacles, together with a minor proportion of echinoids, serpulids and calcareous algae, is therefore accumulating in many parts of this area to form carbonate sands and gravels.

The sediment dispersal system described above is geologically atypical and results from the recent and geologically very rapid Flandrian transgression. In other words, the erosional and depositional agencies have not yet had time to produce a shelf sediment distribution pattern that is in equilibrium with the presently acting marine processes.

18b The near future: a return to normality?

In terms of the geological time scale, the British Isles have only just emerged from the last glaciation. Most of the British landscape and the adjacent shelf sea-bed is characteristically glacial in both geomorphology and sediment distribution. What happens in the near future depends largely on how the climate changes.

If it were assumed that the Quaternary ice age

has come to an end, we might predict that most of these glacial characteristics would be progressively eradicated. Soft glacial sediments would be removed by fluvial and coastal erosion. Lakes would fill with sediment, estuaries would tend to silt up and eventually fluvial sediment would be contributed directly to the sea floor. The sedimentary basins would, presumably, continue to subside and accumulate predominantly marine sediments. All traces of the Quaternary glaciation would eventually be removed from the land, but glacial, glaciomarine and glaciofluvial sediments would be preserved within the marine sequence of the major sedimentary basins. The position of the coastline would depend on the balance between subsidence and sediment availability, which itself would depend on the tectonics of the source area. In the absence of any major changes, a return to the palaeogeography of the Mesozoic might be imagined.

However, it is unlikely that the Quaternary ice age has ended. Mathematical analysis of the climatic curves deduced from deep sea cores shows that these curves can be thought of as the sum of a series of cycles with periodicities of 20 000 – 120 000 years (Hayes *et al.* 1976). As these cycles can be projected forwards, we can predict that, in the absence of interference from man himself, glacial conditions should return in the geologically not too distant future. We appear to be living in an interglacial. If this is the case, the whole spectrum of geological change that can be inferred for, say, the Devensian will be repeated. Eventually the return to 'normality' will come, but we do not know how many glacial advances away that return may be. Predictions about the more immediate future have to allow for the impact of man's activities on climate. World climate became generally warmer following the 'Little Ice Age' of the sixteenth to nineteenth centuries, although in the northern hemisphere this amelioration seems to have reversed since about 1950. Over the last 100 years or so sea level has been rising eustatically at about 1 mm a^{-1}, and Mercer (1978) has predicted that within the next 100 years a catastrophically rapid rise due to melting of part of the Antarctic ice cap could follow a rise in polar temperatures resulting from the increasing concentration of atmospheric carbon dioxide derived from the burning of fossil fuels.

18c　The distant future: analogies with the past

It is a basic tenet of plate tectonics that ocean basins can only expand to a certain critical size; their adjacent passive continental margins are then transformed into active margins, with subduction zones, volcanism and orogenesis. This fate must eventually befall the NW European continental margin. It would be a delusion to think that we could make any sensible predictions about how and when this might happen. Nonetheless, it is useful to consider the difficulties of trying to make any such predictions, because they highlight the problems involved in interpreting the evolution of ancient continental margins such as those of the Caledonian or Hercynian belts.

What would happen if the NW European continental margin became active? Here there is a complex pattern of sedimentary basins of various sizes, shapes and histories bearing no simple geometric relationship to the continental margin. These are separated by uplifted areas, some bounded by faults, others by hingelines, each with its own complex history, in some cases stretching back into the Archaean. What further complications would arise if this whole area were deformed and metamorphosed above a subduction zone? If subduction led to continental collision, what would happen to such microcontinents as the Rockall Bank and the Faroes? Would we in retrospect be able to distinguish between the Rockall Trough and the N. Atlantic proper? Where would the ophiolites, volcanics and mountain belts eventually be found?

The moral of the story is that the major structure of the British Isles has not just evolved in response to the opening of the present Atlantic Ocean; it is the product of the entire span of geological history. If NW Europe were to be involved in a future orogeny, how much of its pre-orogenic history would we be able to unravel looking at the eroded remnants 1000 Ma hence? We offer no answers, but a consideration of the question should prevent us from being too glib in interpreting the evolution of ancient orogenic belts.

18d　Summary

As far as the superficial sediments and morphology are concerned, the British Isles and its adjacent seas have only just emerged from the Devensian glaciation. In the absence of human interference, the area would, after the eventual termination of the Quaternary ice age, probably evolve towards a geography similar to that of preglacial times, with the eradication from the land surface of most of the effects of glaciation. Glacial sediments would be preserved in the marine sequences of the major sedimentary basins. In the long term the NW European continental margin may be caught up in the formation of a new orogenic belt. A consideration of the probable complexities of such a belt gives some insight into why ancient orogenic belts are open to so many alternative interpretations.

References

Note: An asterisk denotes an important and well written contribution which is highly recommended to the reader.

Adams, C. J. D. 1976. Geochronology of the Channel Islands and adjacent French mainland. *J. Geol Soc. Lond.* **132**, 233–50.

Ager, D. V. 1970. The Triassic System in Britain and its stratigraphical nomenclature. *Q. J. Geol Soc. Lond.* **126**, 3–17.

Ager, D. V. 1975. The geological evolution of Europe. *Proc. Geol. Ass.* **86**, 127–54.

Allen, J. R. L. 1960. The Mam Tor sandstones: a 'turbidite' facies of the Namurian deltas of Derbyshire, England. *J. Sedim. Petrol.* **30**, 193–208.

Allen, J. R. L. 1963. Depositional features of Dittonian rocks: Pembrokeshire compared with the Welsh Borderland. *Geol Mag.* **100**, 385–400.

*Allen, J. R. L. 1965. Upper Old Red Sandstone (Farlovian) palaeogeography in South Wales and the Welsh Borderland. *J. Sedim. Petrol.* **35**, 167–95.

*Allen, J. R. L. 1974a. Sedimentology of the Old Red Sandstone (Siluro-Devonian) in the Clee Hills area, Shropshire, England. *Sedim. Geol.* **12**, 73–167.

Allen, J. R. L. 1974b. The Devonian rocks of Wales and the Welsh Borderlands. In *The Upper Palaeozoic and post-Palaeozoic rocks of Wales*, T. R. Owen (ed.), 47–84. Cardiff: University of Wales Press.

Allen, J. R. L. 1975. Source rocks of the lower Old Red Sandstone: Llanishen Conglomerate of the Cardiff area, South Wales. *Proc. Geol. Ass.* **86**, 63–76.

Allen, J. R. L. and P. Kaye 1973. Sedimentary facies of the Forest Marble (Bathonian), Shipton-on-Cherwell Quarry, Oxfordshire. *Geol Mag.* **110**, 153–63.

Allen, J. R. L. and J. Narayan 1964. Cross-stratified units, some with silt bands, in the Folkstone Beds (Lower Greensand) of South-east England. *Geol Mijnb.* **43**, 451–61.

Allen, P. 1967. Origin of the Hastings facies in north-western Europe. *Proc. Geol. Ass.* **78**, 27–105.

Allen, P. 1972. Wealden detrital tourmaline: implications for northwestern Europe. *J. Geol Soc. Lond.* **128**, 273–94.

*Allen, P. 1975. Wealden of the Weald: a new model. *Proc. Geol. Ass.* **86**, 389–438.

Allen, P. 1976. Wealden of the Weald: a new model — reply to discussion. *Proc. Geol. Ass.* **87**, 433–42.

Allen, P., J. Sutton and J. V. Watson 1974. Torridonian tourmaline–quartz pebbles and the Precambrian crust northwest of Britain. *J. Geol Soc. Lond.* **130**, 85–91.

Ali, O. E. 1977. Jurassic hazards to coral growth. *Geol Mag.* **114**, 63–4.

Anderton, R. 1975. Tidal flat and shallow marine sediments from the Craignish Phyllites, middle Dalradian, Argyll, Scotland. *Geol Mag.* **112**, 337–48.

Anderton, R. 1976. Tidal-shelf sedimentation: an example from the Scottish Dalradian. *Sedimentology* **23**, 429–58.

Anhaeusser, C. R. 1973. The evolution of the early Precambrian crust of southern Africa. *Phil Trans R. Soc. Lond.* **273A**, 359–88.

*Arkell, W. J. 1933. *The Jurassic System in Great Britain*, 681 pp. Oxford: Clarendon Press.

*Arkell, W. J. 1956. *Jurassic geology of the world*, 806 pp. Edinburgh: Oliver & Boyd.

Armstrong, M. and I. B. Paterson 1970. *The lower Old Red Sandstone of the Strathmore region.* Rep. Inst. Geol Sci., no. 70/12.

*Arthaud, F. and P. Matte 1977. Late Palaeozoic strike-slip faulting in southern Europe and northern Africa: result of a right-lateral shear zone between the Appalachians and the Urals. *Bull. Geol Soc. Am.* **88**, 1305–20.

*Audley-Charles, M. G. 1970. Triassic palaeogeography of the British Isles. *Q. J. Geol Soc. Lond.* **126**, 49–89.

Badham, J. P. N. 1976. Cornubian geotectonics: lateral thinking. *Proc. Ussh. Soc.* **3**, 448–54.

Badham, J. P. N. and C. Halls 1975. Microplate tectonics, oblique collisions, and the evolution of the Hercynian orogenic system. *Geology* **3**, 373–6.

Bailey, R. J. 1969. Ludlovian sedimentation in south-central Wales. In *The Pre-Cambrian and Lower Palaeozoic rocks of Wales*, A. Wood (ed.), 283–304. Cardiff: University of Wales Press.

Baker, J. W. 1969. Correlation problems of unmetamorphosed Pre-Cambrian rocks in Wales and Southeast Ireland. *Geol Mag.* **106**, 246–59.

Baker, J. W. 1971. The Proterozoic history of southern Britain. *Proc. Geol. Ass.* **82**, 249–66.

*Baker, J. W. 1973. A marginal late Proterozoic ocean basin in the Welsh region. *Geol Mag.* **110**, 447–55.

Bamford, D., K. Nunn, C. Prodehl and B. Jacobs 1977. LISPB – III. Upper crustal structure of northern Britain. *J. Geol Soc. Lond.* **133**, 481–8.

Banham, P. H. 1975. Glacitectonic structures: a general discussion with particular reference to the contorted drift of Norfolk. In *Ice ages: ancient and modern*, A. E. Wright and F. Moseley (eds), 69–94. Liverpool: Seel House Press.

Banks, N. L. 1973. Tide-dominated offshore sedimentation, lower Cambrian, North Norway. *Sedimentology* **20**, 213–28.

Barrow, G. 1893. On an intrusion of muscovite–biotite

gneiss in the south-eastern Highlands of Scotland, and its accompanying metamorphism. *Q. J. Geol Soc. Lond.* **49**, 330−58.

Bassett, M. G., C. H. Holland, R. B. Rickards and P. T. Warren 1975. *The type Wenlock Series.* Rep. Inst. Geol Sci., no. 75/13, 19 pp.

Bath, A. H. 1974. New isotopic data on rocks from the Long Mynd, Shropshire. *J. Geol Soc. Lond.* **130**, 567−74.

Beach, A., M. P. Coward and R. H. Graham 1974. An interpretation of the structural evolution of the Laxford Front, North-west Scotland. *Scott. J. Geol.* **9**, 297−308.

*Bell, J. D. 1976. The Tertiary intrusive complex of Skye. *Proc. Geol. Ass.* **87**, 247−72.

Belt, E. S. 1968. Carboniferous continental sedimentation, Atlantic Provinces, Canada. In *Symposium on Late Paleozoic and Mesozoic continental sedimentation, northeastern North America*, G. de V. Klein (ed.), 127−76. Spec. Pap. Geol Soc. Am, no. 106.

*Belt, E. S., E. C. Freshney and W. A. Read, 1967. Sedimentology of Carboniferous Cementstone facies, British Isles and eastern Canada. *J. Geol.* **75**, 711−21.

Berggren, W. A. 1972. A Cenozoic time-scale: some implications for regional geology and palaeobiology. *Lethaia* **5**, 195−215.

*Bernoulli, D. and H. C. Jenkyns 1974. Alpine, Mediterranean and central Atlantic Mesozoic facies in relation to the early evolution of the Tethys. In *Modern and ancient geosynclinal sedimentation*, R. H. Dott and R. H. Shaver (eds), 129−60. Spec. Publ. Soc. Econ. Palaeont. Mineral. no. 19.

*Berry, W. B. and A. J. Boucot 1973. Glacio-eustatic control of late Ordovician−early Silurian platform sedimentation and faunal changes. *Bull. Geol Soc. Am.* **84**, 275−84.

Bhatt, J. J. 1976. Geochemistry and petrology of the Main Limestone Series (Lower Carboniferous), South Wales, UK. *Sedim. Geol.* **15**, 55−86.

*Bikerman, M., D. R. Bowes and O. van Breeman 1975. Rb−Sr whole rock isotopic studies of Lewisian metasediments and gneisses in the Loch Maree region, Ross-shire. *J. Geol Soc. Lond.* **131**, 237−54.

Binns, P. E., R. McQuillin, N. G. T. Fannin, N. Kenolty and D. A. Ardus 1975. Structure and stratigraphy of sedimentary basins in the Sea of the Hebrides and the Minches. In *Petroleum and the continental shelf of North-west Europe*, A. W. Woodland (ed.), 93−104. London: Applied Science Publishers.

Binns, P. E., R. McQuillin and N. Kenolty 1974. *The geology of the Sea of the Hebrides.* Rep. Inst. Geol. Sci., no. 73/14.

Birkelund, T. and K. P. Perch-Nielsen 1976. Late Palaeozoic−Mesozoic evolution of central East Greenland. In *Geology of Greenland*, A. Escher and W. S. Watt (eds), 305−39. Geological Survey of Greenland.

Birkelund, T., K. Perch-Nielsen, D. Bridgwater and A. K. Higgins 1974. An outline of the geology of the Atlantic coast of Greenland. In *The ocean basins and margins*, A. E. M. Nairn and F. G. Stehli (eds), Vol. 2, 125−59. New York: Plenum Publishing.

Bisat, W. S. 1928. The Carboniferous goniatite zones of England and their continental equivalents. *C.R. Cong. Int. Strat. Géol Carb.*, Heerlen 1927, 117−33.

Bishop, A. C., R. A. Roach and C. J. D. Adams 1975. Precambrian rocks within the Hercynides. In *A correlation of the Precambrian rocks in the British Isles*, A. L. Harris *et al.* (eds), 102−7. Geol Soc. Lond. Spec. Rep., no. 6.

Björlykke, K. 1974. Geochemical and mineralogical influence of Ordovician island arcs on epicontinental clastic sedimentation: a study of Lower Palaeozoic sedimentation in the Oslo region, Norway. *Sedimentology* **21**, 251−72.

Blair, D. G. 1975. Structural styles in North Sea oil and gas fields. In *Petroleum and the continental shelf of North-west Europe*, A. W. Woodland (ed.), 327−38. London: Applied Science Publishers.

Bless, M. J. M., J. Bouckaert, Ph. Bouzet, R. Conil, P. Cornet, M. Fairon-Demaret, E. Groessens, P. J. Longerstaey, J. P. M.Th. Meesen, E. Paproth, H. Pirlet, M. Streel, H. W. J. van Amerom and M. Wolf 1976. Dinantian rocks in the subsurface north of the Brabant and Ardenno-Rhenish Massifs in Belgium, the Netherlands and the Federal Republic of Germany. *Meded. Rijks Geol. Dienst.*, NS **3**, 81−195.

Bloxam, T. W. and J. B. Allen 1960. Glaucophane schists, eclogite, and associated rocks from Knockormal in the Girvan−Ballantrae Complex, south Ayrshire. *Trans R. Soc. Edinb.* **64**, 1−27.

Bluck, B. J. 1967. Deposition of some upper Old Red Sandstone conglomerates in the Clyde area: a study in the significance of bedding. *Scott. J. Geol.* **3**, 139−67.

Bott, M. H. P. 1967. Geophysical investigations of the northern Pennine basement rocks. *Proc. Yorks. Geol Soc.* **36**, 139−68.

*Bowen, D. Q. 1977. Hot and cold climates in prehistoric Britain. *Geogrl Mag.* **49**, 685−98.

*Bowen, J. M. 1975. The Brent oil-field. In *Petroleum and the continental shelf of North-west Europe*, A. W. Woodland (ed.), 353−62. London: Applied Science Publishers.

Bradshaw, M. J. 1975. Origin of montmorillonite bands in the Middle Jurassic of eastern England. *Earth Planet. Sci. Lett.* **26**, 245−52.

Braithwaite, C. J. R. 1967. Carbonate environments in the middle Devonian of south Devon, England. *Sedim. Geol.* **1**, 283−320.

Brenchley, P. J. 1969. The relationship between Caradocian volcanicity and sedimentation in North Wales. In *The Pre-Cambrian and Lower Palaeozoic rocks of Wales*, A. Wood (ed.), 181−202. Cardiff: University of Wales Press.

*Brennand, T. P. 1975. The Triassic of the North Sea. In *Petroleum and the continental shelf of North-west Europe*, A. W. Woodland (ed.), 295−311. London: Applied Science Publishers.

Briden, J. C., W. A. Morris and J. D. A. Piper 1973. Palaeomagnetic Studies in the British Caledonides − VI regional and global implications. *Geophys. J. R. Ast. Soc.* **34**, 107−34.

Bridges, P. H. 1975. The transgression of a hard substrate

shelf: the Llandovery (lower Silurian) of the Welsh Borderland. *J. Sedim. Petrol.* **45**, 79−94.

Bridges, P. H. 1976. Lower Silurian transgressive barrier islands, Southwest Wales. *Sedimentology* **23**, 347−62.

*Briggs, G. (ed.) 1974. *Carboniferous of the southeastern United States*. Spec. pap. Geol Soc. Am., no. 148.

Bristol, H. M. and R. H. Howard 1974. Sub-Pennsylvanian valleys in the Chesterian surface of the Illinois Basin and related Chesterian slump blocks. In *Carboniferous of the southeastern United States*, G. Briggs (ed.), 315−36. Spec. pap. Geol Soc. Am., no. 148.

Bristow, C. R. and F. C. Cox 1973. The Gipping Till: a reappraisal of East Anglian glacial stratigraphy. *J. Geol Soc. Lond.* **129**, 1−37.

Bromley, A. V. 1969. Acid plutonic igneous activity in the Ordovician of North Wales. In *The Pre-Cambrian and Lower Palaeozoic rocks of Wales*, A. Wood (ed.), 387−408. Cardiff: University of Wales Press.

Bromley, A. V. 1976. Granites in mobile belts: the tectonic setting of the Cornubian batholith. *J. Camborne School Mines* **76**, 40−7.

Bromley, R. G. 1967. Some observations on burrows of thalassinidean Crustacea in chalk hardgrounds. *Q. J. Geol Soc. Lond.* **123**, 157−82.

Brook, M., D. Powell and M. S. Brewer 1977. Grenville events in Moine rocks of the northern Highlands, Scotland. *J. Geol Soc. Lond.* **133**, 489−96.

Brooks, C. K. 1973. Rifting and doming in southern East Greenland. *Nature Phys. Sci.* **244**, 23−5.

Brooks, J. R. V. and J. A. Chesher 1975. Review of the offshore Jurassic of the UK northern North Sea. In *Proceedings of the Jurassic northern North Sea symposium*, K. G. Finstad and R. C. Selley (eds), JNNSS 2, 1−24. Stavanger: Norwegian Petroleum Society.

Brooks, M. 1973. Some aspects of the Palaeogene evolution of western Britain in the context of an underlying mantle hot spot. *J. Geol.* **81**, 81−8.

Brück, P. M. 1972. Stratigraphy and sedimentology of the Lower Palaeozoic greywacke formation in Counties Kildare and West Wicklow. *Proc. R. Ir. Acad.* **72B**, 25−53.

Brunstrom, R. G. and P. J. Walmsley 1969. Permian evaporites in the North Sea Basin. *Bull Am. Ass. Petrol. Geol.* **53**, 870−83.

*Burgess, I. C. 1961. Fossil soils of the upper Old Red Sandstone of south Ayrshire. *Trans Geol Soc. Glasg.* **24**, 138−53.

*Burgess, I. C. and M. Mitchell 1976. Viséan lower Yoredale limestones on the Alston and Askrigg blocks, and the base of the D2 zone in northern England. *Proc. Yorks. Geol Soc.* **40**, 613−30.

Burke, K. and J. F. Dewey 1973. Plume-generated triple junctions: key indicators in applying plate tectonics to old rocks. *J. Geol.* **81**, 406−34.

Burne, R. V. 1969. *Sedimentological studies of the Bude Formation.* Unpubl. PhD thesis, University of Oxford.

Burne, R. V. 1973. Palaeogeography of South-west England and Hercynian continental collision. *Nature Phys. Sci.* **241**, 129−31.

Burne, R. V. and L. J. Moore 1971. The Upper Carboniferous rocks of Devon and Cornwall. *Proc. Ussh. Soc.* **2**, 288−98.

Burrett, C. F. 1972. Plate tectonics and the Hercynian Orogeny. *Nature* **239**, 155−7.

Burrett, C. F. 1973. Ordovician biogeography and continental drift. *Palaeogeog., Palaeoclimatol., Palaeoecol.* **13**, 161−201.

Calef, C. E. and N. J. Hancock 1974. Wenlock and Ludlow marine communities. *Palaeontology* **17**, 779−810.

Callaway, C. 1878. On the quartzites of Shropshire. *Q. J. Geol Soc. Lond.* **34**, 754−68.

*Calver, M. A. 1969. Westphalian of Britain. *C.R. 6me Cong. Int. Strat. Géol. Carb.*, Sheffield 1967, **1**, 233−54.

*Casey, R. 1961. The stratigraphical palaeontology of the Lower Greensand. *Palaeontology* **3**, 487−621.

*Casey, R. 1973. The ammonite succession at the Jurassic/Cretaceous boundary in eastern England. In *The Boreal Lower Cretaceous*, R. Casey and P. F. Rawson (eds), 193−266. Liverpool: Seel House Press.

Casey, R. and C. R. Bristow 1964. Notes on some ferruginous strata in Buckinghamshire and Wiltshire. *Geol Mag.* **101**, 116−28.

*Casey, R. and P. F. Rawson (eds) 1973. *The Boreal Lower Cretaceous*, 448 pp. Liverpool: Seel House Press.

Catt, J. A. and L. F. Penny 1966. The Pleistocene deposits of Holderness, east Yorkshire. *Proc. Yorks. Geol Soc.* **35**, 375−420.

Chisholm, J. I. and J. M. Dean 1974. The upper Old Red Sandstone of Fife and Kinross: a fluviatile sequence with evidence of marine incursion. *Scott. J. Geol.* **10**, 1−30.

Christian, H. E. 1969. Some observations on the initiation of salt structures in the southern British North Sea. In *The exploration for petroleum in Europe and North Africa*, P. Hepple (ed.), 231−50. Amsterdam: Elsevier.

Church, W. R. and R. A. Gayer 1973. The Ballantrae ophiolite. *Geol Mag.* **110**, 497−592.

Clarkson, C. M., G. Y. Craig and E. K. Walton 1975. The Silurian rocks bordering Kirkcudbright Bay, South Scotland. *Trans R. Soc. Edinb.* **69**, 313−25.

Clemmey, H. 1975. World's oldest animal traces. *Nature* **261**, 576−8.

Cloos, H. 1939. Hebung−Spaltung−Vulcanismus. *Geol. Rundsch.* **30**, 405−527.

Cloud, P. 1972. A working model of the primitive earth. *Am. J. Sci.* **272**, 537−48.

Cobbold, E. S. 1927. The stratigraphy and geological structure of the Cambrian area of Comley (Shropshire). *Q. J. Geol Soc. Lond.* **83**, 551−73.

Cocks, L. R. M., C. H. Holland, R. B. Rickards, and I. Strachan 1971. A correlation of Silurian rocks in the British Isles. *J. Geol Soc. Lond.* **127**, 103−36.

*Collinson, J. D. 1969. The sedimentology of the Grindslow Shales and the Kinderscout Grit: a deltaic complex in the Namurian of northern England. *J. Sedim. Petrol.* **39**, 194−221.

Collinson, J. D. 1970. Deep channels, massive beds and turbidity current genesis in the central Pennine Basin.

Proc. Yorks. Geol Soc. **37**, 495–519.

Condie, K. C. 1973. Archean magmatism and crustal thickening. *Bull. Geol Soc. Am.* **84**, 2981–92.

*Coope, G. R. 1975. Climatic fluctuations in Northwest Europe since the last interglacial, indicated by fossil assemblages of Coleoptera. In *Ice ages: ancient and modern*, A. E. Wright and F. Moseley (eds), 153–68. Liverpool: Seel House Press.

Cottençon, A., B. Parent and G. Flacelière 1975. Lower Cretaceous gas-fields in Holland. In *Petroleum and the continental shelf of North-west Europe*, A. W. Woodland (ed.), 403–12. London: Applied Science Publishers.

Coward, M. P., P. W. Francis, R. H. Graham, J. S. Myers and J. Watson 1969. Remnants of an early metasedimentary assemblage in the Lewisian Complex of the Outer Hebrides. *Proc. Geol. Ass.* **80**, 387–408.

Cowie, J. W., A. W. A. Rushton and C. J. Stubblefield 1972. *A correlation of Cambrian rocks in the British Isles.* Geol Soc. Lond. Spec. Rep., no. 2, 42, pp.

Cox, A. 1969. Geomagnetic reversals. *Science* **163**, 237–45.

Craig, G. Y. and E. K. Walton 1959. Sequence and structure in the Silurian rocks of Kirkcudbrightshire. *Geol Mag.* **96**, 209–20.

Cribb, S. J. 1975. Rubiduim–strontium ages and strontium isotope ratios from the igneous rocks of Leicestershire. *J. Geol Soc. Lond.* **131**, 203–12.

*Crimes, T. P. 1970a. A facies analysis of the Cambrian of Wales. *Palaeogeog., Palaeoclimatol., Palaeoecol.* **7**, 113–70.

Crimes, T. P. 1970b. A facies analysis of the Arenig of western Lleyn, North Wales. *Proc. Geol. Ass.* **81**, 221–39.

Crimes, T. P. and J. D. Crossley 1968. The stratigraphy, sedimentology, ichnology and structure of the Lower Palaeozoic rocks of part of north-eastern Co. Wexford. *Proc. R. Ir. Acad.* **67B**, 185–215.

Cummins, W. A. 1969. Patterns of sedimentation in the Silurian rocks of Wales. In *The Pre-Cambrian and Lower Palaeozoic rocks of Wales*, A. Wood (ed.), 219–37. Cardiff: University of Wales Press.

Curry, D. 1965. The Palaeogene beds of South-east England. *Proc. Geol. Ass.* **76**, 151–73.

Curry, D. 1966. Problems of correlation in the Anglo-Paris–Belgian Basin. *Proc. Geol. Ass.* **77**, 437–67.

Curtis, C. D. and D. A. Spears 1968. The formation of sedimentary iron minerals. *Econ. Geol.* **63**, 257–70.

Curtis, M. L. K. 1968. The Tremadoc rocks of the Tortworth Inlier, Gloucestershire. *Proc. Geol. Ass.* **79**, 349–62.

Dagger, G. W. 1977. Controls of copper mineralisation at Coniston, English Lake District. *Geol Mag.* **144**, 195–202.

*Daley, B. 1972. Some problems concerning the early Tertiary climate of southern Britain. *Palaeogeog., Palaeoclimatol., Palaeoecol.* **11**, 177–90.

Dalland, A. 1975. The Mesozoic rocks of Andøy, northern Norway. *Norges Geol. Unders.* **316**, 271–87.

Dansgaard, W., S. J. Johnsen, H. B. Clausen and C. C. Langway Jr 1971. Climatic record revealed by the Camp Century ice core. In *The late Cenozoic glacial ages*, K. H. Turekian (ed.), 37–56. New Haven: Yale University Press.

Dansgaard, W. and H. Tauber 1969. Glacier oxygen-18 content and Pleistocene ocean temperatures. *Science* **166**, 499–502.

Davies, F. B. 1974. A layered basic complex in the Lewisian, south of Loch Lexford, Sutherland. *J. Geol Soc. Lond.* **130**, 279–84.

Dearman, W. R. 1971. A general view of the structure of Cornubia. *Proc. Ussh. Soc.* **2**, 220–36.

Deegan, C. E. 1973. Tectonic control of sedimentation at the margin of a Carboniferous depositional basin in Kirkcudbrightshire. *Scott. J. Geol.* **9**, 1–28.

Deer, W. A. 1976. Tertiary igneous rocks between Scoresby Sund and Kap Gustav Holm, East Greenland. In *Geology of Greenland*, A. Escher and W. S. Watt (eds), 404–29. Geological Survey of Greenland.

de Raaf, J. F. M. and J. R. Boersma 1971. Tidal deposits and their sedimentary structures (seven examples from western Europe). *Geol Mijnb.* **50**, 479–504.

de Raaf, J. F. M., H. G. Reading and R. G. Walker 1965. Cyclic sedimentation in the lower Westphalian of north Devon, England. *Sedimentology* **4**, 1–52.

Devereux, I. 1967. Oxygen isotope paleotemperature measurements on New Zealand Tertiary fossils. *NZ J. Sci.* **10**, 988–1011.

*Dewey, J. F. 1969a. Evolution of the Appalachian–Caledonian orogen. *Nature* **222**, 124–8.

*Dewey, J. F. 1969b. Structure and sequence in paratectonic British Caledonides. In *North Atlantic geology and continental drift: a symposium*, M. Kay (ed.), 309–35. Mem. Am. Ass. Petrol. Geol., no. 12.

*Dewey, J. F. 1971. A model for the Lower Palaeozoic evolution of the southern margin of the early Caledonides of Scotland and Ireland. *Scott. J. Geol.* **7**, 219–40.

Dewey, J. F. 1974. Continental margins and ophiolite obduction: Appalachian–Caledonian System. In *The geology of continental margins*, C. A. Burk and C. L. Drake (eds), 933–50. Berlin: Springer-Verlag.

*Dewey, J. F. and K. C. A. Burke 1973. Tibetan, Variscan and Precambrian basement reactivation: products of continental collision. *J. Geol.* **81**, 683–92.

Dike, E. F. 1972. *Ophiomorpha nodosa* (Lundgren): environmental implications in the Lower Greensand of the Isle of Wight. *Proc. Geol. Ass.* **83**, 165–77.

Dimroth, E. and M. M. Kimberley 1976. Precambrian atmospheric oxygen: evidence in the sedimentary distributions of carbon, sulphur, uranium, and iron. *Can J. Earth Sci.* **13**, 1161–85.

Dineley, D. L. 1961. The Devonian System in south Devonshire. *Field Studies* **1**, 121–40.

Dineley, D. L. 1966. The Dartmouth Beds of Bigbury Bay, south Devon. *Q. J. Geol Soc. Lond.* **122**, 187–217.

Dobson, M. R., D. Evans and R. Wittington 1975. The offshore extension of the Loch Gruinart Fault, Islay. *Scott. J. Geol.* **11**, 23–35.

*Dodson, M. H. and D. C. Rex 1971. Potassium–argon

ages of slates and phyllites from South-west England. *J. Geol Soc. Lond.* **126**, 465–99.

*Donovan, R. N. 1975. Devonian lacustrine limestones at the margin of the Orcadian Basin, Scotland. *J. Geol Soc. Lond.* **131**, 489–510.

Donovan, R. N., R. Archer, P. Turner and D. H. Tarling 1976. Devonian palaeogeography of the Orcadian Basin and the Great Glen Fault. *Nature* **259**, 550–1.

Donovan, R. N., R. J. Foster and T. S. Westoll 1974. A stratigraphic revision of the Old Red Sandstone of north-eastern Caithness. *Trans R. Soc. Edinb.* **69**, 167–201.

Downie, C., T. R. Lister, A. L. Harris and D. J. Fettes 1971. *A palynological investigation of the Dalradian rocks of Scotland.* Rep. Inst. Geol Sci., no. 71/9.

Duff, K. L. 1975. Palaeoecology of a bituminous shale: the Lower Oxford Clay of central England. *Palaeontology* **18**, 443–82.

Duff, P. McL.D. and E. K. Walton 1962. Statistical basis for cyclothems: a quantitative study of the sedimentary succession in the east Pennine coalfield. *Sedimentology* **1**, 235–55.

Eames, T. D. 1975. Coal rank and gas source relationships: Rotliegendes reservoirs. In *Petroleum and the continental shelf of North-west Europe*, A. W. Woodland (ed.), 191–204. London: Applied Science Publishers.

Eden, R. A., R. Holmes and N. G. T. Fannin 1977. *Quaternary deposits of the central North Sea, 6: depositional environment of offshore Quaternary deposits of the continental shelf around Scotland.* Rep. Inst. Geol Sci., no. 77/15.

Edmonds, E. A., M. C. McKeown and M. Williams 1975. *South-west England.* In *British regional geology*, 130 pp. London: HMSO.

*Elliott, T. 1974. Abandonment facies of high-construction lobate deltas, with an example from the Yoredale Series. *Proc. Geol. Ass.* **85**, 359–65.

Elliott, T. 1975. The sedimentary history of a delta lobe from a Yoredale (Carboniferous) cyclothem. *Proc. Yorks. Geol Soc.* **30**, 505–36.

Elliott, T. 1976. Upper Carboniferous sedimentary cycles produced by river-dominated, elongate deltas. *J. Geol Soc. Lond.* **132**, 199–208.

Evans, A. L., F. J. Fitch and J. A. Miller 1973. Potassium–argon age determinations on some British Tertiary igneous rocks. *J. Geol Soc. Lond.* **129**, 419–44.

Evans, A. M. 1968. Precambrian rocks, A: Charnwood Forest. In *The geology of the East Midlands*, P. C. Sylvester-Bradley and T. D. Ford (eds), 1–12. Leicester: Leicester University Press.

Evans, C. R. and R. St J. Lambert 1974. The Lewisian of Lochinver, Sutherland: the type area for the Inverian metamorphism. *J. Geol Soc. Lond.* **130**, 125–50.

Falcon, N. L. and P. E. Kent 1960. *Geological results of petroleum exploration in Britain 1945–1957.* Mem. Geol Soc. Lond., no. 2.

Fitch, F. J., J. A. Miller, A. L. Evans, R. L. Grasty and M. Y. Meneisy 1969. Isotopic age determinations on rocks from Wales and the Welsh Borders. In *The Pre-Cambrian and Lower Palaeozoic rocks of Wales*, A. Wood (ed.), 23–45. Cardiff: University of Wales Press.

Fitch, F. J., J. A. Miller and D. B. Thompson 1966. The palaeogeographic significance of isotopic age determinations on detrital micas from the Triassic of the Stockport–Macclesfield district, Cheshire, England. *Palaeogeog., Palaeoclimatol., Palaeoecol.* **2**, 281–312.

*Fitton, J. G. and D. J. Hughes 1970. Volcanism and plate tectonics in the British Ordovician. *Earth Planet. Sci. Lett.* **8**, 223–8.

Flint, R. F. 1971. *Glacial and Quaternary geology.* New York: John Wiley.

*Floyd, P. A. 1972. Geochemistry, origin and tectonic environment of the basic and acidic rocks of Cornubia, England. *Proc. Geol. Ass.* **83**, 385–404.

Floyd, P. A., G. J. Lees and A. Parker 1976. A preliminary geochemical twist to the Lizard's new tale. *Proc. Ussh. Soc.* **3**, 414–25.

Ford, T. D. 1968. Precambrian rocks, B: the Precambrian palaeontology of Charnwood Forest. In *The geology of the East Midlands*, P. C. Sylvester-Bradley and T. D. Ford (eds), 12–14. Leicester: Leicester University Press.

Foster, R. J. 1972. *The solid geology of north-east Caithness.* Unpubl. PhD thesis, University of Newcastle-upon-Tyne.

Francis, E. H. 1967. Review of Carboniferous–Permian volcanicity in Scotland. *Geol. Rundsch.* **57**, 219–46.

Francis, E. H. 1978. The Midland Valley as a rift, seen in connection with the late-Palaeozoic rift system. In *Tectonics geophysics of continental drift*, I. B. Ramberg and E. R. Neumann (eds). Reidal, Dordrecht: Proc. NATO Adv. Study Inst. 2.

Furness, R. R. 1965. The petrography and provenance of the Coniston Grits east of the Lune Valley, Westmorland. *Geol Mag.* **102**, 252–60.

Furness, R. R., P. G. Llewellyn, T. N. Norman and R. B. Rickards 1967. A review of Wenlock and Ludlow stratigraphy and sedimentation in NW England. *Geol Mag.* **104**, 132–47.

Fürsich, F. T. and J. M. Hurst 1974. Environmental factors determining the distribution of brachiopods. *Palaeontology* **17**, 879–900.

Fyfe, W. S. 1973. The granulite facies, partial melting and the Archaean crust. *Phil Trans R. Soc. Lond.* **273A**, 457–61.

Gallois, R. W. 1976. Coccolith blooms in the Kimmeridge Clay and origin of North Sea oil. *Nature* **259**, 473–5.

Garrard, R. A. 1977. The sediments of the south Irish Sea and Nymphe Bank area of the Celtic Sea. In *The Quaternary history of the Irish Sea*, C. Kidson and M. J. Tooley (eds), 69–92. Liverpool: Seel House Press.

*Garrels, R. M., E. A. Perry and F. T. Mackenzie 1973. Genesis of Precambrian iron-formations and the development of atmospheric oxygen. *Econ. Geol.* **68**, 1173–9.

Gastil, G. 1960. The distribution of mineral dates in time and space. *Am. J. Sci.* **258**, 1–35.

Gaunt, G. D. 1976. The Devensian maximum ice limit in the Vale of York. *Proc. Yorks. Geol Soc.* **40**, 631–7.

Gauss, A. G. and M. R. House 1972. The Devonian

successions in the Padstow area, north Cornwall. *J. Geol. Soc. Lond.* **128**, 151–72.

*Geiger, M. E. and C. A. Hopping 1968. Triassic stratigraphy of the southern North Sea Basin. *Phil Trans R. Soc. Lond.* **254B**, 1–36.

*George, T. N. 1958. Lower Carboniferous palaeogeography of the British Isles. *Proc. Yorks. Geol. Soc.* **31**, 227–318.

George, T. N., G. A. L. Johnson, M. Mitchell, J. E. Prentice, W. H. C. Ramsbottom, G. D. Sevastopulo and R. B. Wilson. 1976. *A correlation of Dinantian rocks in the British Isles.* Geol Soc. Lond., Spec. Rep., no. 7.

Gilligan, A. 1920. The petrography of the Millstone Grit of Yorkshire. *Q. J. Geol Soc. Lond.* **75**, 251–94.

Glennie, K. W. 1970. *Desert sedimentary environments.* Amsterdam: Elsevier.

*Glennie, K. W. 1972. Permian Rotliegendes of Northwest Europe interpreted in light of modern desert sedimentation studies. *Bull. Am. Ass. Petrol. Geol.* **56**, 1046–71.

*Goldring, R. 1971. *Shallow water sedimentation as illustrated in the upper Devonian Baggy Beds.* Mem. Geol Soc. Lond., no. 5.

Goldring, R., M. R. House, E. B. Selwood, S. Simpson and R. St J. Lambert 1967. Devonian of southern Britain. In *International symposium on the Devonian System*, D. H. Oswald (ed.), Vol. 1, 1–14. Calgary: Alberta Society of Petroleum Geologists.

*Graham, C. M. 1976. Petrochemistry and tectonic significance of Dalradian metabasaltic rocks of the SW Scottish Highlands. *J. Geol Soc. Lond.* **132**, 61–84.

Graham, J. R. 1975. Analysis of an Upper Palaeozoic transgressive sequence in southwest County Cork, Eire. *Sedim. Geol.* **13**, 267–90.

Gunn, P. J. 1973. Location of the Proto-Atlantic in the British Isles. *Nature* **242**, 111–12.

Hallam, A. 1958. The concept of Jurassic axes of uplift. *Science Prog.* **183**, 441–9.

Hallam, A. 1960. The White Lias of the Devon coast. *Proc. Geol. Ass.* **71**, 47–60.

Hallam, A. 1972. Relation of Palaeogene ridge and basin structures and vulcanicity in the Hebrides and Irish Sea regions of the British Isles to the opening of the North Atlantic. *Earth Planet. Sci. Lett.* **16**, 171–7.

*Hallam, A. 1975. *Jurassic environments.* Cambridge: Cambridge University Press.

Hallam, A. and B. W. Sellwood 1968. Origin of fuller's earth in the Mesozoic of southern England. *Nature* **220**, 1193–5.

*Hallam, A. and B. W. Sellwood 1976. Middle Mesozoic sedimentation in relation to tectonics in the British area. *J. Geol.* **84**, 301–21.

Halstead, L. B. 1975. *The evolution and ecology of the dinosaurs.* Frome: Peter Lowe, Eurobook.

Hamilton, D. 1961. Algal growths in the Rhaetic Cotham Marble of southern England. *Palaeontology* **4**, 324–33.

Hancock, J. M. 1969. Transgression of the Cretaceous sea in South-west England. *Proc. Ussh. Soc.* **2**, 61–83.

*Hancock, J. M. 1975. The petrology of the Chalk. *Proc. Geol. Ass.* **86**, 499–536.

Hancock, J. M. and P. A. Scholle 1975. Chalk of the North Sea. In *Petroleum and the continental shelf of North-west Europe*, A. W. Woodland (ed.), 413–27. London: Applied Science Publishers.

Harland, W. B. and R. A. Gayer 1972. The Arctic Caledonides and earlier oceans. *Geol Mag.* **109**, 289–314.

Harris, A. L., H. J. Bradbury and M. H. McGonigal 1976. The evolution and transport of the Tay Nappe. *Scott. J. Geol.* **12**, 103–13.

*Harris, A. L. and W. S. Pitcher 1975. The Dalradian Supergroup. In *A correlation of the Precambrian rocks in the British Isles*, A. L. Harris *et al.* (eds), 52–75. Geol. Soc. Lond. Spec. Rep., no. 6.

Harris, T. M. 1976. A slender upright plant from Wealden sandstones. *Proc. Geol. Ass.* **87**, 413–22.

Harwood, G. M. 1976. The Staddon Grits – or Meadfoot Beds? *Proc. Ussh. Soc.* **3**, 333–8.

Hawkesworth, C. J., S. Moorbath, R. K. O'Nions and J. F. Wilson 1975. Age relationships between greenstone belts and 'granites' in the Rhodesian Archean craton. *Earth Planet. Sci. Lett.* **25**, 251–62.

Hays, J. D., J. Imbrie and N. J. Shackleton 1976. Variations in the Earth's orbit: pacemaker of the ice ages. *Science* **194**, 1121–32.

Heckel, P. H. 1975. Carbonate buildups in the geologic record: a review. In *Reefs in time and space*, L. F. Laporte (ed.), Spec. Publ. Soc. Econ. Paleont. Miner., no. 18. 90–154.

Hemingway, J. E. 1974. The Jurassic. In *The geology and mineral resources of Yorkshire*, D. H. Rayner and J. E. Hemingway (eds), 161–223. Yorkshire Geological Society.

Henderson, S. M. K. 1935. Ordovician submarine disturbances in the Girvan district. *Trans R. Soc. Edinb.* **63**, 487–509.

Henningsmoen, G. 1973. The Cambro-Ordovician boundary. *Lethaia*, **6**, 423–39.

Heybroek, P. 1975. On the structure of the Dutch part of the central North Sea Graben. In *Petroleum and the continental shelf of North-west Europe*, A. W. Woodland (ed.), 339–51. London: Applied Science Publishers.

*Hickman, A. H. 1975. The stratigraphy of late Precambrian metasediments between Glen Roy and Lismore. *Scott. J. Geol.* **11**, 117–42.

Hodgson, A. V. 1978. Braided river bedforms and related sedimentary structures in the Fell Sandstone Group (Lower Carboniferous) of North Northumberland. *Proc. Yorks. Geol Soc.* **41**, 509–22.

Hoffman, P., J. F. Dewey and K. Burke 1974. Aulacogens and their genetic relation to geosynclines, with a Proterozoic example from Great Slave Lake, Canada. In *Modern and ancient geosynclinal sedimentation*, R. H. Dott and R. H. Shaver (eds), 38–55. Spec. Publ. Soc. Paleont. Miner., no. 19.

Holmes, R. 1977. *Quaternary deposits of the central North Sea, 5: the Quaternary geology of the UK sector of the North Sea between 56° and 58°N*. Rep. Inst. Geol Sci., no. 77/14.

Horne, R. R. 1975. The association of alluvial fan, aeolian and fluviatile facies in the Caherbla Group (Devonian), Dingle Peninsula, Ireland. *J. Sedim. Petrol.* **45**, 535–40.

Horton, A. 1977. The age of the Middle Jurassic 'white sands' of north Oxfordshire. *Proc. Geol. Ass.* **88**, 147–62.

*House, M. R. 1963. Devonian ammonoid successions and facies in Devon and Cornwall. *Q. J. Geol Soc. Lond.* **119**, 1–27.

*House, M. R. 1975a. Facies and time in Devonian tropical areas. *Proc. Yorks. Geol Soc.* **40**, 233–80.

*House, M. R. 1975b. Faunas and time in the marine Devonian. *Proc. Yorks. Geol Soc.* **40**, 459–90.

Howells, M. F. , B. E. Leveridge and C. D. R. Evans 1973. *Ordovician ash-flow tuffs in eastern Snowdonia*. Rep. Inst. Geol Sci., no. 73/3.

Howitt, F., E. R. Aston and M. Jacqué 1975. The occurrence of Jurassic volcanics in the North Sea. In *Petroleum and the continental shelf of North-west Europe*, A. W. Woodland (ed.), 379–87. London: Applied Science Publishers.

Hsü, K. J. 1972. Origin of saline giants: a critical review after the discovery of the Mediterranean evaporite. *Earth Sci. Rev.* **8**, 371–96.

Hubert, J. F. 1969. Late Ordovician sedimentation in the Caledonian geosyncline, southwestern Scotland. In *North Atlantic geology and continental drift: a symposium*, M. Kay (ed.), 309–35. Mem. Am. Ass. Petrol. Geol., no. 12.

Hudson, J. D. 1963. The recognition of salinity-controlled mollusc assemblages in the Great Estuarine Series (middle Jurassic) of the Inner Hebrides. *Palaeontology* **6**, 318–26.

Hudson, J. D. 1965. The petrology of the sandstones of the Great Estuarine Series, and the Jurassic palaeogeography of Scotland. *Proc. Geol. Ass.* **75**, 499–528.

Hudson, J. D. 1970. Algal limestones with pseudomorphs after gypsum from the middle Jurassic of Scotland. *Lethaia* **3**, 11–40.

Hudson, J. D. and N. Morton 1969. Guide for western Scotland. In *International field symposium on the British Jurassic Excursion No. 4*, p. D1–D47. University of Keele.

Hudson, J. D. and T. J. Palmer 1976. A euryhaline oyster from the middle Jurassic and the origin of the true oysters. *Palaeontology* **19**, 79–94.

Hurley, P. M. and J. R. Rand 1969. Pre-drift continental nuclei. *Science* **164**, 1229–42.

Imbrie, J. and N. G. Kipp 1971. A new micropaleontological method for quantitative paleoclimatology: application to a late Pleistocene Caribbean core. In *The late Cenozoic glacial ages*, K. H. Turekian (ed.), 71–181. New Haven: Yale University Press.

Ireland, R. J., J. E. Pollard, R. J. Steel and D. B. Thompson 1978. Intertidal sediments and trace fossils from the Waterstones (Scythian–Anisian?) at Dartsbury, Cheshire. *Proc. Yorks. Geol Soc.* **41**, 399–436.

Ivimey-Cook, H. C. 1974. The Permian and Triassic

deposits of Wales. In *The Upper Palaeozoic and post-Palaeozoic rocks of Wales*, T. R. Owen (ed.), 295–321. Cardiff: University of Wales Press.

Jacqué, M. and J. Thouvenin 1975. Lower Tertiary tuffs and volcanic activity in the North Sea. In *Petroleum and the continental shelf of North-west Europe*, A. W. Woodland (ed.), 455–65. London: Applied Science Publishers.

James, D. M. D. and J. James 1969. The influence of deep fractures on some areas of Ashgillian–Llandoverian sedimentation in Wales. *Geol Mag.* **106**, 562–82.

Jeans, C. V. 1973. The Market Weighton structure: tectonics, sedimentation and diagenesis during the Cretaceous. *Proc. Yorks. Geol Soc.* **39**, 409–44.

Jeans, C. V., R. J. Merriman and J. G. Mitchell 1977. Origin of middle Jurassic and lower Cretaceous fuller's earths in England. *Clay Minerals* **12**, 11–44.

Jeans, P. J. F. 1973. Plate tectonic reconstruction of the southern Caledonides of Great Britain. *Nature Phys. Sci.* **245**, 120–2.

*Johnson, G. A. L. 1973. Closing of the Carboniferous sea in western Europe. In *Implications of continental drift to the earth sciences*, D. K. Tarling and S. K. Runcorn (eds), Vol. 2, 845–50. London: Academic Press.

Johnson, M. R. W. 1975. Morarian Orogeny and Grenville belt in Britain. *Nature* **257**, 301–2.

Johnstone, G. S. 1975. The Moine Succession. In *A correlation of the Precambrian rocks in the British Isles*, A. L. Harris *et al.* (eds), 30–42. Geol. Soc. Lond. Spec. Rep., no. 6.

Jones, E. L., H. P. Raveling and H. R. Taylor 1975. Statfjord field. In *Proceedings of the Jurassic northern North Sea symposium*, K. G. Finstad and R. C. Selley (eds), JNNSS 20, 1–19. Stavanger: Norwegian Petroleum Society.

Jones, O. T. 1938. On the evolution of a geosyncline. *Q. J. Geol Soc. Lond.* **94**, 60–110.

Jones, O. T. 1940. Some Lower Palaeozoic contacts in Pembrokeshire. *Geol Mag.* **77**, 405–9.

Joplin, G. A. 1959. On the origin and occurrence of basic bodies associated with discordant bathyliths. *Geol Mag.* **96**, 361–73.

Kauffman, E. G. 1973. Cretaceous Bivalvia. In *Atlas of palaeobiogeography*, A. Hallam (ed.), 353–83. Amsterdam: Elsevier.

Kauffman, E. G. 1977. Evolutionary rates and biostratigraphy, in *Concepts and methods of biostratigraphy*, E. G. Kauffman and J. E. Hazel (eds), 109–41. Stroudsburg, Penn.: Dowden, Hutchinson and Ross.

Keene, J. B. 1976. *Cherts and porcellanites from the North Pacific, log 32, Deep Sea Drilling Project*. Initial Rep. Deep Sea Drill. Proj., no. 32.

Kelling, G. 1961. The stratigraphy and structure of the Ordovician rocks of the Rhinns of Galloway. *Q. J. Geol Soc. Lond.* **117**, 37–75.

*Kelling, G. 1962. The petrology and sedimentation of upper Ordovician rocks in the Rhinns of Galloway, Southwest Scotland. *Trans R. Soc. Edinb.* **65**, 107–37.

Kelling, G. 1974. Upper Carboniferous sedimentation in South Wales. In *The Upper Palaeozoic and post-Palaeozoic rocks of Wales*, T. A. Owen (ed.), 185–224. Cardiff:

University of Wales Press.

Kelling, G. and M. A. Woollands 1969. The stratigraphy and sedimentation of the Llandoverian rocks of the Rhayader district. In *The Pre-Cambrian and Lower Palaeozoic rocks of Wales*, A. Wood (ed.), 255−82. Cardiff: University of Wales Press.

Kennedy, M. J. 1976. Southeastern margin of the northeastern Appalachians: late Precambrian orogeny on a continental margin. *Bull. Geol Soc. Am.* **87**, 1317−25.

Kennedy, W. J. 1967. Burrows and surface traces from the Chalk of South-east England. *Bull. Br. Mus. Nat. Hist. (Geol.)* **15**, 125−67.

Kennedy, W. J. 1969. The correlation of the lower Chalk of South-east England. *Proc. Geol. Ass.* **80**, 459−560.

Kennedy, W. J. 1970. Trace fossils in the Chalk environment. In *Trace fossils*, T. P. Crimes and J. C. Harper (eds), 263−82. Liverpool: Seel House Press.

Kennedy, W. J. 1978. Cretaceous. In *The ecology of fossils*, W. S. McKerrow (ed.), 280−322. London: Duckworth.

Kennedy, W. J. and W. A. Cobban 1977. The role of ammonites in biostratigraphy, In *Concepts and methods of biostratigraphy*, E. G. Kauffman and J. E. Hazel (eds), 309−20. Stroudsburg, Penn.: Dowden, Hutchinson and Ross.

Kennedy, W. J. and W. A. Cobban 1976. *Aspects of ammonite biology, biogeography and biostratigraphy*. Spec. Pap. Palaeontology, no. 17.

Kennedy, W. J. and R. E. Garrison 1975. Morphology and genesis of nodular chalks and hardgrounds in the upper Cretaceous of southern England. *Sedimentology* **22**, 311−86.

Kent, P. E. 1967. A contour map of the sub-Carboniferous floor in the northeast Midlands. *Proc. Yorks. Geol Soc.* **36**, 175−94.

Kent, P. E. 1970. Problems of the Rhaetic in the East Midlands. *Mercian Geologist* **3**, 361−72.

Kent, P. E. 1975. The tectonic development of Great Britain and the surrounding seas. In *Petroleum and the continental shelf of North-west Europe*, A. W. Woodland (ed.), 3−28. London: Applied Science Publishers.

Knill, J. L. 1963. A sedimentary history of the Dalradian Series. In *The British Caledonides*, M. R. W. Johnson and F. H. Stewart (eds), 99−121. Edinburgh: Oliver & Boyd.

Knox, R. W. O. 1970. The Eller Beck Formation (Bajocian) of the Ravenscar Group of north-east Yorkshire. *Geol. Mag.* **110**, 511−34.

Knox, R. W. O. 1977. Upper Jurassic pyroclastic rocks in Skye, West Scotland. *Nature* **265**, 323−4.

Krantz, R. 1972. Die Sponge-Gravels von Faringdon (England). *N. Jb. Geol. Paläont. Abh.* **140**, 207−31.

Krebs, W. and H. Wachendorf 1973. Proterozoic−Palaeozoic geosynclinal and orogenic evolution of central Europe. *Bull. Geol Soc. Amer.* **84**, 2611−29.

Krylov, I. N. and M. A. Semikhatov 1976. Table of time-ranges of the principal groups of Precambrian stromatolites. In *Stromatolites*, M. R. Walter (ed.), 693−4. Amsterdam: Elsevier.

Kuijpers, E. P. 1971. Transition from fluviatile to tidal marine sediments in the Upper Devonian of Seven Heads Peninsula, south County Cork, Ireland. *Geol Mijnb.* **50**, 443−50.

Kulm, L. D. and G. A. Fowler 1974. Oregon continental margin structure and stratigraphy: a test of the imbricate thrust model. In *The geology of continental margins*, C. A. Burk and C. L. Drake (eds), 261−83. Berlin: Springer-Verlag.

Lambert, J. L. M. 1965. A reinterpretation of the breccias in the meneage crush-zone of the Lizard boundary, South-west England. *Q. J. Geol Soc. Lond.* **121**, 77−102.

Lambert, R. St J. 1969. In discussion of Fitch *et al.* (1969).

Lambert, R. St. J. 1971. *The pre-Pleistocene Phanerozoic time-scale: a review*, 9−31. Spec. Publ. Geol. Soc. Lond., no. 5.

Lambert, R. St J. and W. S. McKerrow 1976. The Grampian Orogeny. *Scott. J. Geol.* **12**, 271−92.

Lambert, R. St J. and D. C. Rex 1966. Isotopic ages of minerals from the Pre-Cambrian complex of the Malverns. *Nature* **209**, 605−6.

Laughton, A. S. 1975. Tectonic evolution of the Northeast Atlantic Ocean: a review. *Norges Geol. Unders.* **316**, 169−93.

Laurent, R. 1972. The Hercynides of South Europe: a model. *24th Int. Geol Congr.* **3**, 363−370.

Lawson, D. E. 1976. Sandstone−boulder conglomerates and a Torridonian cliffed shoreline between Gairloch and Stoer, Northwest Scotland. *Scott. J. Geol.* **12**, 67−88.

le Brecque, J., D. V. Kent and S. C. Cande 1977. Revised magnetic polarity time scale for late Cretaceous and Cenozoic time. *Geology* **5**, 330−5.

Leeder, M. R. 1973. Sedimentology and palaeogeography of the upper Old Red Sandstone in the Scottish Border Basin. *Scott. J. Geol.* **9**, 117−44.

*Leeder, M. R. 1974a. Lower Border Group (Tournaisian) fluvio-deltaic sedimentation and palaeogeography of the Northumberland Basin. *Proc. Yorks. Geol Soc.* **40**, 129−80.

Leeder, M. R. 1974b. Origin of the Northumberland Basin. *Scott. J. Geol.* **10**, 283−96.

Leeder, M. R. 1976a. Palaeogeographic significance of pedogenic carbonates in the topmost upper Old Red Sandstone of the Scottish Border Basin. *Geol J.* **11**, 21−8.

Leeder, M. R. 1976b. Sedimentary facies and the origins of basin subsidence along the northern margin of the supposed Hercynian ocean. *Tectonophysics* **36**, 167−79.

*Lees, A. 1964. The structure and origin of the Waulsortian (lower Carboniferous) 'reefs' of west−central Eire. *Phil Trans R. Soc. Lond.* **247B**, 483−531.

Lewis, H. P. 1926. On *Bolopora undosa* gen. et sp. nov.: a rock-building bryozoan with phosphatised skeleton, from the basal Arenig rocks of Ffestiniog (North Wales). *Q. J. Geol Soc. Lond.* **82**, 411−27.

Long, C. B. and M. D. Max 1977. Metamorphic rocks in the SW Ox Mountains Inlier, Ireland: their structural compartmentation and place in the Caledonian orogen. *J. Geol Soc. Lond.* **133**, 413−32.

Lorenz, V. 1976. Formation of Hercynian subplates: possible causes and consequences. *Nature* **262**, 374−7.

Lowell, J. D. and G. J. Genik 1972. Sea floor spreading and structural evolution of the southern Red Sea. *Bull. Am. Ass. Petrol. Geol.* **52**, 247−59.

McCabe, P. J. 1977. Deep distributary channels and giant bedforms in the upper Carboniferous of the central Pennines, northern England. *Sedimentology* **24**, 271−90.

McCave, N. 1971. Wave effectiveness at the sea bed and its relationship to bed-forms and deposition of mud. *J. Sedim. Petrol.* **41**, 89−96.

*MacDermot, C. V. and G. D. Sevastopulo 1973. Upper Devonian and lower Carboniferous stratigraphical setting of Irish mineralisation. *Bull. Geol Surv. Ireland* **1**, 267−80.

MacDonald, R., J. E. Thomas and S. A. Rizzello 1977. Variations in basalt chemistry with time in the Midland Valley Province during the Carboniferous and Permian. *Scott. J. Geol.* **13**, 11−22.

*McKerrow, W. S. and L. R. M. Cocks 1976. Progressive faunal migration across the Iapetus Ocean. *Nature* **263**, 304−6.

*McKerrow, W. S., J. K. Leggett and M. H. Eales 1977. Imbricate thrust model of the Southern Uplands of Scotland. *Nature* **267**, 237−9.

Madgett, P. A. and J. A. Catt 1978. Petrography, stratigraphy and weathering of Late Pleistocene tills in East Yorkshire, Lincolnshire and north Norfolk. *Proc. Yorks. Geol Soc.* **42**, 55−108.

Maltman, A. J. 1975. Ultramafic rocks in Anglesey: their non-tectonic emplacement. *J. Geol Soc. Lond.* **131**, 593−605.

*Marie, J. P. P. 1975. Rotliegendes stratigraphy and diagenesis. In *Petroleum and the continental shelf of Northwest Europe*, A. W. Woodland (ed), 205−11. London: Applied Science Publishers.

Mathews, S. C. 1974. Exmoor Thrust? Variscan front? *Proc. Ussh. Soc.* **3**, 82−94.

Max, M. D. 1975. Precambrian rocks of Southeast Ireland. In *A correlation of the Precambrian rocks in the British Isles*, A. L. Harris *et al.* (eds), 97−101, Geol. Soc. Lond. Spec. Rep., No. 6.

Max, M. D. 1976. The pre-Palaeozoic basement in south-eastern Scotland and the Southern Uplands Fault. *Nature* **264**, 485−6.

Meischner, D. 1971. Clastic sedimentation in the Variscan geosyncline east of the River Rhine. In *Sedimentology of parts of central Europe*, G. Müller (ed.), 9−44. Field trip Guidebook, 8th Int. sed. Cong., Heidelberg.

Meneisy, M. Y. and J. A. Miller 1963. A geochronological study of the crystalline rocks of Charnwood Forest, England. *Geol Mag.* **100**, 507−23.

Mercer, J. H. 1978. West Antarctic ice sheet and CO_2 greenhouse effect: a threat of disaster. *Nature* **271**, 321−5.

Middleton, G. V. 1960. Spilitic rocks in southeast Devonshire. *Geol Mag.* **97**, 192−207.

*Miller, J. and R. F. Grayson 1972. Origin and structure of the lower Viséan 'reef' limestones near Clitheroe, Lancashire. *Proc. Yorks. Geol Soc.* **38**, 607−38.

Miller, J. A. and D. H. Green 1961. Age determinations of rocks in the Lizard (Cornwall) area. *Nature* **192**, 1175−6.

Mitchell, A. H. G. and W. S. McKerrow 1975. Analogous evolution of the Burma orogen and the Scottish Caledonides. *Bull. Geol Soc. Am.* **86**, 305−15.

Mitchell, G. F., L. F. Penny, F. W. Shotton and R. G. West 1973. *A correlation of Quaternary deposits in the British Isles.* Geol. Soc. Lond. Spec. Rep., No. 4.

Mitchell, J. G., R. M. MacIntyre and I. R. Pringle 1975. K−Ar and Rb−Sr isotopic age studies on the Wolf Rock nosean phonolite, Cornwall. *Geol Mag.* **112**, 55−61.

Mohr, P. A. 1964. Genesis of the Cambrian manganese carbonate rocks of North Wales. *J. Sedim. Petrol.* **34**, 819−29..

Montford, H. M. 1970. The terrestrial environment during Upper Cretaceous and Tertiary times. *Proc. Geol. Ass.* **81**, 181−204.

*Moorbath, S. 1975. The geological significance of early Precambrian rocks. *Proc. Geol. Ass.* **86**, 259−79.

Moorbath, S., J. L. Powell and P. N. Taylor 1975. Isotope evidence for the age and origin of the 'grey gneiss' complex of the southern Outer Hebrides, Scotland. *J. Geol Soc. Lond.* **131**, 213−22.

Moore, D. G., J. R. Curray and F. J. Emmel 1976. Large submarine slide (olistostrome) associated with Sunda Arc subduction zone, Northeast Indian Ocean. *Marine Geol.* **21**, 211−26.

Morgan, A. V. 1973. The Pleistocene geology of the area north and west of Wolverhampton, Staffordshire, England. *Phil Trans R. Soc. Lond.* **265B**, 233−97.

Morris, W. A. 1976. Transcurrent motion determined paleomagnetically in the northern Appalachians and Caledonides and the Acadian Orogeny. *Can J. Earth Sci.* **13**, 1236−43.

Morton, N. 1965. The Bearreraig Sandstone Series (middle Jurassic) of Skye and Raasay. *Scott. J. Geol.* **1**, 189−216.

Murray, J. W. and C. A. Wright 1974. *Palaeogene foraminiferida and palaeoecology, Hampshire and Paris Basins and the English Channel*, Spec. Pap. Palaeontology, No. 14: 171 pp.

Mykura, W. and J. Phemister 1976. *The geology of western Shetland.* Mem. Geol Surv. GB.

Nami, M. 1976. An exhumed Jurassic meander belt from Yorkshire, England. *Geol Mag.* **133**, 47−52.

Nami, M. and M. R. Leeder 1978. Changing channel morphology and magnitude in the Scalby Formation (M. Jurassic) of Yorkshire, England. In *Fluvial sedimentology*, A. D. Miall (ed.), Can Soc. Petrol. Geol. Mem. No. 5.

Narayan, J. 1963. Cross-stratification and palaeogeography of the Lower Greensand of South-east England and Bas-Boulonnois, France. *Nature* **199**, 1246−7.

Narayan, J. 1971. Sedimentary structures in the Lower Greensand of the Weald, England, and Bas-Boulonnois, France. *Sedim. Geol.* **6**, 73−109.

Neves, R. and R. C. Selley 1975. A review of the Jurassic

rocks of North-east Scotland. In *Proceedings of the Jurassic of the northern North Sea symposium*, K. G. Finstad and R. C. Selley (eds), JNNSS 5, 1–29. Stavanger: Norwegian Petroleum Society.

Nicolas, A. 1972. Was the Hercynian orogenic belt of Europe of the Andean type? *Nature* **236**, 221–3.

Oliver, R. L. 1954. Welded tuffs in the Borrowdale Volcanic Series, English Lake District, with a note on similar rocks in Wales. *Geol. Mag.* **91**, 473–83.

Owen, H. G. 1971. The stratigraphy of the Gault in the Thames Estuary and its bearing on the Mesozoic tectonic history of the area. *Proc. Geol. Ass.* **82**, 187–207.

Owen, H. G. 1975. The stratigraphy of the Gault and Upper Greensand of the Weald. *Proc. Geol. Ass.* **86**, 475–98.

Owen, T. R. 1976. *The geological evolution of the British Isles.* Oxford: Pergamon Press.

Owen, T. R., T. W. Bloxam, D. G. Jones, V. G. Walmsley and B. P. J. Williams 1971. Summer (1968) field meeting in Pembrokeshire, South Wales: report by the directors. *Proc. Geol. Ass.* **82**, 17–60.

Palmer, A. R. 1969. Cambrian trilobite distributions in North America and their bearing on the Cambrian palaeogeography of Newfoundland. In *North Atlantic geology and continental drift: a symposium*, M. Kay (ed.), 139–44. Mem. Am. Ass. Petrol. Geol., No. 12.

Palmer, T. J. and H. C. Jenkyns 1975. A carbonate island barrier from the Great Oolite (middle Jurassic) of central England. *Sedimentology* **22**, 125–35.

*Parker, J. R. 1975. Lower Tertiary sand development in the central North Sea. In *Petroleum and the continental shelf of North-west Europe*, A. W. Woodland (ed.), 447–53. London: Applied Science Publishers.

Parker, R. J. 1975. The petrology and origin of some glauconite and glauco-conglomeratic phosphorites from the South African continental margin. *J. Sedim. Petrol.* **45**, 230–42.

Pattison, J. 1970. A review of the marine fossils from the upper Permian rocks of northern Ireland and Northwest England. *Bull. Geol. Surv. GB.* **32**, 123–65.

Pearce, J. A. and J. R. Cann 1973. Tectonic setting of basic volcanic rocks determined using trace element analyses. *Earth Planet. Sci. Lett.* **19**, 290–300.

Penn, I. E. and C. D. R. Evans 1976. *The middle Jurassic (mainly Bathonian) of Cardigan Bay and its palaeogeographical significance.* Rep. Inst. Geol. Sci., no. 76/6.

Perch-Nielsen, K. 1973. Campanian–Maastrichtian coccoliths from Nugssuaq, West Greenland. *Bull. Geol. Soc. Den.* **22**, 79–82.

Perrin, R. M. S., H. Davies and M. D. Fysh 1973. Lithology of the Chalky Boulder Clay. *Nature Phys. Sci.* **245**, 101–4.

Phillips, W. E. A. 1973. The pre-Silurian rocks of Clare Island, Co. Mayo, Ireland, and the age of the metamorphism of the Dalradian in Ireland. *J. Geol Soc. Lond.* **129**, 585–606.

Phillips, W. E. A. 1974. The stratigraphy, sedimentary environments and palaeogeography of the Silurian strata of Clare Island, Co. Mayo, Ireland. *J. Geol Soc.*

Lond. **130**, 19–41.

*Phillips, W. E. A., C. J. Stillman and T. Murphy 1976. A Caledonian plate tectonic model. *J. Geol Soc. Lond.* **132**, 579–609.

Phillips, W. E. A., W. E. G. Taylor and I. S. Sanders 1975. An analysis of the geological history of the Ox Mountains Inlier. *Sci. Proc. R. Dubl. Soc.* **5A**, 311–29.

Piper, D. J. W. 1972. Sedimentary environments and palaeogeography of the late Llandovery and earliest Wenlock of north Connemara, Ireland. *J. Geol Soc. Lond.* **128**, 33–51.

Piper, J. D. A. 1974. Proterozoic crustal distribution, mobile belts and apparent polar movements. *Nature* **251**, 381–4.

Pitman, W. C. and M. Talwani 1972. Seafloor spreading in the North Atlantic. *Bull. Geol. Soc. Am.* **83**, 619–46.

Pomerol, C. 1973. *Ère Cénozoique*, Paris: Doin.

Porrenga, D. H. 1966. Glauconite and chamosite as depth indicators in the marine environment. *Marine Geol.* **5**, 495–501.

Potter, J. F. and J. H. Price 1965. Comparative sections through rocks of Ludlovian–Downtonian age in the Llandovery and Llandeilo districts. *Proc. Geol. Ass.* **76**, 379–402.

*Powell, D. 1974. Stratigraphy and structure of the western Moine and the problem of Moine orogenesis. *J. Geol Soc. Lond.* **130**, 575–93.

Powell, D. W. 1971. A model for the Lower Palaeozoic evolution of the southern margin of the early Caledonides of Scotland and Ireland. *Scott. J. Geol.* **7**, 369–72.

Purser, B. H. (ed.) 1973. *The Persian Gulf: Holocene carbonate sedimentation and diagenesis in a shallow epicontinental sea.* Berlin: Springer-Verlag.

Purser, B. H. 1975. Tidal sediments and their evolution in the Bathonian carbonates of Burgundy, France. In *Tidal deposits*, R. N. Ginsburg (ed.) 335–43. Berlin: Springer-Verlag.

Ramsbottom, W. H. C. 1966. A pictorial diagram of the Namurian rocks of the Pennines. *Trans Leeds Geol Ass.* **7**, 181–4.

Ramsbottom, W. H. C. 1969. The Namurian of Britain. *G.R. 6me Cong. Int. Strat. Géol. Carb.*, Sheffield 1967, **1**, 219–32.

*Ramsbottom, W. H. C. 1973. Transgressions and regressions in the Dinantian: a new synthesis of British Dinantian stratigraphy. *Proc. Yorks. Geol Soc.* **39**, 567–607.

Ramsbottom, W. H. C. 1977. Major cycles of transgression and regression (mesothems) in the Namurian. *Proc. Yorks. Geol Soc.* **41**, 261–91.

Ramsbottom, W. H. C. 1978. Permian. In *The ecology of fossils*, W. S. McKerrow (ed.), 146–83. London: Duckworth.

*Rast, N. 1969. The relationship between Ordovician structure and volcanicity in Wales. In *The Pre-Cambrian and Lower Palaeozoic rocks of Wales*, A. Wood (ed.), 305–38. Cardiff: University of Wales Press.

*Rast, N., B. H. O'Brien and R. J. Wardle 1976. Relationships between Precambrian and Lower Palaeozoic rocks of the Avalon Platform in New Brunswick, the northeast Appalachians and the British Isles. *Tectonophysics* **30**, 315–38.

Raup, D. M. and S. M. Stanley 1978. *Principles of palaeontology*, 2nd edn. San Francisco: W. H. Freeman.

*Rawson, P. F., D. Curry, F. C. Dilley, J. M. Hancock, W. J. Kennedy, J. W. Neale, C. J. Wood and B. C. Worssam 1978. *A correlation of Cretaceous rocks in the British Isles*. Geol Soc. Lond. Spec. Rep., no. 9

*Rayner, D. H. 1963. The Achanarras Limestone of the middle Old Red Sandstone, Caithness, Scotland. *Proc. Yorks. Geol Soc.* **34**, 117–38.

Read, H. H. and J. Watson 1975. *Introduction to geology, 2: Earth history. Later stages of Earth history*. London: Macmillan.

Read, W. A. and S. R. H. Johnson 1967. The sedimentology of sandstone formations within the upper Old Red Sandstone and lowest Calciferous Sandstone Measures west of Stirling, Scotland. *Scott. J. Geol.* **3**, 242–67.

Reading, H. G. 1973. The tectonic environment of Southwest England: discussion of Floyd (1972). *Proc. Geol Ass.* **84**, 237–9.

Reading, H. G. 1975. Strike slip fault systems: an ancient example from the Cantabrians. *9th Int. Sed. Cong.*, Nice, **4**, 287–291.

Reid, R. E. H. 1968. Bathymetric distributions of Calcarea and Hexactinellida in the present and the past. *Geol Mag.* **105**, 546–59.

Renouf, J. T. 1974. The Proterozoic and Palaeozoic development of the Armorican and Cornubian Provinces. *Proc. Ussh. Soc.* **3**, 6–43.

Rhys, G. H. 1974. *A proposed standard lithostratigraphic nomenclature for the southern North Sea and an outline structural nomenclature for the whole of the (UK) North Sea*. Rep. Inst. Geol Sci., no. 74/8.

Richardson, S. W. and R. Powell 1976. Thermal causes of Dalradian metamorphism in the central Highlands of Scotland. *Scott. J. Geol.* **12**, 237–68.

Rider, M. H. 1969. *Sedimentological studies in the West Clare Namurian, Ireland, and the Mississippi River delta*. Unpublished PhD thesis, University of London.

Riding, R. 1974. Model of the Hercynian fold belt. *Earth Planet. Sci Lett.* **24**, 125–35.

Riding, R. 1977. Calcified Plectonema (blue-green algae): a Recent example of *Girvanella* from Aldabra Atoll. *Palaeontology* **20**, 33–46.

Ries, A. C. and R. M. Shackleton 1976. Patterns of strain variation in arcuate fold belts. *Phil Trans R. Soc. Lond.* **283A**, 281–8.

Roberts, B. 1969. The Llwyd Mawr ignimbrite and its associated volcanic rocks. In *The Pre-Cambrian and Lower Palaeozoic rocks of Wales*, A. Wood (ed.), 337–56. Cardiff: University of Wales Press.

Roberts, D. E. 1977. The structure of the Skiddaw Slates in the Blencathra–Mungrisdale area, Cumbria. *Geol J.* **12**, 33–58.

Roberts, D. G. 1974. Structural development of the British Isles, the continental margin, and the Rockall

Plateau. In *The geology of continental margins*, C. A. Burk and C. L. Drake (eds), 343–59. Berlin: Springer-Verlag.

Roberts, D. G. 1975. Tectonic and stratigraphic evolution of the Rockall Plateau and Trough. In *Petroleum and the continental shelf of North-west Europe*, A. W. Woodland (ed.), 77–89. London: Applied Science Publishers.

Roberts, J. L. 1974. The structure of the Dalradian rocks in the SW Highlands of Scotland. *J. Geol Soc. Lond.* **130**, 93–124.

Rohrlich, V., S. E. Calvert and N. B. Price 1969. Chamosite in the Recent sediments of Loch Etive, Scotland. *J. Sedim. Petrol.* **39**, 624–31.

*Rose, J. and P. Allen 1977. Middle Pleistocene stratigraphy in south-east Suffolk. *J. Geol Soc. Lond.* **133**, 83–102.

Rossiter, J. R. 1972. Sea-level observations and their secular variation. *Phil Trans R. Soc. Lond.* **272A**, 131–9.

Russell, M. J. 1972. North–south geofractures in Scotland and Ireland. *Scott. J. Geol.* **8**, 75–84.

Rust, B. R. 1965. The stratigraphy and structure of the Whithorn area of Wigtownshire, Scotland. *Scott. J. Geol.* **1**, 101–33.

*Sadler, P. M. 1973. An interpretation of new stratigraphic evidence from south Cornwall. *Proc. Ussh. Soc.* **2**, 535–50.

*Sadler, P. M. 1974. An appraisal of the 'Lizard–Dodman–Start Thrust' concept. *Proc. Ussh. Soc.* **3**, 71–81.

Sancetta, C., J. Imbrie and N. G. Kipp 1973. Climatic record of the past 130 000 years in North Atlantic deep-sea core V23–82: correlation with the terrestrial record. *Quat. Res.* **3**, 110–16.

*Sanderson, D. J. and W. R. Dearman 1973. Structural zones of the Variscan fold belt in SW England: their location and development. *J. Geol Soc. Lond.* **129**, 527–36.

Schwarzacher, W. 1958. The stratification of the Great Scar Limestone in the Settle district of Yorkshire. *Lpool Manchr Geol J.* **2**, 124–42.

Scoffin, T. P. 1971. The conditions of growth of the Wenlock reefs of Shropshire. *Sedimentology* **17**, 173–219.

Scott, A. C. 1977. A review of the ecology of Upper Carboniferous plant assemblages with new data from Strathclyde. *Palaeontology*, **20**, 447–73.

Scrutton, C. T. 1969. Corals and stromatoporoids from the Chudleigh Limestone. *Proc. Ussh. Soc.* **2**, 102–6.

Scrutton, C. T. 1975. The Devonian limestones of Newton Abbot and their equivalents at Torquay. *Proc. Ussh. Soc.* **3**, 263.

*Scrutton, C. T. 1977. Facies variations in the Devonian limestones of eastern South Devon. *Geol Mag.* **114**, 165–93.

Sedgwick, A. 1835. On the geological relations and internal structure of the Magnesium Limestone and lower portions of the New Red Sandstone Series in their range through Nottinghamshire, Derbyshire, Yorkshire and Durham, to the southern extremity of Northumberland. *Trans Geol Soc. Lond.*, ser. 2, **3**, 37–124.

*Seely, D. R., P. R. Vail and G. G. Walton 1974. Trench

slope model. In *The geology of continental margins*, C. A. Burk and C. L. Drake (eds), 249–60. Berlin: Springer-Verlag.

Selley, R. C. 1976. The habitat of North Sea oil. *Proc. Geol. Ass.* **87**, 359–88.

Sellwood, B. W. 1970. The relation of trace fossils to small scale sedimentary cycles in the British Lias. In *Trace fossils*, T. P. Crimes and J. C. Harper (eds), 489–504. Liverpool: Seel House Press.

Sellwood, B. W. 1972. Regional environmental changes across a lower Jurassic stage-boundary in Britain. *Palaeontology* **15**, 125–57.

Sellwood, B. W. 1975. Lower Jurassic tidal-flat deposits, Bornholm, Denmark. In *Tidal deposits*, R. N. Ginsburg (ed.), 93–102. New York: Springer-Verlag.

Sellwood, B. W. 1978. Jurassic. In *The ecology of fossils*, W. S. McKerrow (ed.), 204–79. London: Duckworth.

Sellwood, B. W., M. K. Durkin and W. J. Kennedy 1970. Field meeting on the Jurassic and Cretaceous rocks of Wessex. *Proc. Geol. Ass.* **81**, 715–32.

Sellwood, B. W. and A. Hallam 1974. Bathonian volcanicity and North Sea rifting. *Nature* **252**, 27.

*Sellwood, B. W. and H. C. Jenkyns 1975. Basins and swells and the evolution of an epeiric sea (Pliensbachian–Bajocian of Great Britain). *J. Geol Soc. Lond.* **131**, 373–88.

Sellwood, B. W. and W. S. McKerrow 1974. Depositional environments in the lower part of the Great Oolite Group of Oxfordshire and north Gloucestershire. *Proc. Geol. Ass.* **85**, 189–210.

Shackleton, N. J. and N. D. Opdyke 1973. Oxygen isotope and palaeomagnetic stratigraphy of equatorial Pacific core V28–238: oxygen isotope temperatures and ice volumes on a 10^5 year and 10^6 year scale. *Quat. Res.* **3**, 39–55.

Shackleton, R. M. 1969. The Pre-Cambrian of North Wales. In *The Pre-Cambrian and Lower Palaeozoic rocks of Wales*, A. Wood (ed.), 1–22. Cardiff: University of Wales Press.

*Shackleton, R. M. 1975. Precambrian rocks of Wales. In *A correlation of the Precambrian rocks in the British Isles*, A. L. Harris *et al.* (eds), 76–82. Geol Soc. Lond. Spec. Rep., no. 6.

Sheraton, J. W., J. Tarney, T. J. Wheatley and A. E. Wright 1973. The structural history of the Assynt district. In *The early Precambrian of Scotland and related rocks of Greenland*, R. G. Park and J. Tarney (eds), 31–43. University of Keele.

Shotton, F. W. 1956. Some aspects of the New Red desert in Britain. *Lpool Machr Geol J.* **1**, 450–65.

Sibuet, J. C. 1973. South Armorican shear zone and continental fit before the opening of the Bay of Biscay. *Earth Planet. Sci. Lett.* **18**, 153–7.

Simpson, A. 1967. The stratigraphy and tectonics of the Skiddaw Slates and the relationship of the overlying Borrowdale Volcanic Series in part of the Lake District. *Geol J.* **5**, 391–418.

*Smith, A. G. and J. C. Briden 1977. *Mesozoic and Cenozoic palaeocontinental maps*. Cambridge: Cambridge University Press.

Smith, A. G., J. C. Briden and G. E. Drewry 1973. Phanerozoic world maps. In *Organisms and continents through time*, N. F. Hughes (ed.), 1–42. Palaeont. Ass. Spec. Pap., no. 12.

Smith, D. B. 1958. Observations on the Magnesian Limestone reefs of north-east Durham. *Bull. Geol Surv. GB.* **15**, 71–84.

*Smith, D. B., R. G. W. Brunstrom, P. I. Manning, S. Simpson and F. W. Shotton 1974. A correlation of Permian rocks in the British Isles. *J. Geol Soc. Lond.* **130**, 1–45.

Smith, E. G. and G. Warrington 1971. The age and relationships of the Triassic rocks assigned to the lower part of the Keuper in north Nottinghamshire, north-west Lincolnshire and south Yorkshire. *Proc. Yorks. Geol Soc.* **38**, 201–27.

*Spencer, A. M. 1971. *Late Pre-Cambrian glaciation in Scotland*. Mem. Geol Soc. Lond., no. 6.

*Spencer, A. M. 1975. Late Precambrian glaciation in the North Atlantic region. In *Ice ages: ancient and modern*, A. E. Wright and F. Moseley (eds), 217–40. Liverpool: Seel House Press.

Spencer, A. M. and M. O. Spencer 1972. The late Precambrian/lower Cambrian Bonahaven Dolomite of Islay and its stromatolites. *Scott, J. Geol.* **8**, 269–82.

Stanley, S. M. 1968. Post-Palaeozoic adaptive radiation of infaunal bivalve molluscs: a consequence of mantle fusion and siphon formation. *J. Palaeont.* **42**, 214–29.

Steel, R. J. 1974. Continental sedimentation in the New Red Sandstone of the Hebridean Province, Scotland. *J. Sedim. Petrol.* **44**, 336–57.

*Steel, R. J., R. Nicholson and L. Kalander 1975. Triassic sedimentation and palaeogeography in central Skye. *Scott. J. Geol.* **11**, 1–13.

Stewart, A. D. 1969. Torridonian rocks of Scotland reviewed. In *North Atlantic geology and continental drift: a symposium*, M. Kay (ed.), 595–608. Mem. Am. Ass. Petrol. Geol., no. 12.

*Stewart, A. D. 1975. 'Torridonian' rocks of western Scotland. In *A correlation of the Precambrian rocks in the British Isles*, A. L. Harris *et al.* (eds), 43–51. Geol Soc. Lond. Spec. Rep., no. 6.

*Stillman, C. J., K. Downes and E. J. Schiener 1974. Caradocian volcanic activity in East and South-east Ireland. *Sci. Proc. R. Dubl. Soc.* **5A**, 87–98.

Stockwell, C. H. 1972. *Revised Precambrian time scale for the Canadian Shield*. Geol Surv. Can. Pap., no. 72–52.

Surlyk, F. 1975. Block faulting and associated marine sedimentation at the Jurassic/Cretaceous boundary, East Greenland. In *Proceedings of the Jurassic northern North Sea symposium*, K. G. Finstad and R. C. Selley (eds), JNNSS 7, 1–31. Stavanger: Norwegian Petroleum Society.

Surlyk, F. 1978. *Submarine fan sedimentation along fault scarps on tilted fault blocks (Jurassic/Cretaceous boundary, East Greenland)*. Bull. Grønlands Geologiske Undersøgelse no. 128, 108 pp.

Sutton, J. and J. V. Watson 1951. The pre-Torridonian

metamorphic history of the Loch Torridon and Scourie areas in the north-west Highlands and its bearing on the chronological classification of the Lewisian. *Q. J. Geol Soc. Lond.* **106**, 241–307.

*Swett, K. and D. E. Smit 1972. Palaeogeography and depositional environments of the Cambro-Ordovician shallow marine facies of the North Atlantic. *Bull. Geol Soc. Amer.* **83**, 3223–48.

Sykes, R. M. 1975. The stratigraphy of the Callovian and Oxfordian in northern Scotland. *Scott. J. Geol.* **11**, 1–28.

Sylvester-Bradley, P. C. 1972. An evolutionary model for the origin of life. In *Understanding the earth*, I. G. Gass, P. J. Smith and R. C. L. Wilson (eds), 123–41. Horsham, Sussex: Artemis Press.

Talbot, M. R. 1973. Major sedimentary cycles in the Corallian Beds. *Palaeogeog., Palaeoclimatol., Palaeoecol.* **14**, 293–317.

Tarling, D. H. 1974. A palaeomagnetic study of Eocambrian tillites in Scotland. *J. Geol Soc. Lond.* **130**, 163–77.

*Taylor, J. C. M. and V. S. Colter 1975. Zechstein of the English sector of the southern North Sea Basin. In *Petroleum and the continental shelf of North-west Europe*, A. W. Woodland (ed.), 249–63. London: Applied Science Publishers.

Taylor, J. H. 1949. *Petrology of the Northampton Sand Ironstone Formation.* Mem. Geol Surv. GB.

Thomas, A. N., P. J. Walmsley and D. A. L. Jenkins 1974. Forties field, North Sea. *Bull. Am. Ass. Petrol. Geol.* **58**, 396–406.

Thompson, M. E. and R. A. Eden 1977. *Quaternary deposits of the central North Sea, 3: the Quaternary sequence in the west-central North Sea.* Rep. Inst. Geol Sci., no. 77/12.

*Thorpe, R. S. 1974. Aspects of magmatism and plate tectonics in the Precambrian of England and Wales. *Geol J.* **9**, 115–36.

Townson, W. G. 1975. Lithostratigraphy and deposition of the type Portlandian. *J. Geol Soc. Lond.* **131**, 619–38.

Trümpy, R. 1973. The timing of orogenic events in the central and western Alps. In *Gravity and tectonics*, K. A. de Jong and R. Scholten (eds), 229–52. New York: John Wiley.

Tucker, M. E. 1969. Crinoidal turbidites from the Devonian of Cornwall and their palaeogeographical significance. *Sedimentology* **13**, 281–90.

Tucker, M. E. 1975. Exfoliated pebbles and sheeting in the Triassic. *Nature* **252**, 375–6.

*Tucker, M. E. 1977. The marginal Triassic deposits of South Wales: continental facies and palaeogeography. *Geol J.* **12**, 169–88.

Tucker, M. E. 1978. Triassic lacustrine sediments from South Wales: shore zone clastics, evaporites and carbonates. In *Modern and ancient lake sediments*, A. Matter and M. E. Tucker (eds). Oxford: Blackwell Sci. Publications.

Tucker, M. E. and P. van Straaten 1970. Conodonts and facies on the Chudleigh schwelle. *Proc. Ussh. Soc.* **2**, 160–70.

Tunbridge, I. P. 1976. Notes on the Hangman Sandstones

(middle Devonian) of north Devon. *Proc. Ussh. Soc.* **3**, 339.

Turner, P. and D. H. Tarling 1975. Implications of new palaeomagnetic results from the Carboniferous system of Britain. *J. Geol Soc. Lond.* **131**, 469–88.

Upton, B. G. J., P. Aspen, A. Graham and N. A. Chapman 1976. Pre-Palaeozoic basement of the Scottish Midland Valley. *Nature* **260**, 517–18.

Vail, P. R. 1977. Seismic stratigraphy and global changes in sea level, part 6: stratigraphic interpretation of seismic reflection patterns in depositional systems. In *Seismic stratigraphy: applications to hydrocarbon exploration*, 117–33, Mem. Am. Ass. Petrol. Geol., no. 26.

Valentine, J. W. 1973. *Evolutionary palaeoecology of the marine biosphere.* Englewood Cliffs (NJ): Prentice-Hall.

van Breeman, O., M. Aftalion and R. T. Pidgeon 1971. The age of the granitic injection complex of Harris, Outer Hebrides. *Scott. J. Geol.* **7**, 139–52.

Van der Voo, R. and R. B. French 1974. Apparent polar wandering for the Atlantic-bordering continents: late Carboniferous to Eocene. *Earth Sci. Rev.* **10**, 99–119.

van Straaten, P. and M. E. Tucker 1972. The upper Devonian Saltern Cove goniatite bed is an intraformational slump. *Palaeontology* **15**, 430–8.

Walkden, G. M. 1974. Palaeokarst surfaces in upper Viséan (Carboniferous) limestones of the Derbyshire block, England. *J. Sedim. Petrol.* **44**, 1232–47.

Walker, A. D. 1969. The reptile fauna of the 'lower Keuper' sandstone. *Geol Mag.* **106**, 470–6.

Walker, R. G. 1966. Shale Grit and Grindslow Shales: transition from turbidite to shallow water sediments in the upper Carboniferous of northern England. *J. Sedim. Petrol.* **36**, 90–114.

Walker, R. G. 1970. Deposition of turbidites and agitated-water siltstones: a study of the upper Carboniferous Westward Ho! Formation, north Devon. *Proc. Geol. Ass.* **81**, 43–68.

Walker, T. R. 1976. Diagenetic origin of continental red beds. In *The continental Permian in Central, West and South Europe*, H. Falke (ed.), 240–82. Dordrecht, Holland: Reidel.

Walmsley, V. G. and M. G. Bassett 1976. Biostratigraphy and correlation of the Coralliferous Group and Gray Sandstone Group (Silurian) of Pembrokeshire, Wales. *Proc. Geol. Ass.* **87**, 191–220.

Walton, E. K. 1963. Sedimentation and structure in the Southern Uplands. In *The British Caledonides*, F. H. Stewart and M. R. W. Johnson (eds), 71–97. Edinburgh: Oliver & Boyd.

Walton, E. K. 1965. Lower Palaeozoic rocks: stratigraphy. In *The geology of Scotland*, G. Y. Craig (ed.), 161–200. Edinburgh: Oliver & Boyd.

Warrington, G. 1970. The stratigraphy and palaeontology of the 'Keuper' Series of the central Midlands of England. *Q. J. Geol Soc. Lond.* **126**, 183–223.

Watson, J. V. 1973. Effects of reworking on high-grade gneiss complexes. *Phil Trans R. Soc. Lond.* **273A**, 443–55.

*Watson, J. V. 1975. The Lewisian Complex. In *A correla-*

tion of the Precambrian rocks in the British Isles, A. L. Harris *et al.* (eds), 15–29. Geol. Soc. Lond. Spec. Rep., no. 6.

Watts, N. L. 1976. Palaeopedogenic palygorskite from the basal Permo–Triassic of Northwest Scotland. *Am. Miner.* **61**, 299–302.

*Webby, B. D. 1966. Middle–upper Devonian palaeogeography of north Devon and west Somerset, England. *Palaeogeog., Palaeoclimatol., Palaeoecol.* **2**, 27–46.

Weir, J. A. 1973. Lower Palaeozoic graptolitic facies in Ireland and Scotland: review, correlation and palaeogeography. *Sci. Proc. R. Dubl. Soc.* **1A**, 439–60.

Weir, J. A. 1975. Palaeogeographical implications of two Silurian shelly faunas from the Arra Mountains and Cratloe Hills, Ireland. *Palaeontology* **18**, 343–50.

West, I. M. 1975. Evaporites and associated sediments of the basal Purbeck Formation (upper Jurassic) of Dorset. *Proc. Geol. Ass.* **86**, 205–26.

West, R. G. 1968. *Pleistocene geology and biology*. London: Longman.

West, R. G. and P. E. P. Norton 1974. The Icenian Crag of southeast Suffolk. *Phil Trans R. Soc. Lond.* **269B**, 1–28.

Whitaker, J. H. McD. 1962. The geology of the area around Leintwardine, Herefordshire. *Q. J. Geol Soc. Lond.* **118**, 319–51.

Whiteman, A. J., D. Naylor, R. Pegrum and G. Rees 1975. North Sea troughs and plate tectonics. *Tectonophysics* **26**, 39–54.

Whittington, H. B. 1972. Scotland. In *A correlation of Ordovician rocks in the British Isles*, A. Williams *et al.* (eds), 49–53. Geol. Soc. Lond. Spec. Rep., no. 3.

Wilkinson, J. M. and J. R. Cann 1974. Trace elements and tectonic relationships of basaltic rocks in the Ballantrae igneous complex, Ayrshire. *Geol Mag.* **111**, 35–41.

Williams, A. 1962. *The Barr and Lower Ardmillan Series (Caradoc) of the Girvan district, south-western Ayrshire, with descriptions of the Brachiopoda*. Mem. Geol. Soc. Lond., no. 3.

Williams, A. 1969a. Ordovician faunal provinces with reference to brachiopod distribution. In *The Precambrian and Lower Palaeozoic rocks of Wales*, A. Wood (ed.), 117–54. Cardiff: University of Wales Press.

Williams, A. 1969b. Ordovician of the British Isles. In *North Atlantic geology and continental drift: a symposium*, M. Kay (ed.) 236–64. Mem. Am. Ass. Petrol. Geol., no. 12.

Williams, A. 1976. Plate tectonics and biofacies evolution as factors in Ordovician correlation. In *The Ordovician System: proceedings of a Palaeontological Association symposium, September 1974*, M. G. Bassett (ed.), 29–66. Cardiff: University of Wales Press.

Williams, A., I. Strachan, D. A. Bassett, W. T. Dean, J. K. Ingham, A. D. Wright and H. B. Whittington 1972. *A correlation of Ordovician rocks in the British Isles*. Geol Soc. Spec. Rep., no. 3.

Williams, B. P. J. 1971. Sedimentary features of the Old Red Sandstone and Lower Limestone shales of south Pembrokeshire, south of the Ritec Fault. In *Geological*

excursions in South Wales and the Forest of Dean*, D. A. and M. G. Bassett (eds), 222–39. London: Geologists' Association.

Williams, G. E. 1966. Palaeogeography of the Torridonian Applecross Group. *Nature* **209**, 1303–6.

Williams, G. E. 1969. Characteristics of a Precambrian pediment. *J. Geol.* **77**, 183–207.

Williams, G. E. 1975. Late Precambrian glacial climate and the Earth's obliquity. *Geol Mag.* **112**, 441–65.

*Williams, J. J., D. C. Conner and K. E. Peterson 1975. The Piper oil-field, UK North Sea: a fault block structure with upper Jurassic beach-bar reservoir sands. In *Petroleum and the continental shelf of North-west Europe*, A. W. Woodland (ed.), 363–77. London: Applied Science Publishers.

Wills, L. J. 1973. *A palaeogeological map of the Palaeozoic floor below the Permian and Mesozoic formations in England and Wales*. Mem. Geol Soc. Lond., no. 17.

Wilson, A. C. 1971. *Lower Devonian sedimentation in the NW Midland Valley of Scotland*. Unpubl. PhD thesis, University of Glasgow.

Wilson, J. L. 1975. *Carbonate facies in geologic history*. Berlin: Springer-Verlag.

Wilson, J. T. 1966. Did the Atlantic close and then re-open? *Nature* **211**, 676–81.

Wilson, R. C. L. 1968. Upper Oxfordian palaeogeography of southern England. *Palaeogeog., Palaeoclimatol., Palaeoecol.* **4**, 5–28.

*Winchester, J. A. 1974. The zonal pattern of regional metamorphism in the Scottish Caledonides. *J. Geol Soc. Lond.* **130**, 509–24.

*Windley, B. F. 1973. Crustal development in the Precambrian. *Phil Trans R. Soc. Lond.* **273A**, 321–41.

Wise, S. W. and F. W. Weaver 1974. Chertification of oceanic sediments. In *Pelagic sediments on land and under the sea*, K. J. Hsü and H. C. Jenkyns (eds), 301–26. Oxford: Blackwell Sci. Publications.

Woods, D. S. 1974. Ophiolites, mélanges, blueschists and ignimbrites: early Caledonian subduction in Wales. In *Modern and ancient geosyncline sedimentation*, R. H. Dott and R. H. Shaver (eds), 334–44. Spec. Publ. Soc. Econ. Paleont. Miner., no. 19.

Woodcock, N. H. 1976. Structural style in slump sheets: Ludlow Series, Powys, Wales. *J. Geol Soc. Lond.* **132**, 399–415.

Woodrow, D. L., F. W. Fletcher and W. F. Ahrnsbrak 1973. Palaeogeography and palaeoclimate at the deposition sites of the Devonian Catskill and Old Red facies. *Bull. Geol Soc. Amer.* **84**, 3051–64.

Woollands, M. A. 1970. *Sedimentation and stratigraphy of the lower Silurian of the Garth and Rhayader districts, Wales*. Unpubl. PhD thesis, University of London.

Wright, A. E. 1969. Precambrian rocks of England, Wales and South-east Ireland. In *North Atlantic geology and continental drift: a symposium*, M. Kay (ed.), 93–109. Mem. Am. Ass. Petrol. Geol., no. 12.

Wright, A. E. 1976. Alternating subduction direction and the evolution of the Atlantic Caledonides. *Nature* **264**, 156–60.

Zagwijn, W. H. 1974. The palaeogeographical evolution of the Netherlands during the Quaternary. *Geol. Mijnb.* **53**, 369–85.

Zankl, H. 1971. Upper Triassic carbonate facies in the northern Limestone Alps. In *Sedimentology of parts of central Europe*, G. Müller (ed.) 147–85, Fieldtrip Guidebook, 8th Int. Sed. Cong. Heidelberg.

Ziegler, A. M. 1970. Geosynclinal development of the British Isles during the Silurian period. *J. Geol.* **78**, 445–79.

Ziegler, A. M., L. R. M. Cocks and W. S. McKerrow 1968. The Llandovery transgression of the Welsh Borderland. *Palaeontology* **11**, 736–82.

Ziegler, A. M. and W. S. McKerrow 1975. Silurian marine red beds. *Am. J. Sci.* **275**, 31–56.

*Ziegler, P. A. 1975. North Sea Basin history in the tectonic framework of north-western Europe. In *Petroleum and the continental shelf of North-west Europe*, A. W. Woodland (ed.), 131–49. London: Applied Science Publishers.

Zwart, H. J. 1967. The duality of orogenic belts *Geol Mijnb.* **46**, 283–98.

INDEX

References to figures appear in *italic type*.
References to sections appear in **bold type**.